HIGH-T_C SUPERCONDUCTORS: MAGNETIC INTERACTIONS

SERIES ON PROGRESS IN HIGH TEMPERATURE SUPERCONDUCTIVITY

Published

Vol. 1 — Proceedings of the Adriatico Research Conference on High Temperature Superconductors
(*eds. S. Lundqvist, E. Tosatti, M. Tosi and Yu Lu*)

Vol. 2 — Proceedings of the Beijing International Workshop on High Temperature Superconductivity
(*eds. Z. Z. Gan, G. J. Cui, G. Z. Yang and Q. S. Yang*)

Vol. 3 — Proceedings of the Drexel International Conference on High Temperature Superconductivity
(*eds. S. Bose and S. Tyagi*)

Vol. 4 — Proceedings of the 2nd Soviet-Italian Symposium on Weak Superconductivity
(*eds. A. Barone and A. Larkin*)

Vol. 5 — Proceedings of the IX Winter Meeting on Low Temperature Physics — High Temperature Superconductors
(*eds. J. L. Heiras, R. A. Barrio, T. Akachi and J. Tagüeña*)

Vol. 7 — Chemical and Structural Aspects of High Temperature Superconductors
(*ed. C. N. R. Rao*)

Vol. 8 — World Congress on Superconductivity
(*eds. C. G. Burnham and R. Kane*)

Vol. 9 — First Latin-American Conference on High Temperature Superconductivity
(*eds. R. Nicolsky, R. A. Barrio, R. Escudero and O. F. de Lima*)

Vol. 12 — High Temperature Superconductivity and Other Related Topics — 1st Asia-Pacific Conference on Condensed Matter Physics
(*eds. C. K. Chew et al.*)

Vol. 14 — Proceedings of the Adriatico Research Conference and Workshop on Towards the Theoretical Understanding of High Temperature Superconductors
(*eds. S. Lundqvist, E. Tosatti, M. Tosi and Yu Lu*)

Vol. 16 — Proceedings of the Srinagar Workshop on High Temperature Superconductivity
(*eds. C. N. R. Rao et al.*)

Forthcoming

Vol. 10 — Macroscopic Theories of Superfluidity and Superconductivity
(*ed. A. A. Sobyanin*)

Vol. 11 — High Temperature Superconductivity from Russia
(*eds. A. Larkin and N. Zavaritski*)

Vol. 13 — Applications of High Temperature Superconductors
(*ed. C. Y. Huang*)

Vol. 15 — International Symposium on New Developments in Applied Superconductivity
(*ed. Y. Murakami*)

Vol. 18 — Proceedings of the Tokai University International Symposium on the Science of Superconductivity and New Materials
(*ed. S. Nakajima*)

Vol. 19 — Superconductivity and Applications — Taiwan International Symposium on Superconductivity
(*eds. H. C. Ku and P. T. Wu*)

Vol. 20 — Proceedings of the X Winter Meeting on Low Temperature Physics — High Temperature Superconductors
(*eds. T. Akachi, J. A. Cogordán and A. A. Valladares*)

Progress in High Temperature Superconductivity — Vol. 17

PROCEEDINGS OF HIGH-T_C SUPERCONDUCTORS: MAGNETIC INTERACTIONS

11-13 October 1988, Gaithersburg, Maryland, USA

Editors

L. H. Bennett
Y. Flom
G. C. Vezzoli

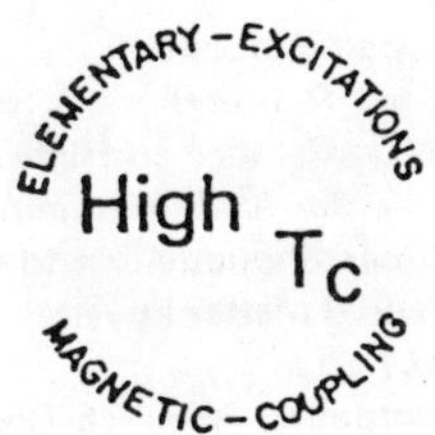

World Scientific
Singapore • New Jersey • London • Hong Kong

Published by

World Scientific Publishing Co. Pte. Ltd.
P O Box 128, Farrer Road, Singapore 9128

USA office: World Scientific Publishing Co., Inc.
687 Hartwell Street, Teaneck, NJ 07666, USA

UK office: World Scientific Publishing Co. Pte. Ltd.
73 Lynton Mead, Totteridge, London N20 8DH, England

PROCEEDINGS OF HIGH-T_c SUPERCONDUCTORS: MAGNETIC INTERACTIONS

ISBN 9971-50-820-6

Printed in Singapore by JBW Printers & Binders Pte. Ltd.

– IN MEMORIUM –

. . . . Dedicated to Dr. B.T. Matthias of The University of California and Dr. Issai Lefkowitz of The Army Research Office . . . .

. . . . only wishing that they could have been among us to see the realization of high-T_c superconductivity for which they strived and in which they believed . . . .

G. C. Vezzoli

FOREWORD

This three-day workshop, sponsored by the National Institute of Standards and Technology (formerly NBS) and the Goddard Space Flight Center, was held at NIST in Gaithersburg, MD, on October 11 – 13. Those persons who attended this workshop/conference with its subtitle "Elementary Excitations and Magnetic Coupling" certainly will acknowledge that the mission which we had in mind when we planned this event has indeed been accomplished. We sought to bring together invited experts, which included both world class solid-state theorists and experimentalists, to assess the role of magnetic phenomena as related to the great discovery of superconductivity near and above liquid-nitrogen temperature, as well as to provide a forum for interaction between communities representing fundamental research and application of HTSC materials. Truly, we were not intent on affirming or denying whether magnetic interactions were at the heart of the mechanism of high-T_c superconductivity, even though such an over-ambitious goal seemed timely based on recently announced theories. To wit, we were quite satisfied to endeavor to prove or disprove the involvement of magnetic coupling, magnetic interactions, and magnetic phenomena as an intrinsic property of the high-T_c superconducting materials. This objective, the one-hundred conference attendees and speakers, in truth, achieved. Magnetic interactions are indeed inescapably present and may provide a contribution toward the Cooper-pairing interaction energy generating the high-T_c state.

Toward our objectives, we felt very strongly that prior to the presentation of almost forty state-of-the-art theoretical and experimental papers, we should set the stage of the experimentally confirmed facts as constraints on any theory that intends to explain high-T_c superconductivity with a high level of credibility and timeless validity. Therefore we asked a senior scientist, who here in the United States had been the first champion of high-T_c superconductivity due to a non-phonon mechanism, to be our keynote speaker. Much to our delight Professor Bill Little of Stanford University accepted our invitation. Beyond even anyone's greatest expectations, Dr. Little accomplished the objective of establishing the rightful point of departure to commence this workshop. It seemed also proper and correct to afford this honor and responsibility to the scientist who almost a quarter of a century ago had virtually stood alone as an advocate of high-T_c superconductivity and who first introduced an excitonic mechanism thereof. When one thinks of all of the very great names associated with the remarkable and almost thought-defying phenomenon of superconductivity such as Onnes, Froilich, Meissner, Ochsenfeld, London, Onsager, Bardeen, Josephson, Giaver, Matthias, Anderson, Abrikosov, Gorkov, Freeman, Kresin, Müeller, and Chu (to name but a few), it is most interesting to note that this phenomenon was one of the few in physics not predicted by theorists. It is equally noteworthy to be able to state that high-T_c superconductivity was indeed foreseen in a rich theme of scientific literature in the mid 1960's in the Soviet Union, and here in the United States by Professor Bill Little.

Despite the fact that the most recent experiments have narrowed down the number of possible explanations of the phenomenon, a unified theory of the coupling mechanism is yet to be developed. No predictions were made of an increase in T_c or critical current in future high-T_c materials. It was stated several times during the presentations that the quality of experimental data has greatly improved in the last several months.

The tone of the discussions at the Workshop lacked the aspects of speculation and unfounded excitement which were so characteristic of previous meetings on high-T_c superconductivity. Interestingly enough, the report from Taiwanese researchers of a Tl compound with a critical temperature of 162 K and the latest theoretical work of Dr. W. Goddard from Caltech (which provides directions to experimentalist on how to achieve a high T_c) were not even mentioned at the conference which indicates the caution of the scientific community in unconfirmed reports.

A synopsis of the theoretical advances and critical experiments discussed at this workshop are given in the concluding remarks by Dr. V. Z. Kresin which appear in these proceedings. A summary of the participants' ideas on the elementary excitations important to high-T_c superconductivity is given in the panel report provided by Prof. J. Ashkenazi. The talk given by Dr. Susan Breon on space applications of high T_c superconductors was favorably received by the attendees and led to some interesting questions and discussions; some of these are reviewed in two panel summaries provided in these proceedings.

One feature of the Workshop were demonstrations. Among the things demonstrated during poster sessions were Bi-compound fibers grown at Stanford University, Y-Ba-Cu-O wires fashioned at Argonne National Laboratories, and superconducting composites fabricated at Oregon Graduate Center using shock compression techniques. Also demonstrated was a very intersting property of Y-Ba-Cu-O compounds doped with silver; these compounds exhibit strong flux trapping which was shown by levitating them above or below a magnet. A considerable effort at NIST in the synthesis of high-T_c superconductors via aqueous mixing of metal acetates and the fabrication of polymer/HTSC ceramic composites was displayed, along with the results of various measurements. A poster describing proposed applications of high-T_c superconducting materials at NASA-Goddard also attracted considerable attention.

L.H. Bennett
Gaithersburg, MD

G. C. Vezzoli
Watertown, MA

Y. Flom
Greenbelt, MD

A TRIBUTE TO LEF

D. Heiman
Francis Bitter National Magnet Laboratory
Massachusetts Institute of Technology

ONLY ONCE or twice in one's life does someone like Issai Lefkowitz come along. He caused a profound change in the direction of my life, from small-town auto mechanic to a career in condensed matter physics. After completing a B.S. degree at the local state university, I was drafted into the Army in 1968. At that time, about three-quarters of the inductees went into the infantry, and their chances of coming back intact were about fifty-fifty. Hoping to beat these odds, during orientation at boot camp I checked off a box marked "physical science assistant" on a job preference form. By sheer chance, this extremely unlikely job category opened up at the time I applied, and was assigned to replace someone leaving Lef's lab at Frankford Arsenal in Philadelphia. Lef headed a small group that was investigating perovskite ferroelectrics, including barium titanate piezoelectric detonators. I remember doing rather poorly in the lab, but through Lef's enormous enthusiasm for the scientific process and his constant reassurances about my not-too-obvious talents I was forced to live up to his estimation.

In a similar respect, he had an intuitive eye for clearly seeing someone else's dream, sometimes even before they saw it, and would use his vast resources to help make their dream come true. He kept insisting that I not go back to a job in auto repair in my small hometown in California, but instead should continue my education in a Ph.D. program. One of his suggestions was to work with Professor Ushioda at the University of California at Irvine. He made a few phone calls to Katsu and I found myself in graduate school doing optical spectroscopy. Without Lef's constant encouragement and devoted support I might still be doing valve jobs.

In the early 1980s, Lef became a strong believer in high-T_c superconductors. He would send me papers, then telephone to hold forth on his latest results with CuCl. These crazy notions of Cu-based high-T_c superconductors, also shared by Paul Chu, Gary Vezzoli and others, were essentially heretical to most of the scientific community at that time. Unfortunately, Lef's death on November 29, 1983 was just short of the recent discovery in La- Ba-Cu-O. I'm sure that his long career in oxide perovskites, when coupled with his fervent belief in Cu-based superconductors would have placed him at the forefront of the recent developments in high-T_c materials.

I consider myself extremely lucky to have had a mentor with so much energy, inspiration, optimism and foresight. I have never stopped being thankful.

HISTORICAL REMARKS

I would like to ask the reader's kind indulgence to allow me to track the history of how the present workshop came into being and my personal observations on some of the less-known background regarding the road to high-T_C superconductivity.

My own involvement in high-T_C superconductivity started in 1978-1979 while employed as a solid-state physicist/materials scientist at the US Army Picatinny Arsenal in north central New Jersey. Having conducted many years of experimental physics studies of insulator-metal phase transformations and threshold switching in chalcogenide glasses and transition metal oxides, as well as being strongly involved in high-pressure physics and crystal chemistry, I was deeply intrigued by the high-pressure cuprous chloride work reported by Chu and Russakov in the late 1970's and hypothetically suggesting effects akin to superconductivity at liquid nitrogen temperatures or somewhat above. This work motivated me to read the papers of Abrikosov, Gorkov, Brandt, Little, and Ginzburg, and to try to duplicate the experimental findings of Chu and Russakov. After many failures, I was finally able to reproduce those data using a Bridgman Anvil and diamond cell apparatus but quickly realized that I was not simply studying CuCl, however, I was studying the system $Cu^{+}Cl$-Cu^{0}-$Cu^{2+}Cl_2$ due to pressure-induced micro (and macro) disproportionation. Although I could <u>not</u> reproduce the Chu data on an experiment-by-experiment basis, nonetheless, my interest was irreversibly ignited. These first experiments allowed me to continue those studies via internal funding referred to as In-Laboratory-Independent-Research (ILIR) and granted through the recommendation of Dr. Yvon Carrignan and Dr. Thomas Davidson. Throughout this year of study I was most frequently encouraged in my work by Dr. Issai Lefkowitz of the Army Research Office in Research Triangle Park, North Carolina, a strong and cooperative proponent of high-T_C.

By the end of 1981 it had become apparent to me that it was necessary to switch from studies of copper halides to studies of

copper oxides if we were to ever achieve some kind of convincing consistency related to observed
effects akin to a collective effect like superconductivity. In this regard I noted early encouraging findings in interface samples of Cu_2O-CuCl at high pressure. I published this as an 'Note Added in Proof' to my October 1982 paper in Phys. Rev. B. It was also apparent to me at that time that excitons were a powerful candidate as a high-T_C mediator, also noted in that and an earlier paper of mine.

During the early 1980's the Army sponsored two conferences in high-pressure physics at the Benet Laboratories of Watervliet Arsenal near Albany, New York, and at the Rennsellaerville Conference Center near Troy, New York. Although I personally felt that copper oxide-copper chloride should be my next target study, nonetheless administrative supervision earnestly encouraged me to pursue the cadmium sulfide (chlorinated) work that was reported by Watervliet scientists Clark Homan and Dave Kendal and RPI's Professor Bob MacCrone. At the two conferences in the Albany-Troy area the findings related to CuCl and CdS were presented and discussed and as expected, generated enormous controversy related to whether these data were suggestive of superconductivity and related to the purity of the crystals. These studies, including my own, were also presented at two international conferences of The Association of International Research at Pressure and Temperature (AIRAPT) in Le Creusot, France, and Uppsalla, Sweden during this period. Very detailed high-pressure x-ray diffraction studies conducted by respected groups at the then National Bureau of Standards and The Naval -search Laboratory were presented on the copper chloride and cadmium sulfide systems at several of these conferences.

If indeed we were dealing with Cu^+Cl-Cu^0-$Cu^{2+}Cl_2$ in the high pressure
cuprous chloride work, we soon realized that in the CdS studies we were addressing interfacial mixtures of the sphalerite and wurzite phases in pressure-quenched cadmium sulfide. With the help of Stan Block and Gasper Piermarini of the Bureau and a 1964 Penn State paper by Miller, Dachille and Roy, I also soon realized that the role of the

chlorine impurity was to allow the rocksalt structure to indeed be quenchable to one atmosphere and then in mixed-phase form be addressable by the low temperature studies which revealed an Meissner effect.

Notwithstanding the embryo of significant findings in these systems, the clearcut disinterest of many major institutions and the inability to produce a true high-T_C superconductor caused the rather modest channels of funding to come to an abrupt halt. Even a few scattered startling findings by Chu et al. and at The Pennsylvannia State University's Materials Research Laboratory (as well as in Europe, The Soviet Union, and Japan) failed to reverse the decision to virtually eliminate almost all sources of high-T_C funding. My own experimental work in high-T_C thus came to a temporary halt in late 1982.

In early 1985, an unusual irony of fate took place. I, who like many others, had grown accustomed to many years of famine-funding suddenly became in charge of a well-funded program in the area of opto-electronic physics related to signal processing. In February of 1985, I placed a telephone call to my old friend, C.W. Paul Chu, who I had only seen a few times since I was a senior undergraduate at Fordham College in New York City, where Paul Chu was a Masters candidate at that time working in superconductivity with Professor Joseph Budnick. At the <u>very same time</u> after all of these years Paul Chu was atempting to telephone <u>me!</u> Our lines became coincidentally cross-connected. My intention was to ask Paul if he were interested in studying the effect of high pressure on excitonic absorption in a GaAs-$Ga_{1-x}Al_xAs$ multiple quantum-well device (patterned after the design of Dr. D.A.B. Miller of Bell Telephone Laboratories and grown for us at Cornell University by Dr. Doran Smith of Fort Monmouth's Electronics Technology and Devices Laboratory). Paul, on the other hand, wanted to ask me if I were interested in participating in a program aimed at growing by Molecular Beam Epitaxy ultra-pure thin films of hard-to-grow materials in the ultra high vacuum of space remote to the US Space Shuttle (a program later to be funded by NASA as the Space Vacuum Epitaxy Center (SVEC) located at the University of

Houston and directed originally by Chu and presently by Dr. Alex Ignatiev). Both Paul and myself answered each other's questions in the affirmative!

In March of 1986, I traveled to the University of Houston to join other scientists for the purpose of assessing the feasibility of thin film growth in space. It was here where I had the wonderful pleasure of first meeting Arthur Freeman of Northwestern University and Carl Rau of Rice who have become in these two subsequent years two of my closest colleagues, both being then very open-minded about superconductivity at high-T_C.

For the next eight months, I travelled extensively throughout the United States to recruit first class scientists to participate in my opto-electronic physics basic research program, and eventually I funded over seventy such scientists at over fifty distinguished institutions. On the occasion of each of my visits I presented a seminar on the subject of Army needs in the areas of electro-optics and magneto-optics. Toward the end of each of these seminars I devoted about fifteen minutes or so to state the case for continued, even unfunded, research in high-T_C superconductivity. To assist me in this "plea", so to speak, I asked Paul Chu to join me on several of these trips, notably to the Francis Bitter National Magnet Laboratory at MIT under the invitation of Peter Wolf and August Witt, to the University of Michigan under the invitation of Palab Bhattacharya and Roy Clark, to the 3M Company as invited by Chuck Callo, and to the University of Minnesota under the invitation of Alan Goldman. I also asked my long-standing collaborator Peter Walsh of Fairleigh-Dickinson University and Dr. Mark Mentzer (an expert in electro-optics) to join me at these particular meetings.

During the spring and summer of 1986 it was becoming apparent to me, almost intuitively, that a breakthrough in high-T_C was on the horizon. A 1978 paper authored by J. A. Wilson (at that time working at Bell Laboratories) and published in The Philosophical Magazine had suggested the importance of the interface between two different polyhedral building blocks (in the same unit cell) to create large internal inhomogeneous fields and possibly lead to a exciton high-T_C

in a copper compound. Wilson cited interfaces between building block polyhedra of octahedral and tetrahedral symmetry. This was interestingly exactly what existed in the pressure-quenched mixed phase material in our earlier chlorinated CdS samples! (Recently at the March 1988 meeting of the American Physical Society in New Orleans, over a year after the announcements of high-T_C, Wilson presented an excellent extension of his arguments). In March of 1985, Bernard Raveau et al. from the University of Cannes published the now famous paper on the early lanthanum cuprates which Raveau, as a brillant crystal chemist, synthesized for the purpose of employing different polyhedral building blocks to achieve materials <u>of anisotropic thermal conductivity.</u> Scientists who may have had the good fortune to read both the Wilson and the Raveau paper were now armed with the ingredients necessary to complete the road to high-T_C! My preaching to pursue the high-T_C research regardless of funding thus intensified!

Finally under the invitation of Dr. Albert Feldman of the then National Bureau of Standards, the time arrived that I came to Caithersberg to discuss mutual interest in diamond particles and films (both grown from methane) for use as optical windows and hard semi-conductors. I again presented a seminar and again discussed high-T_C. Present in the audience was Dr. Bob Shull of the Bureau who told Dr. L. H. Bennett of my copper chloride disproportion-ation and cadmium sulfide mixed phase pressure-quench, both indicating a Meissner efffect near liquid nitrogen temperature. Dr. Bennett decided immediately to reanalyze the data related to the cuprous chloride system.

In December of 1986, I assembled an Investigator's Meeting at Picatinny Arsenal to bring together the seventy researchers who I had funded to participate in a 3-day conference. I invited Paul Chu to be the guest speaker at the Conference banquet scheduled for December 16th (my birthday) at the Parsippany Hilton Hotel. It was at this meeting where Paul Chu announced T_C in excess of 55K achieved in the La-Sr-Cu-O system at moderately high pressure. The audience conferees were overwhelmed! Doubt was of course mixed with elation. Some sneered and some marvelled. Within less than a month the data were confirmed at Argonne, Bell Laboratories, and MIT and Chu's superconductivity had by then even soared far above 77K into the 90's.

During subsequent months, Dr. Larry Bennett and myself kept in contact to exchange ideas and data. By June of 1987 I transferred to the US Army Materials Technology Laboratory (formerly Watertown Arsenal AMMRC) in Watertown, Massachusetts, to work with Dr. R. N. Katz and also accepted a Visiting Scientist position at The Massachusetts Institute of Technology (sponsored by August Witt) where I commenced J_C, Hall effect, and H_{C2} studies on $Y_1Ba_2Cu_3O_7$ assisted by B. M. Moon and Terence Burke, my graduate students from Rutgers (where I hold an Adjunct Professorship). In the Autumn of 1987 I invited Dr. Bennett to our laboratory at the Ceramics Research Branch to review some very interesting data obtained on the MIT SQUID magnetometer at very low magnetic field which was suggestive of a Néel Temperature of about 390K in $Y_1Ba_2Cu_3O_{6.9}$. It was at this meeting where Dr. Bennett and I cemented the foundations for the conference that is documented in these Proceedings. Soon afterwards a T_N of ~400K was reported by Tranquada et al. in $Y_1Ba_2Cu_3O_{6.15}$ using neutron diffraction. Dr. Bennett had also recently achieved some very fascinating and valuable research findings on the high-T_C superconductors (as related to magnetic interactions) at the Bureau. We both realized that the area of magnetic coupling in high-T_C materials needed unification, discussion, debate, and dialogue. Theories were beginning to emerge embodying magnons and spinons, holes in the oxygen, and the virtual spin on such holes. It was clear that the field of magnetic correlations in high-T_C materials was becoming a respected discipline but we believed the need for perhaps a year or so to sufficiently mature the field such that a major conference would be valuable to all. The present conference, therefore, was then planned, and came into being at The National Institute of Standards and Technology (formerly The National Bureau of Standards) in Gaithersberg, Maryland on October 11th to 13th 1988 and was co-sponsored by The National Aeronautics and Space Agency through the cooperation of Dr. Yury Flom. I am indebted to both Dr. Larry Bennett and Dr. Flom for their great effort toward making this conference a superb success.

Dr. Gary Vezzoli
Watertown, MA

WORKSHOP ON THE MATERIALS SCIENCE OF HIGH T_C SUPERCONDUCTORS: MAGNETIC INTERACTIONS

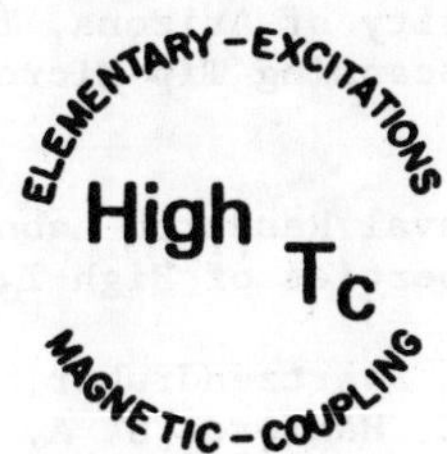

October 11–13, 1988

National Institute of Standards and Technology
Gaithersburg, MD 20899

TUESDAY, 11 October

CHAIR: L.H. BENNETT, National Institute of Standards and Technology, Gaithersburg, MD

8:30 **L.H. SCHWARTZ, Director, Institute for Materials Science and Engineering, NIST**
Welcome to the Workshop

8:45 **KEYNOTE SPEECH: W.A. Little, Stanford University**
Experimental Constraints on Theories of High T_C Superconductors

9:30 **S.-C. ZHANG, Institute for Theoretical Physics, UCSB, Santa Barbara, CA**
The Spin Bag Mechanism of High Temperature Superconductivity
(Coworkers: J.R. Schrieffer, X.-G. Wen)

10:00 **L. KRUSIN-ELBAUM, IBM, Yorktown Heights, NY**
Magnetic Penetration Depth and Lower Critical Fields in YBaCuO Single Crystals
(Coworkers: A.P. Malozemoff, R.L. Greene, Y. Yeshurun, F. Holtzberg)

10:30 ----------------------COFFEE BREAK----------------------------

CHAIR: G. VEZZOLI, Army Materials Technology Laboratory Watertown, MA and M.I.T., Cambridge, MA

11:00 R.E. COHEN, Naval Research Laboratory, Washington, DC
First-Principles Phonon Calculations for La_2CuO_4
(Coworkers: W.E. Pickett, H. Krakauer)

11:30 D. SARID, University of Arizona, Tucson, AZ
Applications of Scanning Tip Microscopy to Superconducting Technology

12:00 M. RUBINSTEIN, Naval Research Laboratory, Washington, DC
Normal State Properties of High T_C Compounds: Magnetism and Conductivity
(Coworkers: L. J. Swartzendruber, L. H. Bennett, T. K. Chaki, M. Z. Chaki, M. Z. Harford, S. A. Wolf, A. K. Singh, A. E. Edelstein)

Tuesday, 11 October

12:30 --------------------------LUNCH------------------------------

CHAIR: B.B. RATH, NRL, Washington, DC

2:00 C.M. VARMA, AT&T Bell Laboratories, Murray Hill, NJ
Issues and Problems in a Magnetic Mechanism for High Temperature Superconductivity

2:30 T. R. THURSTON, MIT, Cambridge, MA and Brookhaven National Laboratory, Upton, NY
Neutron Scattering Studies on $La_{2-x}Sr_xCuO_4$ Single Crystals

3:00 S. BERKO, Brandeis University, Waltham, MA
Is There a Fermi Surface in High T_c Superconductors? --The present Status of Positron Experiments.

3:30 A. MORRISH, University of Manitoba, Winnipeg, Canada
^{57}Fe Mössbauer Study of a Tl-Ca-Ba-Cu-O High T_C Superconductor
(Coworkers: X.Z. Zhou, Y.L. Luo)

4:00 ----------------------COFFEE BREAK--------------------------

CHAIR: R.D. SHULL, NIST, Gaithersburg, MD

4:30 G. VEZZOLI, Army Materials Technology Laboratory, Watertown, MA and M.I.T., Cambridge, MA
Role of Substituted Rare Earths in High T_C $Y_1Ba_2Cu_3O_{7-\delta}$ Superconductors: Polarizing of Fluctuations from

Antiferromagnetism; Breakdown of Cooper Pairing at Close Proximity; Hall Effect
(Coworkers: B.M. Moon, B. Lalevic, A. Safari)

5:00 C.-P.S. WANG, University of Maryland, College Park, MD
Magnetic Interactions in High-T_C Superconductors

5:30 D. TANNER, University of Florida, Gainsville, FL
Optical Excitations in Thin Film $YBa_2Cu_3O_7$
(Coworkers: K. Kamaras, S.L. Herr, C.D. Porter, S. Etemad, S.W. Chan)

5:50 M.B. SALAMON, University of Illinois, Urbana, IL
(Present address: NIST, Gaithersburg, MD)
Magnetic-Field Effects and Fluctuations at T_C in $YBa_2Cu_3O_{7-\delta}$
(Coworkers: S.E. Inderhees, M. Howson, D.M. Ginsberg, J.P. Rice, B.G. Pazol)

6:30 BUS TO MARRIOTT

7:00 Cash Bar - Marriott

8:00 BANQUET - Marriott
After-dinner speaker - **JACOB RABINOW**
"Is Invention an Art Form?"

WEDNESDAY, 12 October

CHAIR: A.J. FREEMAN, Northwestern University, Evanston, IL

8:30 J. B. GOODENOUGH, University of Texas, Austin, TX
Mechanism in High-T_C Superconductivity

9:00 J. J. RHYNE, JR., NIST, Gaithersburg, MD
Atomic Site Substitutions and Magnetism in 1-2-3 Superconductors
(Coworkers: J.K. Stalick, D.A. Neumann, C.L. Chien, G. Xiao, M.Z. Cieplak, D. Musser, A. Gavrin, F.H. Streitz, P.F. Miceli, J.M. Tarascon)

9:30 J. CROW, Temple University, Philadelphia, PA
Effect of Rare Earth and Transition Metal Substitutions on the Superconductivity of 1-2-3

9:50 B. R. COOPER, University of West Virginia, WV
Magnetism vs. High Temperature Superconductivity - The Role of Competing Hybridization

10:10 P. J. WALSH, Fairleigh Dickenson University, Teaneck, NJ
Analysis of Microwave Surface Resistance in High T_C Superconductors

10:30 --------------------COFFEE BREAK--------------------

CHAIR: L. COHEN, Institute for Defense Analyses, Alexandria, VA

11:00 K. MOORJANI, Johns Hopkins University Applied Physics Laboratory
Magnetically Modulated Microwave Absorption Method for Studying Superconductors
(Coworkers: B.F. Kim, F.J. Adrian and J. Bohandy)

11:30 C. VITTORIA, Northeastern University, Boston, MA
Low Field Dependence of the Microwave Absorption in Superconducting $YBa_2Cu_3O_{7-x}$ near T_C
(Coworkers: R. Karim, S. Oliver)

12:00 V. Z. KRESIN, Lawrence Berkeley Laboratory, Berkeley, CA
Normal Properties of the Cuprates and Their Influence on the Mechanisms of High T_C

12:30 --------------------LUNCH--------------------

CHAIR: D. BUTRYMOWICZ, NIST, Gaithersburg, MD

2:00 J. RUVALDS, University of Virginia, Charlottesville, VA
Why Are Some Cu-O Planes Favorable for Superconductivity?

2:30 R. C. O'HANDLEY, MIT, Cambridge, MA
Flux Flow, Pinning and Critical Currents in High T_c Superconductors

Wednesday, 12 October

3:00 J. GENOSSAR, Technion, Haifa, Israel
The Normal State Resistivity in $YBa_2Cu_3O_x$: A Master Curve
(Coworkers: B. Fisher, I.O. Lelong, J. Ashkenazi, L. Patlegan)

3:20 F. ADRIAN, Johns Hopkins University Applied Physics Laboratory, Laurel, MD
Crystal-Field-Model Interpretation of Magnetic Resonance Data in High T_c Superconductors

3:40 L. H. BENNETT, NIST, Gaithersburg, MD
Magnetic Properties of a Chemically Synthesized Bi(Pb)SrCaCuO Superconductor
(Coworkers: D. Lundy, J. Ritter, L.J. Swartzendruber, R.D. Shull)

4:00 --------------------COFFEE BREAK--------------------------

CHAIR: Y. FLOM, NASA Goddard Space Flight Center, Greenbelt, MD

4:30 A. J. FREEMAN, Northwestern University, Evanston, IL
Electronic Structure and Excitonic Superconductivity in the High T_c Copper Oxides

5:00 J.I. BUDNICK, University of Connecticut, Storrs, CT
Muon Spin Rotation Studies (μSR) of Magnetic Ordering and Magnetic Interactions in High T_C Superconducting Oxide Systems
(Coworkers: B. Chamberland, M. Filipkowski, Z. Tan, R. O'Connor, G. German and X. Ling)

5:30 S. R. BREON, NASA-Goddard Space Flight Center, Beltsville, MD
High Temperature Superconductor Applications in Space
(Coworkers: S. H. Castios and Y. Flom)

6:00 R. D. Shull, NIST, Gaithersburg, MD
RAPPORTEUR for Posters and Demonstrations

6:45 BUS TO MARRIOTT

THURSDAY, 13 October

<u>Posters, Demonstrations, Tours, Panels</u>

POSTERS AND DEMONSTRATIONS
Chair: R.D. SHULL, NIST, Gaithersburg, MD

11:00-12:30 and 1:30-2:30

1. T. BURKE, Rutgers University, New Burnswick, NJ
Critical Current vs. Temperature in the High-T_c $Y_1Ba_2Cu_3O_7$-Superconductor
(Coworkers: G.C. Vezzoli, M. Moon, B. Lalevic, A. Saffari, A. Sundar)

2. A. DeREGGI, NIST, Gaithersburg, MD
Magnetic Measurements on Polymer-High T_c Superconductor Composites
(Coworker: C.K. Chiang, L.J. Swartzendruber, G.T. Davis)

3. N. G. EROR, Oregon Graduate Center, Beaverton, OR
(1) Synthesis of High Temperature Superconducting Y-Ba-Cu-O Powders
(2) Shock Compression Fabrication of High Temperature (Y-Ba-Cu-O) Superconductor/Metal Composite Monoliths
DEMONSTRATION

4. Y. FLOM, NASA Goddard Space Flight Center, Greenbelt, MD
Requirements for High Temperature Superconductors in Certain Space Applications
(Coworkers: S.R. Breon, S.H. Castles, J. Brasunas)

5. C. GALLO, Superconix, Inc., St. Paul, MN
Measurement of Magnetic Properties in a Melt Cast Bi-Ca-Sr-Cu-O Superconductor
(Coworkers: L.J. Swartzendruber and L. H. Bennett)

6. D. GAZIT, Center for Materials Research and Stanford University, Stanford, CA
The Growth of Very Thin High Temperature Superconducting Wires
(Coworker: R. S. Feigelson)

7. A. GOULD, University of Maryland, College Park, MD
Microwave Absorption, DC Magnetization, and AC Susceptibility of High Temperature Y, Bi and Tl Based Superconductors
(Coworkers: E.M. Jackson, S.D. Murphy, G. Shaw, R. Crittenden, Z.Y. Li, A.M. Stewart, S.M. Bhagat)

8. R.H. ONO, NIST, Boulder, CO
Switching Noise in $YBa_2Cu_3O_x$ Microbridges
(Coworkers: J.A. Beall, M.W. Cromar, W. Skocpol, P.M. Mankiewich, R.E. Howard)

9. R. PETERSON, NIST, Boulder, CO
Flux Pinning and Critical Currents in High T_c Superconductors
(Coworker: J. W. Ekin)

10. R. B. POEPPEL, Argonne National Laboratory, Argonne, IL
A Meissner Motor Using Yttrium Barium Copper Oxide High-T_c Superconductor
DEMONSTRATION

11. J. RITTER, NIST, Gaithersburg, MD
The Hydroxycarbonate Precursor as an Approach to $YBa_2Cu_3O_{7-x}$ High T_c Superconductor Materials with High Transport Current Densities
(Coworkers: J.F. Kelly and S. A. Soulen)

12. R. STOCKBAUER, NIST, Gaithersburg, MD
Studies of Electronic Interactions and Material Degradation of High Temperature Superconductors
(Coworkers: R.L. Kurtz, D. Mueller, A. Shih, L.E. Toth, M. Osofsky, S.A. Wolf)

13. E. VENTURINI, Sandia National Laboratory, Alburquerque, NM
Magnetization, Transport and Structural Studies of Tl-Ca-Ba-Cu-O Single Crystals, Polycrystalline Thin Films and Bulk Ceramics
(Coworkers: J.F. Kwak, D.S. Ginley, B. Morosin, R.J. Baughman)

14. M. K. WU, University of Alabama, Huntsville, AL
Magnetic Levitation by Attraction in Some High T_c Superconductors
(Coworkers: R.D. Shull, L.J. Swartzendruber, C.K. Chiang, P.N. Peters, C.Y. Huang)
DEMONSTRATION

PANELS

Chair: L.J. Swartzendruber, NIST, Gaithersburg, MD

9:00 - 10:30

IA SPACE APPLICATIONS
Moderator: Y. FLOM, NASA Goddard Space Flight Center, Greenbelt, MD

IIA THEORY - ELEMENTARY EXCITATIONS
Moderator: J. ASHKENAZI, University of Miami, Coral Gables, FL

IIIA HIGHER T_C?
Moderators: E. VENTURINI, Sandia National Laboratory, Albuquerque, NM,
AND S. BHAGAT, University of Maryland, College Park, MD

IVA MEASUREMENT METHODS
Moderator: N. DALAL, West Virginia University, Morgantown, WV

PANELS
11:00 - 12:30

IB THIN FILM APPLICATIONS
Moderator: J. BRASUNAS, NASA Goddard Space Flight Center, Greenbelt, MD

IIB THEORY - CRITICAL EXPERIMENTS
Moderator: Z. TESANOVIC, Johns Hopkins University, Baltimore, MD

IIIB PATENTS
Moderator: G. GOLDBERG, Patent and Trademark Office, Crystal City, VA

IVB SUBSTITUTIONS OF MAGNETIC IONS (TM AND RE)
Moderator: J. J. RHYNE, JR., NIST, Gaithersburg, MD

TABLE OF CONTENTS

EXPERIMENTAL AND APPLICATIONS

PANEL SUMMARIES

CONCLUDING REMARKS

KEYNOTE ADDRESS

EXPERIMENTAL CONSTRAINTS ON THEORIES OF HIGH T_C SUPERCONDUCTORS

W.A. Little
Physics Department
Stanford University, Stanford, CA 94305

Recent experiments on the high T_c superconductors have begun to narrow the possible theoretical explanations of the phenomenon. Experimental evidence on the size, structure and symmetry of the charge carriers will be reviewed; evidence for and against strong coupling; and, recent results on a search for direct evidence of magnetic signature in the coupling mechanism will be presented. We will show how these experiments impose strong constraints on the theories of these superconductors. A new type of experiment will also be discussed which appears capable of identifying the true nature of the coupling mechanism if the superconductors prove to be BCS-like in nature.

The essence of the lecture can be found in the paper "Experimental Constraints on Theories of High-Transition Temperature Superconductors", Science 242, 1390 (1988).

ELEMENTARY EXCITATIONS

MAGNETISM VERSUS HIGH TEMPERATURE SUPERCONDUCTIVITY - THE ROLE OF COMPETING HYBRIDIZATION

BERNARD R. COOPER
Department of Physics, West Virginia University
Morgantown, WV 26506

ABSTRACT

Recent experiments point to the possibility of competing hybridization mechanisms determining the choice between magnetic and superconducting behavior for 1-2-3 copper-oxide-type materials that can show high-temperature superconductivity. Experiments on $Pr_xY_{1-x}Ba_2Cu_3O_7$ show the disappearance of superconductivity and the appearance of Pr magnetic ordering at somewhat over 60% Pr. The crystal structure remains orthorhombic throughout. For 100% Pr the magnetic ordering temperature is about 17K and the ordered moment at low temperature is quite small, about $0.25\mu_B$ per Pr. We address the question, "*what does this say about the mechanism giving superconductivity and the mechanism determining the choice between superconductivity versus magnetic ordering behavior?*" Our working hypothesis is that the essential feature governing the magnetic and superconducting behavior of the copper-oxide-type systems is a cooperative valence fluctuation effect of the copper ions mediated through hybridization effects dominated by the oxygen p electrons. Then our picture of Pr [which could be either $Pr^{4+}(f^1)$ or $Pr^{3+}(f^2)$] as another hybridizer competing with the copper could give a strong change in the hybridization-mediated copper-copper interactions, killing superconductivity with increasing Pr concentration. Therefore we discuss whether *orbitally-driven* praseodymium-praseodymium magnetic coupling (resulting from the hybridization of Pr f electrons with oxygen-p-derived band electrons) can explain the high magnetic ordering temperature of the 100% Pr system and the low-ordered moment. On the basis of knowledge gained over the past decade for the behavior of correlated f electron systems where cooperative hybridization of the quasi-ionic correlated f electrons with band electrons causes a variety of unusual highly anisotropic orbitally-driven magnetic behavior, we conclude that the observed magnetic behavior indicates an inherently extraordinarily strong hybridization; and we suggest further work to elucidate our picture.

1. INTRODUCTION AND DESCRIPTION OF EXPERIMENTAL BEHAVIOR

The key features of experiments substituting Pr for Y in 1-2-3 high T_c superconductors, i.e., in $Pr_xY_{1-x}Ba_2Cu_3O_7$ are as follows:[1-4]

(1) Superconductivity disappears at somewhat over 60% Pr.

(2) For Pr concentrations beyond the disappearance of superconductivity, the Pr orders magnetically.

(3) The crystal structure remains orthorhombic at all Pr concentrations up to 100%.

(4) For 100% Pr ($PrBa_2Cu_3O_7$) magnetic ordering occurs at about 17K. (The Gd magnetic ordering temperature in $GdBa_2Cu_3O_7$ is only 2.2K, and that for Dy, Er and Nd is even lower.)

(5) Both neutron diffraction and entropy removal as measured by specific heat measurements indicates a very small moment on Pr in $PrBa_2Cu_3O_7$. Neutron diffraction gives a moment of $0.25\mu_B$. Interestingly the magnetic structure indicated by the neutron diffraction is the same as for $GdBa_2Cu_3O_7$.

(6) In the oxygen stoichiometry range where copper orders magnetically at high temperature (and the crystal structure is tetragonal), for 100% Pr substitution the magnetic ordering temperature of the Pr appears to be lowered from 17K to 12K.

What does this say about the mechanism giving superconductivity and the mechanism determining the choice between superconductivity versus magnetic ordering behavior? *The behavior listed above points to the existence of competing hybridization mechanisms as one substitutes Pr for Y.* My working hypothesis is that the essential feature governing the magnetic and superconducting behavior of the copper-oxide-type systems is a cooperative valence fluctuation effect of the copper ions mediated through hybridization effects dominated by the oxygen p electrons. Basically the role of hybridization then is to provide very strong Cu-O-Cu coupling, generically similar to the role of a very strong superexchange. However, there is a very important difference from superexchange and the consequent picture of a magnetic pairing mechanism. Here the magnetic coupling is between *orbital* moments, and there is no real distinction between charge effects and magnetic moment effects -- i.e., *charge shaping and moment formation are the same thing.*[5-7] We can see whether our picture of $Pr^{4+}(f^1)$ or $Pr^{3+}(f^2)$ as another hybridizer competing with the copper gives a strong change in the hybridization-mediated copper-copper interactions, and whether orbitally-driven praseodymium-praseodymium coupling explains the strong magnetic ordering of the praseodymium system.

The question of whether the Pr is $Pr^{4+}(f^1)$ [analagous to $Ce^{3+}(f^1)$] or $Pr^{3+}(f^2)$ is not central to the existence of the mechanisms discussed in this paper, but would only effect quantitative details. The susceptibility behavior above the Pr Neel temperature provides[1] an effective moment much closer to that of Pr^{4+} than to that of Pr^{3+}, but chemical arguments[8] favor Pr^{3+}. The most important difference between Pr^{4+} and Pr^{3+} is that for Pr^{3+} the single-ion crystal-field ground state can be a singlet, and this can provide a moment reduction mechanism in addition to the hybridization-driven ionic state mixing mechanism discussed below.

2. BEHAVIOR EXPECTED FOR HYBRIDIZATION-MEDIATED ORBITALLY-DRIVEN MAGNETISM

Over the past decade, we have developed phenomenological theory for the behavior of "well-ordered" magnetic states of moderately delocalized partially-filled transition-shell systems, characteristically obtaining unusual *anisotropic magnetism* in agreement with experiment.[5-7] The key element of that theory is the treatment of hybridization by transforming the physical mixing to resonant scattering of band electrons off partially-delocalized transition-shell electrons.[9] This resonant scattering corresponds to a virtual configuration transition for the transition-shell ion; and magnetic ordering arises when these virtual configuration transitions become cooperative.

The crucial point of the phenomenology is that partially-delocalized transition-shell systems have *orbitally-driven magnetism.*[5-7] The dominant magnetic coupling is between orbital moments via transition-shell ion electron -- band electron hybridization (mixing). *This gives a parallel channel to conventional spin-driven magnetism, and this orbitally-driven channel dominates the magnetic behavior as the transition-shell electrons move from localized (gadolinium-like) behavior to partially delocalized (cerium-like) behavior. This orbitally-driven magnetism can give magnetic ordering temperatures much higher than expected on the basis of conventionally spin-driven magnetism.* If one pictures the interaction between two cooperatively hybridizing $Pr^{4+}(f^1)$ ions, the f charge clouds wants to take on the shape of elongated discs pointing along the interionic axis, and thus with orbital moments aligned parallel to each other and perpendicular to the interionic axis. It is this cooperative combined charge cloud shaping and orbital moment alignment that gives the strongly anisotropic magnetic behavior. Furthermore one can expect sensitivity to detailed chemical environment once the close relationship between f-electron charge shaping and magnetic behavior is recognized.

The magnetic ordering behavior of the rare earth monopnictides of NaCl-structure, as shown[5,6] in Figure 1, provides an excellent example of the way orbitally-driven magnetism can give magnetic ordering temperatures much higher than expected on the basis of conventionally spin-driven magnetism. Figure 1 shows the experimental magnetic ordering temperatures of the rare earth monopnictides. For comparison we also show a curve giving S(S+l) dependence normalized at gadolinium as might be expected to hold true for conventional spin-driven magnetic interactions. [Crystal-field effects, or exchange dependent on an order of spin higher than bilinear could cause the dependence to fall off faster than S(S+l).] As can be seen from Fig. 1, the rare earths heavier than gadolinium follow expectations for normal spin-driven behavior well; however, *the cerium and neodymium compounds depart greatly from the S(S+l) curve and, especially for the antimony and bismuth compounds, there is clearly no correlation with the change in S(S+l) on going from gadolinium to neodymium to cerium. This is because the magnetic interactions are orbitally driven. That is, the band electrons in acting as a coupling medium interact via hybridization with the orbital moments on two rare earth ions and in effect partially lock together the orbital moment alignment of those two ions. The spin moments are only pulled along parisitically via spin-orbit coupling. It is important to emphasize that the orbitally-driven interactions represent a separate magnetic interaction mechanism that is additive to (in parallel with) the standard spin-driven Ruderman-Kittel-Kasuya-Yoshida (RKKY) interaction. As the f-electron ionic charge cloud becomes more diffuse in the transition regime between localized and delocalized behavior, the orbitally-driven (hybridization-mediated) interactions become dominant.*

In Table 1, we show the S(S+l) values and Neel temperatures[1,10-13] for the Pr, Nd, Gd, Dy and Er 100%-substituted 1-2-3 compounds at the oxygen rich stoichiometry end where the superconducting critical temperatures are about 90K. The Neel temperature of $PrBa_2Cu_3O_7$ is about 160 times the expected value for spin-driven magnetic coupling. (This ratio is for $Pr^{4+}(f^1)$. If the Pr is $Pr^{3+}(f^2)$, then the Neel temperature is still about 61 times the

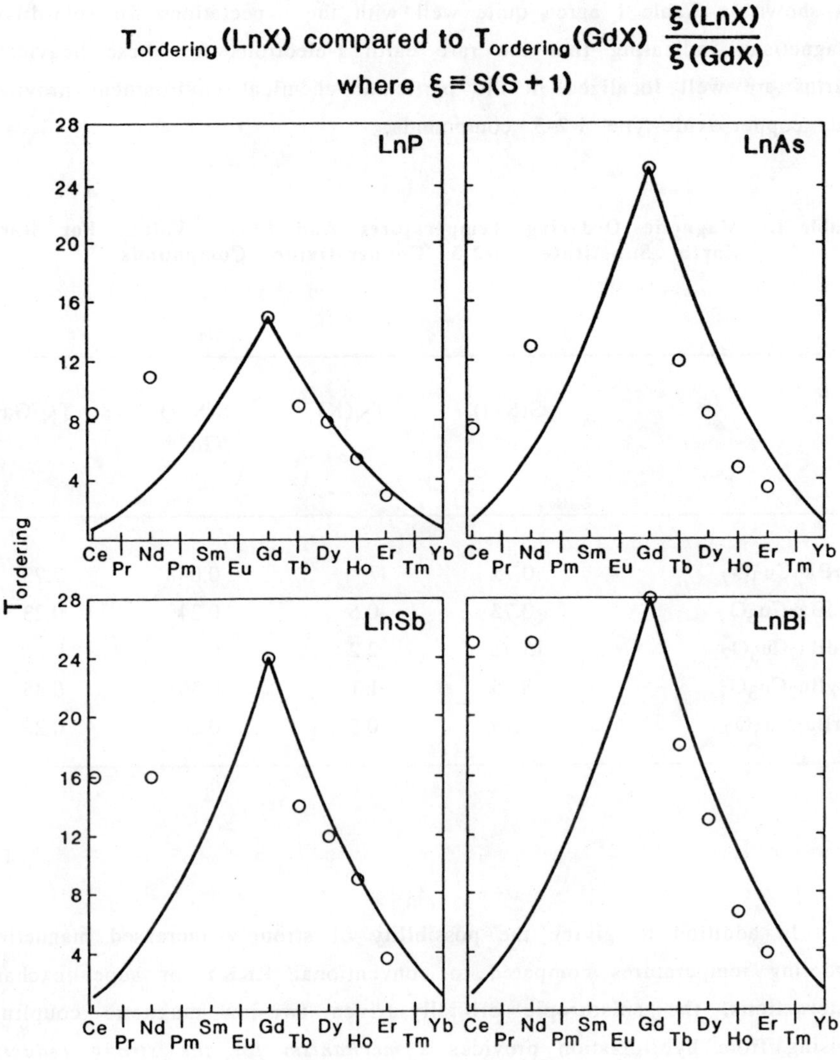

Tordering (LnX) compared to Tordering (GdX) ξ (LnX) / ξ (GdX)
where ξ ≡ S(S+1)
LnP
LnAs
LnSb
LnBi
Tordering
28
24
20
16
12
8
4
Ce Pr Nd Pm Sm Eu Gd Tb Dy Ho Er Tm Yb

Fig. 1

expected value for spin-driven magnetic coupling.) On the other hand, the magnetic ordering behavior for the rare earths heavier than praseodymium as shown in Table 1 agree quite well with the expectations for spin-driven magnetism, indicating that the rare earth f-electrons for these heavier rare earths are well localized in the particular chemical environment provided by the copper-oxide-type 1-2-3 compounds.

Table 1: Magnetic Ordering Temperatures And S(S+1) Values For Rare Earth Substituted 1-2-3 Copper-Oxide Compounds

	S(S+1)	T_N(K)	S(S+1) /Gd^{3+}	T_N/Gd^{3+}
$PrBa_2Cu_3O_7$	0.75	17	0.048	7.73
$NdBa_2Cu_3O_7$	3.75	0.5	0.24	0.23
$GdBa_2Cu_3O_7$	15.75	2.2	1	1
$DyBa_2Cu_3O_7$	8.75	1.0	0.56	0.45
$ErBa_2Cu_3O_7$	3.75	0.5	0.24	0.23

In addition to giving the possibility of strongly increased magnetic-ordering temperatures compared to conventional RKKY or super exchange interactions, the anisotropic orbitally-driven two-ion magnetic coupling arising from hybridization provides a *mechanism for the strong reduction, or indeed total elimination, of rare earth ionic magnetic moments.* This is because the strong coupling of rare earth ions giving the high magnetic ordering temperatures results from each rare earth ion contributing part of its f-electron spectral weight to increasing the strength of coupling, i.e., *as the moments couple more strongly, they are being destroyed.* From our calculations at this point, it appears that the peak value of magnetic ordering temperature occurs before there is significant destruction of magnetic moment with increasing hybridization. In the particular formalism we use to treat magnetic ordering for hybridization-mediated systems, the reduction of ionic magnetic moments comes via state mixing, i.e., of transition-shell ionic states.

Our theory and methodology for dealing with the phenomenology of orbitally-driven magnetic ordering has been developed using resonant scattering type concepts and technique[9,14] based on earlier work of Coqblin and Schrieffer[15] and of Schrieffer and Wolff.[16] In the past four or five years we have placed this theory on an absolute (materially-predictive) basis by performing an absolute evaluation of the Anderson lattice Hamiltonian parameters by use of a non-conventional first principles self-consistent electronic structure calculational technique.[17,18] To give some idea of the predictive power of this technique, I want to briefly describe how we were able to explain[18] the dramatic change in magnetic behavior on going from CeSb to CeTe. Both of these materials have NaCl-structure, and there is very little change in lattice parameter. The change from Sb to Te involves adding one valence p electron. CeSb is an unusual highly anisotropic but *well-ordered* antiferromagnet with T_N=16.8K and an ordered moment of 2.06μ_B (close to the Ce^{3+} free-ion saturation value); while CeTe has a low T_N of 2.4K and very small ordered moment of 0.20μ_B, and indeed is thought to be possibly an incipient heavy fermion system. Our theory predicts such behavior, and we explain the change from CeSb to CeTe through the increase in hybridization corresponding to a greatly increased d-electron density of states at the Fermi energy. Adding a p electron causes the Fermi energy for CeTe to move into a region of greatly enhanced d-electron density of states (increase from 1.46 to 11.43 d-electron states per Ry on going from CeSb to CeTe). When we incorporate the corresponding absolute evaluation of parameters into the phenomenological theory, we find that CeSb has a free energy at zero temperature associated with magnetic ordering (relative to a paramagnetic state) of -3.8K per ion and a magnetic moment close to the free-ion saturation value; while CeTe hovers at the borderline between magnetic ordering and nonmagnetic behavior, barely ordering or being nonmagnetic depending on the level of approximation used.

It is particularly impressive that $PrBa_2Cu_3O_7$ seems to combine a high magnetic ordering temperature (17K) with a strongly reduced Pr ordered moment ($0.25\mu_B$). *This implies an extraordinarily strong hybridization-mediated two-ion interaction if one extrapolates backwards to an ideal material having the peak value of magnetic ordering temperature associated with an almost saturated free-ion moment.*

3. SUGGESTIONS FOR EXPERIMENTAL WORK

In our past work, we have been able to understand a number of strongly anisotropic magnetic phenomena that are characteristic of the hybridizing (partially delocalized) regime of transition-shell electronic behavior. These include:

(1) Marked anisotropy in correlation lengths just above the ordering temperature (near-critical correlation behavior).

(2) Extreme anisotropy in the high field magnetization.

(3) Magnetic structural peculiarities.

(4) Unusual anisotropy and/or extreme damping in the magnetic excitation behavior.[19)]

(5) Large changes in the observed crystal-field splitting from values expected from the behavior of the isostructural heavy rare earth compounds.

(6) Giant magneto-optic effects.

Of these effects, I would especially emphasize the interest in item (4), especially the characteristic strong damping (broadening) associated with orbitally-driven (hybridization-dominated) magnetic behavior.[19)] [Interestingly, this remark was made at the Workshop at which this paper was presented. The following week, in a conversation in which I discussed my views with Dr. Lynda Soderholm of Argonne National Laboratory who has been performing neutron inelastic scattering studies, she informed me of results in experiments for $NdBa_2Cu_3O_7$ and $PrBa_2Cu_3O_7$. While sharp excitation levels were observed for the Nd material at 15K, for the Pr material there were no such sharp excitations.]

As a strong recommendation for experimental work, it would be interesting to do a comparative theoretical/experimental study in 1-2-3 materials where the oxygen stoichiometry is such that the copper lattice orders magnetically,[20] and to follow the praseodymium and copper magnetic ordering changes simultaneously as the stoichiometry is varied. An important possibility is that because of nonlinear hybridization effects associated with magnetic ordering, the band structure entering into the hybridization is apt to be quite different from that expected using conventional electronic structure theory techniques.

ACKNOWLEDGMENTS: This research has been supported by the U.S. Department of Energy through Grant No. DE-FG05-84-ER45134. I benefited greatly from ongoing descriptions of the experimental work provided by J.E. Crow.

REFERENCES

1. J.E. Crow, preceding talk this Workshop; C.S. Jee, A. Kebede, D. Nichols, J.E. Crow, T. Mihalism, G.H. Myer, I. Perez, R.E. Salomon, and P. Schlottmann, to be published (1988); W.H. Li, J.W. Lynn, S. Skanthakumar, T.W. Clinton, A. Kebede, C.S. Jee, J. Crow, and T. Mihalisin, to be published (1988).
2. L. Soderholm, K. Zhang, D.G. Hinks, M.A. Beno, J.D. Jorgensen, C.U. Segre, and I.K. Schuller, Nature *328*, 604 (1987).
3. J.K. Liang, X.T. Xu, S.S. Xie, G.H. Rao, X.Y. Shao, and Z.G. Duan, Z. Phys. B - Condensed Matter *69*, 137 (1987).
4. Y. Dalichaouch, M.S. Torikachvili, E.A. Early, B.W. Lee, C.L. Seaman, K.N. Yang, H. Zhou, and M.B. Maple, Solid State Commun. *65*, 1001 (1988).
5. B.R. Cooper, J. Less-Common Metals *133*, 31 (1987).
6. B.R. Cooper, G.J. Hu, N. Kioussis, and J.M. Wills, J. Magn. Magn. Mat. *63-64*, 121 (1987).
7. B.R. Cooper, J.M. Wills, N. Kioussis, and Q.G. Sheng, to appear in J. Appl. Phys. (1988), to appear in Les Editions de Physique-Colloques (1988).
8. L. Soderholm, private communication, October 20, 1988.

9. G.J. Hu, N. Kioussis, A. Banerjea, and B.R. Cooper, Phys. Rev. B *38*, 2639 (1988).
10. B.W. Lee, J.M. Ferreira, Y. Dalichaouch, M.S. Torachvili, K.N. Yang, and M.B. Maple, Phys. Rev. B *37*, 2368 (1988).
11. D. McK. Paul, H.A. Mook, A.W. Hewat, B.C. Sales, L.A. Boatner, J.R. Thompson, and M. Mostoller, Phys. Rev. B *37*, 2341 (1988).
12. J.W. Lynn, W.H. Li, Q. Li, H.C. Ku, H.D. Yang, and R.N. Shelton, Phys. Rev. B *36*, 2374 (1987).
13. A.I. Goldman, B.X. Yang, J. Tranquada, J.E. Crow, and C.S. Jee, Phys. Rev. B *36*, 7234 (1987).
14. B.R. Cooper, R. Siemann, D. Yang, P. Thayamballi, and A. Banerjea in *Handbook on the Physics and Chemistry of the Actinides*, edited by. A.J. Freeman and G.H. Lander (North-Holland, Amsterdam, 1985), Vol. 2, Chap. 6, pp. 435-500.
15. B. Coqblin and J.R. Schrieffer, Phys. Rev. *185*, 847 (1969).
16. J.R. Schrieffer and P.A. Wolff, Phys. Rev. *149*, 491 (1966).
17. J.M. Wills and B.R. Cooper, Phys. Rev. B *36*, 3809 (1987).
18. N. Kioussis, B.R. Cooper, and J.M. Wills, J. Appl. Phys. *63*, 3683 (1988).
19. G.J. Hu and B.R. Cooper, J. Appl. Phys. *63*, 3826 (1988) and to appear (1988); G.J. Hu, B.R. Cooper and G.H. Lander, to appear in Physica B (1988).
20. J.M. Tranquada, D.E. Cox, W. Kunnmann, H. Moudden, G. Shirane, M. Suenaga, P. Zolliker, D. Vaknin, S.K. Sinha, M.S. Alvarez, A.J. Jacobson, and D.C. Johnston, Phys. Rev. Lett. *60*, 156 (1988).

LOCAL DENSITY ELECTRONIC STRUCTURE AND EXCITONIC SUPERCONDUCTIVITY IN THE HIGH T_c Cu-OXIDES

A. J. Freeman, Jaejun Yu and S. Massidda
Department of Physics and Astronomy
Northwestern University, Evanston, IL 60208, U.S.A.

ABSTRACT

Results of highly precise all-electron local density calculations of the electronic structure for the high T_c Cu-oxide superconductors are presented and the confirmation of these predictions, especially by positron annilihation and recent low temperature UPS experiments, which substantiate the Fermi liquid nature of these metals, are reviewed. Common features of all the four different materials are described and evaluated to indicate the possible role of an excitonic mechanism for the observed high T_c.

I. INTRODUCTION

As is clear to all at this symposium, the discovery of the high T_c superconductors $La_{2-x}M_xCuO_4$[1] and $YBa_2Cu_3O_{7-\delta}$[2] has generated excitement among scientists and technologists on an unprecedented scale. The recent discoveries[3-5] of superconductivity above 85 K in Bi-Sr-Ca-Cu-O and above 120 K in Tl-Ba-Ca-Cu-O, which do not have a rare-earth element, have added a new dimension to the important subject of high T_c. A particularly exciting aspect of having added a third and fourth oxide superconducting material lies in the opportunity for seeking out common features in all four materials which may be relevant to determining the mechanism of their high T_c. One of the starting points is certainly a detailed picture of the electronic structure of the compound, a goal which is achievable by present day supercomputers in combination with highly precise numerical methods to solve the local density functional (LDF) Kohn-Sham equations in a self-consistent way. Even today, the origin of superconductivity in the new metallic oxides remains a challenge despite some intriguing hints obtained from experiment and electronic structure calculations. Detailed high resolution LDF band structure results have served to demonstrate what has been our major emphasis, namely the close relation of the physics (band structure) and chemistry (bonds and valences) to the structural arrangements of the constituent atoms; they may also provide insight into the basic mechanism of their superconductivity. The successes of these LDF studies include excellent agreement of their predictions of their anisotropic Fermi surface[6], and transport and thermopower properties[7-8] with experiment.

For the electronic structure calculations, we used the highly precise full-potential linearized augmented plane wave method (FLAPW)[9-10] within the local density approximation and the Hedin-Lundqvist form for the exchange correlation potential. In the FLAPW approach no shape approximations are made to either the charge density or the potential. Results obtained on the systems we have studied - La_2CuO_4, $YBa_2Cu_3O_7$, $GdBa_2Cu_3O_7$, $Bi_2Sr_2CaCu_2O_8$, $Tl_2Ba_2CaCu_2O_8$ and $Tl_2Ba_2Ca_2Cu_3O_{10}$ - indicate a number of common chemical and physical features, especially the role of intercalated layers such as the CuO chains, and Bi_2O_2 and Tl_2O_2 rock-salt type layers. In this paper, we provide a brief summary of the results on the detailed electronic structures of the $La_{2-x}M_xCuO_4$, $RBa_2Cu_3O_{7-\delta}$ (R = Y and Gd), $Bi_2Sr_2CaCu_2O_8$ and $Tl_2Ba_2CaCu_2O_8$ ($Tl_2Ba_2Ca_2Cu_3O_{10}$) systems, compare them, and point out their relations to an excitonic mechanism of high T_c superconductivity.

II. ELECTRONIC STRUCTURE OF THE NORMAL METALLIC STATES

A. $La_{2-x}M_xCuO_4$

Early on, the results of our highly precise all-electron local density full potential linearized augmented plane wave[9-10] (FLAPW) calculations of the energy band structure, charge densities, Fermi surface, etc., for $La_{2-x}M_xCuO_4$ (M = Sr, Ba)[11-14] demonstrated: (i) that the material consisted of metallic Cu-O(1) planes separated by insulating (dielectric) La-O(2) planes and (ii) that this 2D character and alternating metal/insulator planes would have, as some of their most important consequences, strongly anisotropic (transport, magnetic, etc.) properties. Thus, the calculated band structure along high symmetry directions in the Brillouin zone shows only flat bands, i.e., almost no dispersion, along the c axis, demonstrating that the interactions between the Cu, O(2) and La atoms are quite weak. However, along the basal plane directions there are very strong interactions between the Cu-O(1) atoms leading to large dispersions and a very wide bandwidth (~9 eV).

The band structure near E_F has a number of interesting features[11-12]. What is especially striking is that, in contrast to the complexity of its structure, only a single free electron-like band crosses E_F and gives rise to a simple Fermi surface[13]. Since this band originates from the Cu $d_{x^2-y^2}$-O(1) $p_{x,y}$ orbitals confined within the Cu-O(1) layer, it exhibits clearly all the characteristics of a two dimensional electron system. Particularly striking is the occurrence of a van Hove saddle point singularity (SPS). Such an SPS is expected, and found to contribute strongly, via a singular feature, to the density of states (DOS). Interestingly, the variation of T_c with x, in metallic $La_{2-x}Sr_xCuO_4$ shown esperimentally by Torrance et al.[15] is very consistent with the variation of the DOS at E_F on x, $N(E_F; x)$, and can be explained by our calculations[12]. This dominance of the DOS near E_F by the SPS contribution is responsible for many of the striking properties of this material with divalent ion (M_x) additions (including variations of T_c and other properties).

B. $YBa_2Cu_3O_{7-\delta}$

For the 90K superconductor $YBa_2Cu_3O_{7-\delta}$, discovered by Chu et al.[2], we presented[16-17] detailed high resolution results on the electronic band structure and density of states derived properties as obtained from the same highly precise state-of-the-art local density approach. These results demonstrated the close relation of the band structure to the structural arrangements of the constituent atoms and have helped to provide an integrated chemical and physical picture of the interactions.

The important structural features of the $YBa_2Cu_3O_{7-\delta}$ compounds arise from the fact that $(2+\delta)$ oxygen atoms are missing from the perfect triple perovskite, $YCuO_3(BaCuO_3)_2$. The O vacancies in the Cu plane (between two BaO planes) give rise to the formation of a linear chain of Cu and O ions (labelled Cu1-O1-Cu1). The total absence of O ions in the Y plane leads to the two Cu ions (called Cu2) in five-coordinated positions - as shown in Fig. 1. The double layers of Cu2-O planes in $YBa_2Cu_3O_7$ yield a 2D structure, corresponding to the single CuO_2 plane in $La_{2-x}M_xCuO_4$.

The calculated band structure of stoichiometric $YBa_2Cu_3O_7$ along high symmetry directions in the bottom ($k_z = 0$) plane of the orthorhombic BZ is shown in Fig. 2. The very close similarity in the band structure for the $k_z = 0$ and $k_z = \pi/c$ planes[16] indicates the highly 2D nature of the band structure. It is seen from Fig. 2, that as in the case of La_2CuO_4, a remarkably simple band structure near E_F emerges from this complex set of 36 bands (originating from three Cu (3d) and seven O (2p) atoms). Four bands - two consisting of Cu2(3d)-O2(p)-O3(p) orbitals and two consisting of Cu1(d)-O1(p)-O4(p) orbitals - cross E_F. Two strongly dispersed bands C (S_1' and S_4 in Fig. 2; the labelling is given by their character at S) consist of Cu2($d_{x^2-y^2}$)-O2(p_x)-O3(p_y) combinations and have the 2D character which proved so important for the properties of $La_{2-x}M_xCuO_4$. Significantly, the Cu1($d_{z^2-y^2}$)-O1(p_y)-O4(p_z) anti-bonding band A (S_1 in Fig. 2) shows the (large) 1D dispersion expected from the Cu1-O1-Cu1 linear chains but is almost entirely unoccupied. This band is in sharp contrast to the π-bonding band B (formed from the Cu1 (d_{zy})-O1(p_z)-O4(p_y) orbitals) which is almost entirely occupied in the stoichiometric ($\delta = 0$) compound.

We have predicted the Fermi surfaces (FS) of $YBa_2Cu_3O_7$ determined from our band structure (c.f., Fig. 3). Two 2D Cu-O dpσ bands yield two rounded square FS's (C in Fig. 3) centered around S. These 2D FS show strong nesting features along (100) and (010) directions. In addition, the 1D electronic structure also gives a 1D FS (A in Fig. 3) with possible nesting features along the (010) direction. There are two additional hole pockets (A in Fig. 3) around Y(T) and S(R) which come from the flat dpπ bands at E_F discussed before. Our predictions of the FS for $YBa_2Cu_3O_7$ have been confirmed recently by positron annihilation experiments[6-7]. The dot-dashed lines in Fig. 3 correspond to the experimentally observed FS by Smedskjaer et al. It is important to note that confirmation of the FS results has significant impact on several theories (e.g., the so-called resonant valence band or RVB theory)[18] which deny the Fermi liquid nature of the normal ground state in the Cu-oxide superconductors. Significantly, P.W. Anderson has stated[19] that the proven existence of a Fermi surface would necessitate "withdrawal" of his RVB theory.

Another important confirmation of the Fermi liquid nature of the normal ground state was given recently by Arko et al.'s[20] low temperature photoemission (UPS) experiments. These experiments on single crystals of $EuBa_2Cu_3O_{6.7}$ showed that only when cleaved and measured at 20 K a stable Fermi edge (larger than that in Cu metal) appear - thus demonstrating metallic behavior in agreement with our band calculations[21]. In addition, these experiments provide evidence for both Cu-d and O-2p occupation at E_F as predicted by our calculations. This result is at variance with other theoretical models which assume either the dominance of Cu-3d or O-2p at E_F. Moreover, the predictions of transport properties (Hall coefficient and thermopower) of $YBa_2Cu_3O_7$ by Allen et al.[7,8] with the use of the LDF energy band results show good agreement with experiment[22]. These successful verifications of the predictions made by the LDF band theory confirms the Fermi liquid nature of the normal metallic ground state. These facts reinforce the use of the LDF ground state as a reasonable starting point for the investigation of the origin of the superconductivity in these materials.

Here too, charge density calculations[16-17] reflect the structural properties of the material. Charge density plots for the individual states near E_F demonstrate the 2D nature of Cu2-O2-O3 dpσ bands and the 1D nature of the Cu1-O1-O4 dpσ bands. The ionic Y (or R = rare earth) atoms act as electron donors and do not otherwise participate. Also, the partial DOS at E_F for Y give extremely low values for the conduction electrons (the same is true for Gd). These results give an immediate explanation for the observed[23] coexistence of the high T_c superconductivity and magnetic ordering in the $RBa_2Cu_3O_{7-\delta}$ structures. The lack of conduction electron density around the R-site[24] means that the unpaired rare-earth f-electrons are decoupled from the Cooper pairs (i.e., magnetic isolation) and so cannot pair-break.

However, there are several experimental observations[25-26] of the antiferromagnetic insulators of $YBa_2Cu_3O_6$ as well as La_2CuO_4. These lead to the question whether the antiferromagnetic insulating ground state can be described by a (charge-) spin-density wave state within a band picture. Such a failure of LDF is well known for the case of CoO[27], for example. Later on, there were several reports of a stable magnetic ground state being found (in a band picture), but most of them are not convincing. Although there are still unresolved problems as to the relation of the O vacancies to the anti-ferromagnetic ordering in $La_2CuO_{4-\delta}$ and $YBa_2Cu_3O_{6+x}$, there is now a large effort to overcome this short-coming of the LSDA (local spin-density approximation). We believe that this part of the phase diagram has no relation to the superconductivity observed for the metallic phases.

C. $Bi_2Sr_2CaCu_2O_8$

For the new high T_c superconductor Bi-Sr-Ca-Cu-O, we presented[28] results of a highly precise local density determination of the electronic structure (energy bands, densities of states, Fermi surface, and charge densities). As in the case of the other high T_c Cu-O superconductors, we found a relatively simple band structure near E_F and strongly anisotropic highly 2D properties. One of the interesting points in the Bi-Sr-Ca-Cu-O system is that the Bi-O planes contribute substantially to $N(E_F)$ and to the transport properties.

A proposed structure of $Bi_2Sr_2CaCu_2O_8$ by Sunshine et al.[29] shows the presence of two CuO_2 layers (separated by a Ca layer) and of rock-salt type Bi_2O_2 layers; the $(CuO_2){\cdot}Ca{\cdot}(CuO_2)$ layers are separated by single SrO layer from the Bi_2O_2 layers. It is striking that this new system has no rare-earth elements; instead, it has Bi atoms replacing those strongly electro-positive trivalent ions.

The calculated band structure of $Bi_2Sr_2CaCu_2O_8$ has many points in common with those of the other high T_c Cu-oxide compounds[11,16-17]: above a set of fully occupied bands (in this case 48) with predominant Cu d-O p character, we find a relatively simple band structure at E_F, which in this compound consists of only three bands crossing E_F. Two (almost degenerate) bands with strong Cu-O dpσ character cross E_F and have two dimensional character. They do not cross E_F at the midpoint of the Γ-Z direction because of the existence of the Bi-O band which also crosses E_F. Their quasi-degeneracy proves the weakness of the interplane interactions for these states.

At energies (mostly) above E_F we find a set of six bands corresponding to the antibonding hybrids of the p orbitals of the two Bi atoms in the unit cell with the O2 and O3 p states. These bands form electron pockets near the L point and at the midpoint between Γ and Z (which will be referred to as $\bar{M}$). Their dispersion across the BZ is quite different from that of the Cu-O dpσ bands, as a consequence of the different bonding character (ppσ versus dpσ) and local coordination (rock-salt versus perovskite-like). The doubly periodic dispersion of Bi-O ppσ bands can be understood on the basis of simple tight-binding arguments.

The total density of states at the Fermi level, $N(E_F)$, is 3.03 states/(eV-cell). Large contributions to $N(E_F)$ come from both the Cu-O1 and the Bi-O2 layers. The Bi-O2 contributions are from the pσ bands which create small electron pockets around L (and $\bar{M}$). Therefore (and significantly), both Cu-O1 and Bi-O2 layers provide conduction electrons in this material. This result contrasts with the case of $La_{2-x}M_xCuO_4$ and of $YBa_2Cu_3O_{7-\delta}$, where the cations do not contribute to $N(E_F)$ but give rise to conduction bands which lie 2-3 eV above E_F.

D. $Tl_2Ba_2Ca_{n-1}Cu_nO_x$

New high T_c superconductors of the Tl-Ba-Ca-Cu-O system have been discovered[5] and found to have two different but related superconducting phases[30-31], with compositions $Tl_2Ba_2CaCu_2O_8$ (which we refer to as "Tl/2212") and $Tl_2Ba_2Ca_2Cu_3O_{10}$ ("Tl/2223"), with T_c ~112 K and ~125 K, respectively. For both Tl/2212 and Tl/2223, we presented[32] results of highly precise local density calculations of the electronic structure. A relatively simple band structure is found near E_F and strong 2D properties are predicted - again as in the case of the other high T_c materials.

The crystal structures of Tl/2212 and Tl/2223 determined by Subramanian et al.[31] show essentially the same features as that of $Bi_2Sr_2CaCu_2O_8$. The structure of Tl/2212 consists of two CuO_2 layers (separated by a Ca layer) and of rock-salt type Tl_2O_2 layers, where the (CuO_2)-Ca-(CuO_2) layers are separated by single BaO layers from the Tl_2O_2 layers. Similarly, the Tl/2223 structure is related to the Tl/2212 structure by an addition of extra Ca and CuO_2 layers, where the (CuO_2)-Ca-(CuO_2)-Ca-(CuO_2) layers are separated by single BaO layers from the Tl_2O_2 layers.

The calculated energy bands of Tl/2212 (in an extended zone scheme) are shown in Fig. 4. These bands present, as one would expect, strong similarities with those of all the other high T_c Cu-oxide superconductors[11,13-14,16-17,27]. As in $Bi_2Sr_2CaCu_2O_8$, we have in Tl/2212 two Cu-O dpσ bands (one per Cu-O sheet) crossing E_F, while three Cu-O dpσ bands crossing E_F are present in the Tl/2223 compound.

Despite these common features, the Tl systems present some interesting new points. In both the Tl/2212 and Tl/2223 compounds, there exists the presence of electron pockets around the Γ and Z points. A careful analysis of the character of these states, however, reveals important differences with respect to the $Bi_2Sr_2CaCu_2O_8$ case. While the Bi-O bands at E_F in $Bi_2Sr_2CaCu_2O_8$ originate mainly from the in-plane ppσ Bi-O hybrid, the Tl-O bands at E_F in Tl/2212 and Tl/2223 are mostly from oxygen p states hybridized (anti-bonding) with the Tl orbitals. In fact, the major Tl 6s bands are located at about 7 eV below E_F.

We have found that the 6s electrons of the Tl ions in $Tl_2Ba_2CaCu_2O_8$ are covalently bonded to the out-of-plane oxygens, O2 and O3; similarly, the Bi p orbitals in $Bi_2Sr_2CaCu_2O_8$ form weak covalent bonds with the in-plane O2 oxygens. This result is in contrast to the case of the $La_{2-x}M_xCuO_4$ and $YBa_2Cu_3O_7$ systems where the presence of strongly electro-positive 3+ ions (e.g., La^{3+}, Y^{3+}) is essential. We shall see that this has a significant effect on the electronic structure and may be relevent to understanding the superconducting mechanism.

One of the significant effects of the strong hybridization of the Tl s (d_{z^2}) with O2 and O3 p_z states (discussed above) on the pDOS structure of Tl/2212 and Tl/2223 is the existence of a gap between the non-bonding $p_{x,y}$ bands of O2 and O3 and the anti-bonding Tl s(d_{z^2}) - O2 p_z bands. (This gap, ~2.1 eV wide in Tl/2212, is reduced to $\lesssim$1 eV in Tl/2223 as a consequence of a Madelung shift of the non-bonding O2 states.) These systems are therefore seen to realize alternating metal/semiconductor superstructures, with the metal Fermi level slightly above the conduction band bottom of the semiconductor, a situation reminiscent of the Allender, Bray, and Bardeen[33] model for excitonic superconductivity (for a critical evaluation of a possible shortcoming of this model, see Ref. 34) which we will discuss later.

The calculated $N(E_F)$ for Tl/2212 and Tl/2223 are 2.82 states/eV-cell and 3.80 states/eV-cell, respectively. Thus, the additional CuO_2

sheet increases $N(E_F)$ by 1 state/eV-cell while the other components of $N(E_F)$ change by only 10-20%. Consistent with this is the fact that when we subtract the contribution from the Tl-O3-O2 bands, the $N(E_F)$ per Cu-atom is reduced to ~1.0 states/(eV-Cu atom), which is about the same as in[28] $Bi_2Sr_2CaCu_2O_8$.

The Fermi surfaces (FS) of $Tl_2Ba_2CaCu_2O_8$ (Tl/2212) are shown in Fig. 5 in an extended zone scheme. The electron pockets c and d centered around Γ and Z, respectively, are due to the Tl-O2-O3 bands. The Cu-O dpσ bands produce the two FS indicated by a and b in Fig. 5 (there is a third such surface for Tl/2223 lying between the two shown). These surfaces have a rounded-square shape centered around X. Fermi surface a especially shows striking nesting features along the (100) and (010) directions, with spanning vectors which are not commensurate. This high degree of FS nesting is expected to give rise to singularities in the generalized susceptibility, $\chi(\vec{q})$, of this highly 2D system, and may therefore have important consequences as possible electronically-driven instabilities (e.g., incommensurate charge density waves).

The simple FS of the 2D Cu-O bands in Tl/2212 shown in Fig. 5 as a and b should have a simple origin when looked at from the usual tight binding point of view. In a 2D square lattice, the simple tight-binding band is described by:

$$\varepsilon(\vec{k}) = \varepsilon_0 - 2t_1(\cos k_x a + \cos k_y a) + 4t_2 \cos k_x a \cdot \cos k_y a \quad (1)$$

where t_1 represents the nearest neighbor (n.n.) interaction and t_2 the next-nearest-neighbor (n.n.n.) interaction. From a comparison of the tight-binding bands and the dpσ anti-bonding bands of Tl/2212 and Tl/2223, we showed that the Cu-O1 dpσ anti-bonding bands crossing E_F cannot be properly fitted with a n.n. only tight-binding model while they can be reasonably well described by including the n.n.n. (most likely to be O-O) interactions. We therefore expect that the correct Fermi surface can only be obtained from the fuller tight-binding treatment and not from a simple n.n. tight binding interaction. In fact, the inclusion of the n.n.n. term in Eq. 1 yields a FS which is substantially different from the FS of a simple tight-binding band with only n.n. interactions. As shown in Fig. 6, the square centered at X with perfect nesting along the (110) direction (for n.n. only) has been transformed dramatically (by adding n.n.n.) into a rounded square with strong nesting features along the (100) and (010) directions which closely resembles the actual FS of Tl/2212 (and Tl/2223).

Finally, it is important to note that the same result is also true for the $YBa_2Cu_3O_7$ system,[17] where the FS of the 2D dpσ bands at E_F (see Fig. 3) are rounded squares centered at S with nesting along (100) and (010) directions. This result implies that the commonly used tight-binding model Hamiltonian with only n.n. interactions is not sufficient to describe the anti-bonding bands crossing E_F in Tl/2212 as well as $YBa_2Cu_3O_7$ in that it yields incorrect results. This has important consequences for all such model Hamiltonian descriptions used for explaining the high T_c. Thus, for example, the inclusion of the n.n.n. interactions leads to the reduction of the effective on-site Coulomb repulsion due to the delocalization of the Wannier states.

Again, it is important to emphasize that the good agreement between LDF band theory predictions and experiment establishes strong evidence for the Fermi liquid nature of the metallic ground state of the high T_c Cu-oxide superconductors, and justifies the use of LDF band theory as an excellent starting point for describing the transition to the superconducting state of the high T_c Cu-oxide systems.

III. EXCITONIC MECHANISM OF HIGH T_c SUPERCONDUCTIVITY

We have made crude estimates of the electron-phonon interaction in the Cu-oxide superconductors, $La_{2-x}M_xCuO_4$, $YBa_2Cu_3O_7$, $Bi_2Sr_2CaCu_2O_8$ and $Tl_2Ba_2CaCu_2O_8$, using the rigid muffin-tin approximation (RMTA)[35] to calculate the McMillan-Hopfield constant η and the electron-phonon coupling constant, λ. For all the Cu-oxide superconductors, the largest contributions to η come from the Cu and O ions in the CuO_2 planes, indicating the important role played by the "metallized" oxygens. As a crude approximation - and assuming the most favorable conditions, e.g., strong phonon softening $\theta_D \lesssim 100$ K - we estimated the T_c of these systems by using the strong coupling formula of Allen and Dynes[36]. The highest calculated T_c is found to be 36 K. Even though the T_c for the $La_{2-x}M_xCuO_4$ is close to the values found in the RMTA calculations, it is unlikely that a purely electron-phonon interaction is responsible for its high T_c because these are most favorable (unrealistic) estimates and the corresponding λ values are much larger than the experimental values[12]. For the other systems ($YBa_2Cu_3O_7$, $Bi_2Sr_2CaCu_2O_8$, and $Tl_2Ba_2CaCu_2O_8$), the estimates of T_c are so far off (more than a factor of three) that despite the crudeness of the RMTA approach, they cast doubt on a purely electron-phonon explanation of the observed high T_c. These results suggest the possibility and importance of a non-phonon mechanism of high T_c superconductivity.

Many authors have discussed the excitonic mechanism[37-38] of superconductivity, in which the effective attractive interaction between conduction electrons originates from virtual excitations of excitons rather than phonons. The basic idea of the models proposed is that conduction electrons residing on the conducting filament (or plane) induce electronic transitions on nearby easily polarizable molecules (or complexes), which result in an effective attractive interaction between conduction electrons. As perhaps a striking realization of the excitonic mechanism of superconductivity, $YBa_2Cu_3O_{7-\delta}$ has two 2D conduction bands and additional highly polarizable 1D electronic structure between the two conduction planes.

We have previously discussed[17] the importance of the 1D feature in the electronic structure near E_F, pointing out the possible role played by charge transfer excitations ("excitons") of occupied (localized) Cu1-O dpπ orbitals into their empty (itinerant) Cu1-O dpσ anti-bonding partners. As shown schematically in Fig. 7, we can characterize the 1D electronic structure with two types of electronic states in it, one free-electron-like (the well-dispersed dpσ band) and the other localized (the almost flat dpπ state). When the localized hole is created by the excitation, a strong attractive correlation between the hole and excited electron may lead to an electron-hole bound state ("exciton"). Hence, this excitation of the localized dpπ to the extended dpσ with the electron-hole correlation in the 1D electronic structure will give rise to a strong polarization in the 1D chains between two conduction planes and couple to the 2D conduction electrons, which carry most of the superconductivity.

In comparing the electronic states of the four oxide superconductors, a number of common features emerges which supports the excitonic model of superconductivity: In all materials, the 2D Cu-O dpσ bands dominate the electronic structure near E_F. These bands consist of anti-bonding combinations of Cu $d_{x^2-y^2}$ and in-plane O $p_{x,y}$ orbitals of the CuO_2 planes, which give rise to the strong two-dimensionality of the bands. The remarkable 2D nature of the electronic structure of La_2CuO_4 leads to a simple picture of the conductivity confined essentially to the metallic CuO_2 planes separated by ionic (insulating) planes of the rock-salt type La_2O_2 layers. We note that the slab (LaO)-(CuO_2)-(LaO) has the correct stoichiometry and is charge neutral, where the ionic $La^{3+}O^{2+}$ layers provide residual charge to the CuO_2 layers. Indeed, the (LaO)-(CuO_2)-(LaO) slab becomes a basic building block (with moderate modifications) for the other high T_c Cu-oxide superconductors.

We have seen that in the 90 K superconductor $YBa_2Cu_3O_{7-\delta}$, the building block was modified by introducing the oxygen deficient Y layer between the CuO_2 layers. The new building block for $YBa_2Cu_3O_{7-\delta}$ thus becomes (BaO)-(CuO_2)-(Y)-(CuO_2)-(BaO), where the middle three layers (CuO_2)-(Y)-(CuO_2) correspond to the single CuO_2 layers for La_2CuO_4. Similarly, in $Bi_2Sr_2CaCu_2O_8$, a common building block would be (SrO)-(CuO_2)-(Ca)-(CuO_2)-(SrO) and in $Tl_2Ba_2CaCu_2O_8$, the corresponding one becomes (BaO)-(CuO_2)-(Ca)-(CuO_2)-(BaO). Finally, the one in

$Tl_2Ba_2Ca_2Cu_3O_{10}$ is a mere extension of the one in $Tl_2Ba_2CaCu_2O_8$, i.e., $(BaO)-(CuO_2)-(Ca)-(CuO_2)-(Ca)-(CuO_2)-(BaO)$.

For all of these compounds, the La, Ba, Y, Sr, and Ca atoms are purely ionic and supply extra charges to each CuO_2 layer. In contrast to the strong ionic contribution of the La, Ba, Y, Sr, and Ca atoms, the 2D CuO_2 planes become metallic and give rise to the well dispersed Cu-O dpσ bands at E_F, which are essentially confined within each CuO_2 layer. These 2D Cu-O dpσ bands are essential for all the high T_c Cu-oxide superconductors. In addition, the common structural feature of the layered Cu-oxides superconductor suggests the (intercalated) layer structure as another essential element in the high T_c Cu-oxides. Once we regard the CuO_2 planes as major conduction layers and the Cu-O chains, Bi_2O_2, and Tl_2O_2 as intercalated semi-metallic or insulating layers, then the role of CuO chains, Bi_2O_2, and Tl_2O_2 layers must be to enhance superconductivity.

We have discussed some details of the additional electronic structure induced by these intercalated layers. What all of these electronic structures due to the intercalated layers have in common, is almost empty bands having strong covalent (anti-bonding) character. Furthermore, we also find the existence of occupied localized flat bands or non-bonding bands connected to the anti-bonding bands above E_F. This local electronic structure arising from the intercalated layers can be viewed simply as shown diagramatically in Fig. 7.

As discussed above, in $YBa_2Cu_3O_7$, we have proposed charge transfer excitations of occupied (localized) dpπ states, to empty dpσ bands as a representation of the interband interactions. In $Tl_2Ba_2CaCu_2O_8$ (and similarly in $Bi_2Sr_2CaCu_2O_8$) it becomes clear that the interband interactions between the non-bonding O p states and the almost empty Tl-O sp(dpσ) bands will lead to virtual excitations which couple to the conduction electrons in the CuO_2 planes and may play an important role in their high T_c superconductivity. Indeed, we can consider the role of CuO chains, Bi_2O_2, and Tl_2O_2 layers as providing the low lying charge excitations which couple the conduction electrons in the 2D CuO_2 planes.

We are in the process of quantifying this picture of charge transfer excitations. Such an approach requires detailed calculations of the full dielectric tensor $\varepsilon(\vec{Q}, \vec{Q}')$, including the (important) Umklapp processes, using our band structure results as the starting point.

ACKNOWLEDGMENTS

Work supported by the National Science Foundation (through the Northwestern University Materials Research Center, Grant No. DMR85-20280) and the Office of Naval Research (Grant No. N00014-81-K-0438). We are grateful to NASA Ames and Kirkland Air Force Base personnel for help with the use of their Cray 2. We thank C.L. Fu, D.D. Koelling, T.J. Watson-Yang and J.H. Xu for collaboration on the early aspects of this work.

REFERENCES

1. Bednorz, J.G. and Müller, K.A., Z. Phys. B64, 189 (1986).
2. Wu, M.K., Ashburn, J.R., Torng, C.J., Hor, P.H., Meng, R.L., Gao,

L., Huang, Z.J., Wang, Y.Q. and Chu, C.W., Phys. Rev. Lett. 58, 908 (1987).

3. Maeda, H., Tanaka, Y., Fukutomi, M. and Asano, T., Jpn. J. Appl. Phys. 27, in press (1988).

4. Chu, C.W., et al., Phys. Rev. Lett. 60, 941 (1988).

5. Sheng, Z.Z., Hermann, A.M., El Ali, A., Almason, C., Estrada, J., Datta, T. and Matson, R.J., Phys. Rev. Lett. 60, 937 (1988).

6. Manuel, A.A., Peter, M. and Walker, E. Europhys. Lett. 6, 61 (1987); Smedskaer, L., et al., Physica C156, 269 (1988).

7. Allen, P.B., Pickett, W.E. and Krakauer, H., Phys. Rev. B36, 3926 (1987).

8. Allen, P.B., Pickett, W.E., and Krakaure, H., Phys. Rev. B37, 7482 (1987).

9. Jansen, H.J.F. and Freeman, A.J., Phys. Rev. B30, 561 (1984).

10. Wimmer, E., et al., Phys. Rev. B24, 864 (1981).

11. Yu, J., Freeman, A.J. and Xu, J.-H. Phys. Rev. Lett. 58, 1035 (1987).

12. Freeman, A.J., Yu, J. and Fu, C.L., Phys. Rev. B36, 7111 (1987).

13. Xu, J.-H., Watson-Yang, T.J., Yu, J. and Freeman, A.J., Phys. Lett. A120, 489 (1987).

14. Fu, C.L. and Freeman, A.J., Phys. Rev. B35, 8861 (1987).

15. Torrance, J.B., Tokura, Y., Nazzal, A.I., Bezinge, A., Huang, T.C. and Parkin, S.S.P., Phys. Rev. Lett. 61, 1127 (1988).

16. Massidda, S., Yu, J., Freeman, A.J. and Koelling, D.D., Phys. Lett. 122, 198 (1987).

17. Yu, J., Massidda, S., Freeman, A.J. and Koelling, D.D., Phys.

Lett. 122, 203 (1987).

18. Anderson, P.W., Science 235, 1196 (1987).

19. Anderson, P.W., Bull. Am. Phys. Soc. 33, 459 (1988).

20. Arko, A.J., List, R.S., Fisk, Z., Cheong, S.-W., Thompson, J.D., and O'Rourke, J.A., JMMM 75, L1 (1988).

21. Redinger, J., Freeman, A.J., Yu, J., Massidda, S., Phys. Lett. 124A, 469 (1987).

22. Tozer, S.W., Kleinsasser, A.W., Penney, T., Kaiser, D. and Holtzberg, F., Phys. Rev. Lett. 59, 1768 (1987).

23. Willis, J.O., Fisk, Z., Thompson, J. D., Cheong, S.-W., Aikin, R.M., Smith, J.L. and Zirngiebl, E., J. Magn. Matls. 67, L139 (1987).

24. Yu, J. and Freeman, A.J. (to be published).

25. Vaknin, D., et al., Phys. Rev. Lett. 58, 2802 (1987).

26. Brewer, J.A., et al., Phys. Rev. Lett. 60, 1073 (1988).

27. Terakura, K., et al., Phys. Rev. B30, 4734 (1984).

28. Massidda, S., Yu, J. and Freeman, A.J., Physica C:Superconductivity 152, 251 (1988).

29. Sunshine, S.A., Siegrist, T., Schneemeyer, L.F., Murphy, D.W., Cava, R.J., Batlogg, B., van Dover, R. B., Fleming, R.M., Glarum, S.H., Nakahara, S., Farrow, R., Krajewski, J.J., Zahurak, S.M., Waszczah, J.V., Marshall, J.H., Marsh, P., Rupp, Jr., L. W. and Peck, W.F., Phys. Rev. B38, 893 (1988).

30. Hazen, R.M., Finger, L.W., Angel, R.J., Prewitt, C.T., Ross, N.L., Hadidiacos, C.G., Heaney, P.J., Veblen, D.R., Shen, Z.Z., El Ali, A. and Hermann, A.M., Phys. Rev. Lett. 60, 1657 (1988).

31. Subramanian, M.A., Torardi, C.C., Calabrese, J.C., Gopalakrishnan, J., Morrissey, K.J., Askew, T.R., Flippen, R.B., Chowdhry, U. and Sleight, A.W., Science 239, 1015 (1988).

32. Yu, J., Massidda, S., and Freeman, A.J., Physica C:Superconductivity 152, 273 (1988).

33. Allender, D.W., Bray, J.M. and Bardeen, J., Phys. Rev. B7, 1020 (1973).

34. Cohen, M.L. and Louie, S.G., Superconductivity in d- and f-band Metals, Douglass, D.H., Ed. (Plenum, New York, 1976).

35. Gaspari, G.D. and Gyoffry, B.L., Phys. Rev. Lett. 28, 801 (1972).

36. Allen, P.B. and Dynes, R.C., Phys. Rev. B12, 905 (1975).

37. Little, W.A., Phys. Rev. 134, A1416 (1964).

38. Ginzburg, V.L., JETP 46, 397 (1964).

FIGURE CAPTIONS

Figure 1. A local environment for the Cu1 and Cu2 atoms in $YBa_2Cu_3O_7$, following the Y-Cu2-Ba-Cu1-Ba-Cu2-Y ordering along z.

Figure 2. Band structure of $YBa_2Cu_3O_7$ along symmetry directions in the k_z = 0 plane of the orthorhombic Brillouin zone.

Figure 3. Calculated Fermi Surface for $YBa_2Cu_3O_7$. The dot-dashed lines are experimentally measured Fermi surfaces (Smedskaer et al.)

Figure 4. Energy bands of $Tl_2Ba_2CaCu_2O_8$ along the main symmetry lines of the body-centered tetragonal extended Brillouin zone. (Notation from Reference 11.)

Figure 5. Fermi surfaces of $Tl_2Ba_2CaCu_2O_8$ in an extended zone scheme.

Figure 6. Fermi surfaces of the tight-binding bands for (a) t_2/t_1 = 0.0 (with the n.n. interactions only) and (b) t_2/t_1 = 0.45 (with n.n.n. interactions included). (See text for details.)

Figure 7. Schematic drawing of the 1D electronic structure in $YBa_2Cu_3O_7$.

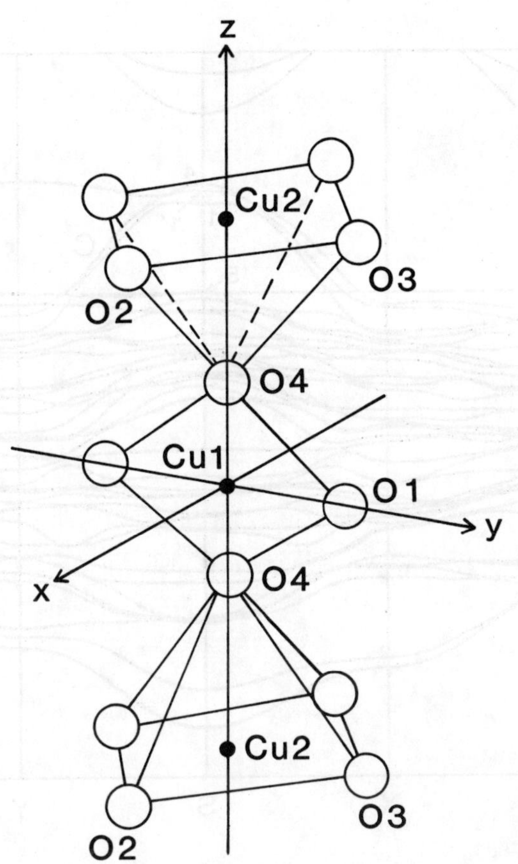

Fig. 1

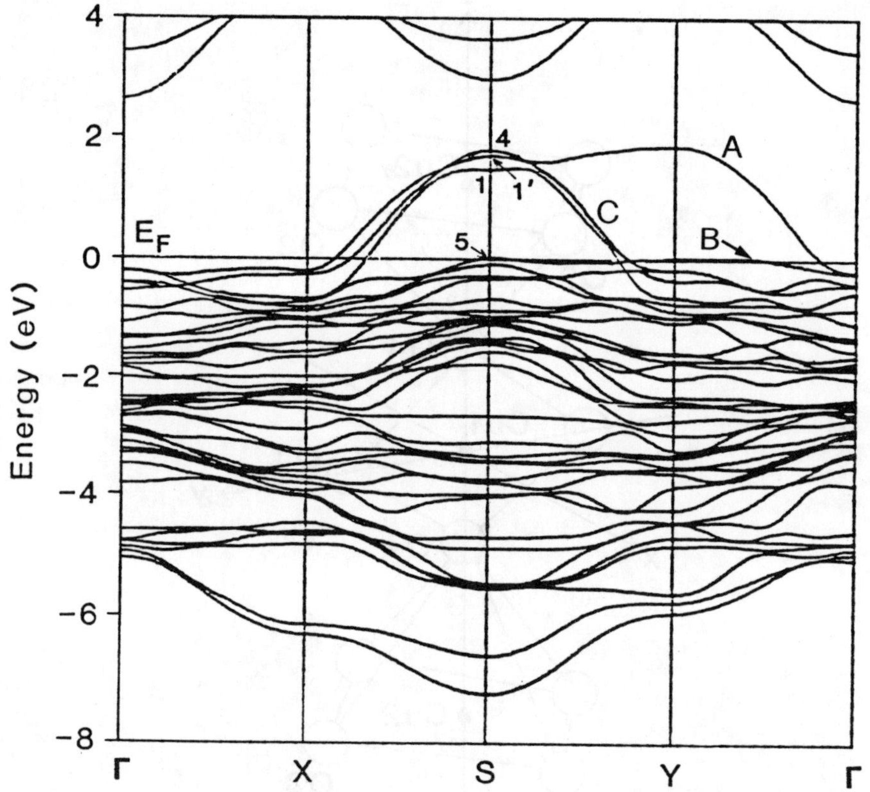

Fig. 2

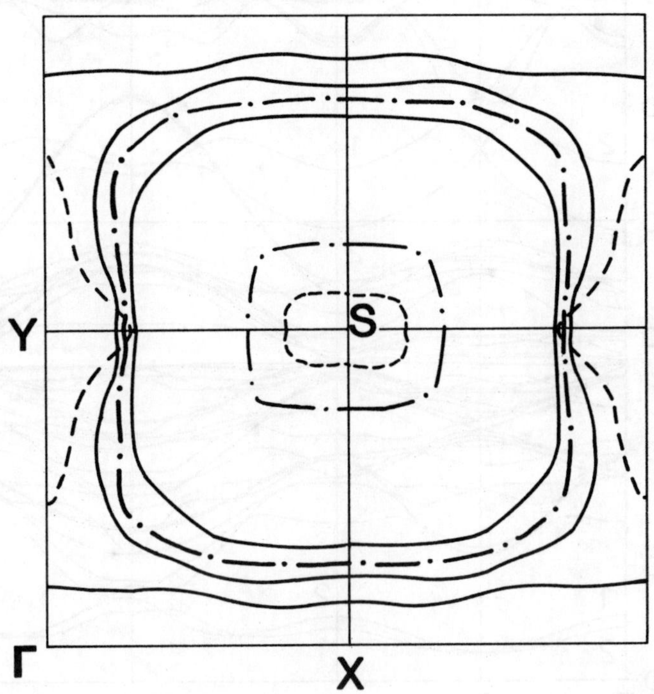

Fig. 3

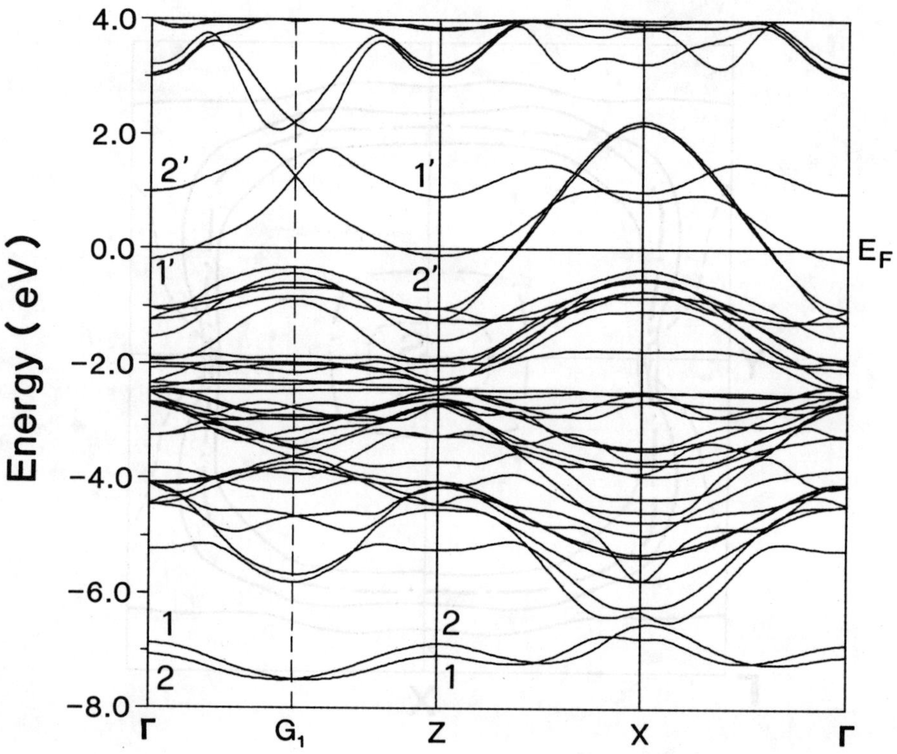

Fig. 4

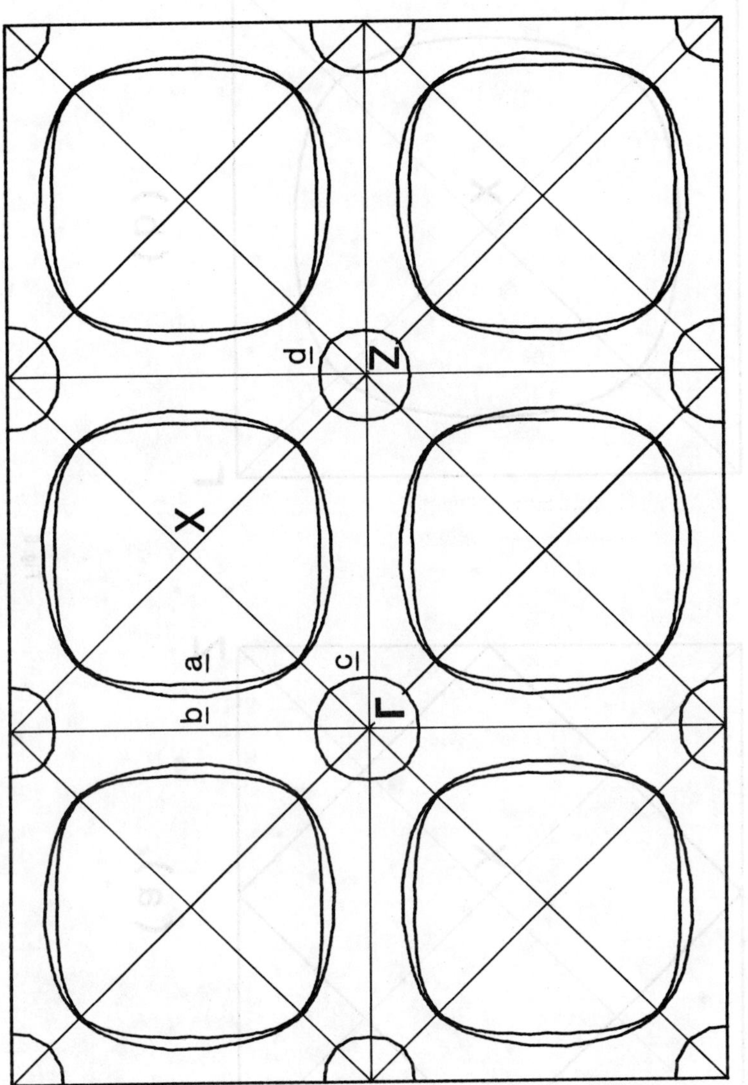

Fig. 5

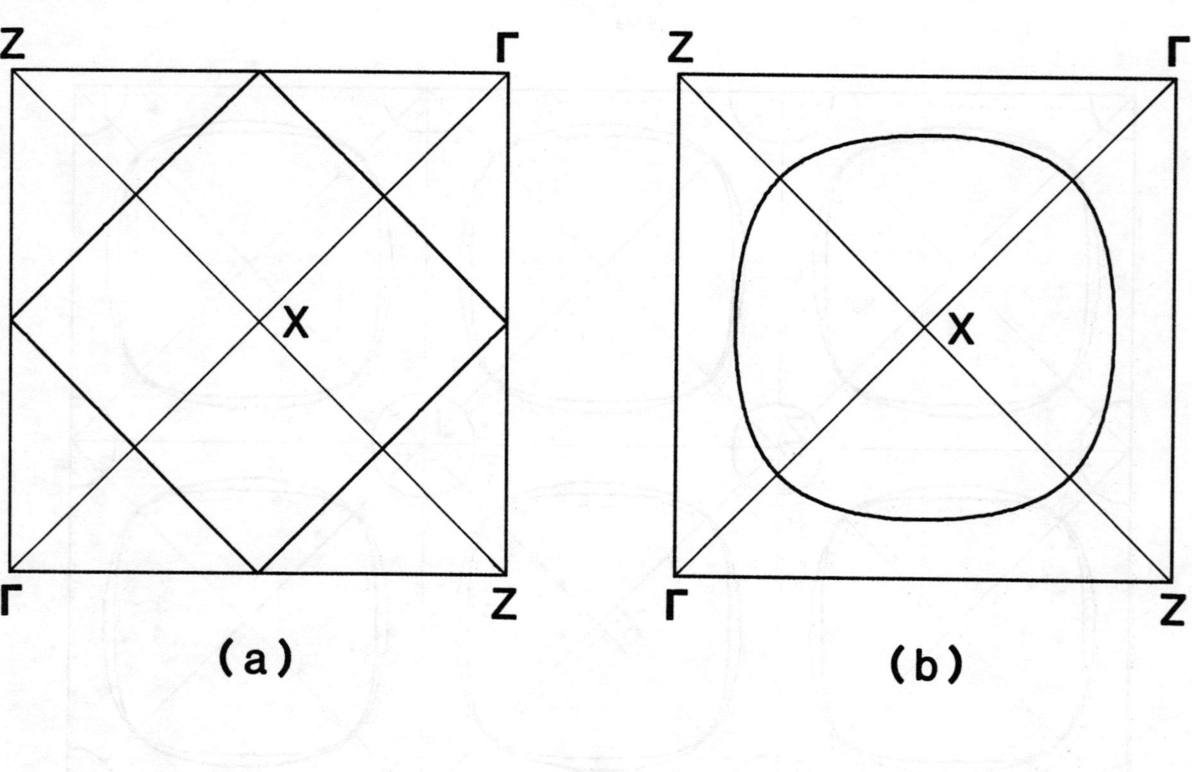

Fig. 6

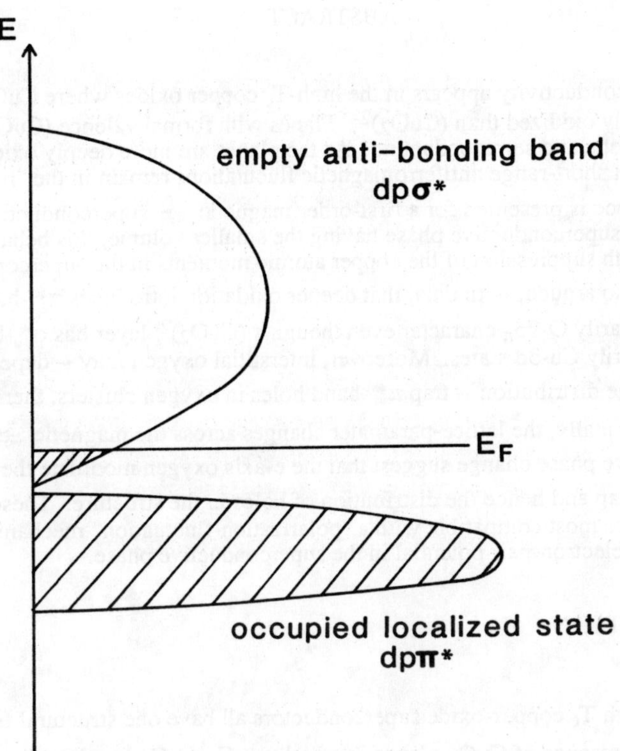
E
empty anti-bonding band
dpσ*
E_F
occupied localized state
dpπ*

Fig. 7

MECHANISM IN HIGH-T_c SUPERCONDUCTIVITY

John B. Goodenough
Center for Materials Science and Engineering, ETC 5.160
University of Texas at Austin, Austin, TX 78712-1084

ABSTRACT

Superconductivity appears in the high-T_c copper oxides where CuO_2 planes are more deeply oxidized than $(CuO_2)^{2-}$. Planes with formal valence $(CuO_2)^{2-}$ are antiferromagnetic and semiconducting. As the planes are more deeply oxidized, T_N disappears, but short-range antiferromagnetic fluctuations remain in the "magnetic" phase. Evidence is presented for a first-order magnetic $\rightleftharpoons$ superconductive phase transition, the superconductive phase having the smaller volume; this behavior is compatible with suppression of the copper atomic moments in the superconductive phase. It is also argued, with data, that deeper oxidation introduces π^* -band holes that have primarily O-$2p_\pi$ character even though a $(CuO_2)^{2-}$ layer has σ^* -band holes in primarily Cu-3d states. Moreover, interstitial oxygen may -- depending on the local charge distribution -- trap π^* -band holes in oxygen clusters, thereby lowering T_c. Finally, the lattice-paramater changes across the magnetic $\rightleftharpoons$ superconductive phase change suggest that the c-axis oxygen modulate the σ^* and π^* -band overlap and hence the distribution of holes in the structure. These several observations are most compatible with a "polarization-fluctuation" mechanism for enhancing the electron-pair potential in the superconductive phase.

INTRODUCTION

The high-T_c copper-oxide superconductors all have one structural feature in common, the presence of CuO_2 planes in which the Cu-O-Cu bonding does not deviate significantly from 180°. Perpendicular to the planes the copper may have no, one, or two oxygen neighbors; any Cu-O distances perpendicular to the plane are longer than those in the plane. Moreover, where the formal oxidation state of the plane is $(CuO_2)^{2-}$, the planes are semiconducting and the copper atoms carry an atomic moment characteristic of a formal copper valence Cu^{2+}; in this situation the planes become antiferromagnetic below a Néel temperature T_N. On the other hand, where the planes become further oxidized to $(CuO_2)^{(2-\delta)-}$, $\delta \gtrsim 0.05$, the compounds become superconductive with a high T_c.

These observations raise four questions that need to be addressed:

- Is the antiferromagnetic $\rightleftharpoons$ superconductive transition first- or second-order?
- Do the copper atoms retain a magnetic moment in the superconductive phase?
- Where are the mobile holes in the superconductive phase?
- What, if any, is the role of the c-axis oxygen?

$La_2CuO_{4+\delta}$

Orthorhombic, stoichiometric $La_2CuO_{4+\delta}$ is an antiferromagnetic semi-conductor[1)], but a filamentary superconductive phase -- corresponding to about 0.04% of the material according to the Meissner effect -- has been noted in some preparations[2-4)]. The observation of a filamentary superconductivity indicates the presence of some CuO_2 layers that are oxidized beyond the formal valence state $(CuO_2)^{2-}$ either through a lanthanum deficiency, $La_{2-x}CuO_4$, or through the introduction of interstitial oxygen, $La_2CuO_{4+\delta}$. The existence of both $La_{2-x}CuO_4$ and $La_2CuO_{4+\delta}$ as superconductive phases has now been established. Of particular interest for the present discussion is the phase $La_2CuO_{4+\delta}$, which was first obtained under 3 kbar of oxygen pressure at 600 ^oC[5,6)].

These initial experiments on $La_2CuO_{4+\delta}$ established a La/Cu ratio of 2.00 ± 0.01 and a trend from orthorhombic toward tetragonal symmetry with increasing oxygen content. Samples with the highest oxygen content were superconductive with a zero resistance near 30 K; Meissner measurements indicated that at least 30% of the material was superconductive. Thermogravimetric analysis (TGA) indicated a $\delta = 0.13 \pm 0.02$ whereas iodometric titration corresponded to a bulk $\delta = 0.032 \pm 0.005$. In an attempt to reconcile this apparent discrepancy, it was postulated that the excess oxygen enter the lattice interstitially as the superoxide ion $(O_2)^-$, and XPS evidence for this species was subsequently reported[7].

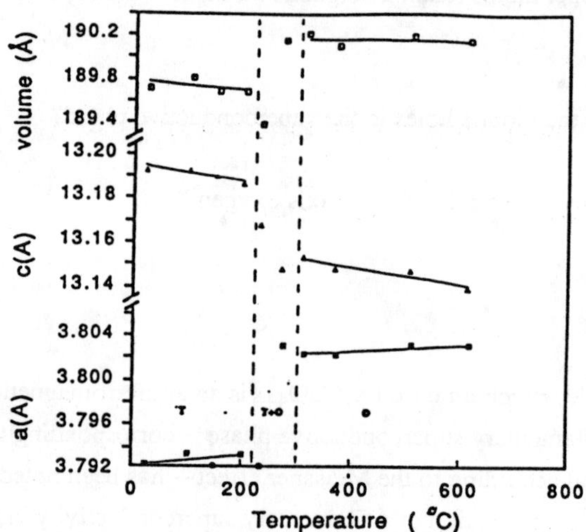

FIG. 1. Lattice parameters and volume versus two-hour air-anneal temperature of initially tetragonal $La_2CuO_{4+\delta}$. Orthorhombic $(a_o+b_o)/2$ shown for comparison with tetragonal a axis.

We[8] have prepared $La_2CuO_{4+\delta}$ at 23 kbar and 800 °C under a high pressure of oxygen supplied by CrO_3 in a separate chamber within a sealed gold crucible. The samples were nearly tetragonal (a = 3.492(3), c = 13.195(4) Å), and the measured excess oxygen corresponded to $\delta = 0.05 \pm 0.01$ by iodometric titration and $\delta = 0.143$ by TGA. The TGA curves showed a continuous loss of oxygen on heating in N_2 up to 700 °C with a superposed abrupt loss of bulk oxygen equivalent to $\delta = 0.05$ in the temperature interval $225 < T < 300$ °C; and differential scanning calorimetry (DSC) indicated a phase change in this interval.

Fig. 1 shows the change in lattice parameter with annealing temperature for $La_2CuO_{4+\delta}$ annealed in air for two hours followed by quenching to room temperature. In the interval $225 < T < 300$ °C there is a first-order phase change for the tetragonal, superconductive phase to the orthorhombic, antiferromagnetic phase. Samples quenched from 350 °C showed filamentary superconductivity; but heating to 800 °C suppressed all superconductivity, and the samples remained semiconductive to lowest temperatures.

Three points deserve emphasis:

- The superconductor $\rightleftharpoons$ antiferromagnetic phase change is first-order.

- The volume of the superconductive phase is smaller even though it contains interstitial oxygen.

- The axial ratio c/a is larger in the superconductive phase primarily as a result of a remarkable contraction of the Cu-O separation in the CuO_2 planes.

Although these data do not provide direct evidence for the suppression of a localized atomic moment on the copper in the superconductive phase, a pronounced broadening of the σ^*_{x2-y2} bands is clearly indicated. Since the σ^*_{x2-y2} bandwidth in the antiferromagnetic phase approaches the correlation splitting $U \approx 6$ eV, the first-order contraction of the CuO_2 planes is probably associated with a delocalization of the σ^*_{x2-y2} -band holes that suppresses the copper atomic moments.

Examination of the tetragonal $La_2CuO_{4+\delta}$ structure, Fig. 2, would place the excess oxygen in the tetrahedral sites of the La_2O_2 layers. In view of the local formal charges of the intergrowth layers, $(La_2O_2)^{2+}$ and $(CuO_2)^{2-}$, such a location would attract electrons from the CuO_2 layers to create interstitial O^{2-} ions within the La_2O_2 layers and oxidize the CuO_2 planes. Moreover, the TGA data are consistent with $\delta = 0.05$ bulk interstitial oxygen present as O^{2-} and $\delta \approx 0.09$ excess oxygen present as superficial oxygen species. The formation of peroxide and superoxide species on the surface can be expected in view of the high oxidizing power of an oxidized $(CuO_2)^{(2-\delta)-}$ layer (see below for further discussion). The XPS evidence for superoxide ions[7)] would thus appear to be associated with superficial oxygen rather than bulk oxygen. Note that location of the interstitial oxygen in the tetrahedral sites of the La_2O_2 layers would force the c-axis oxygen toward the CuO_2 planes, thus further increasing the total Cu-O covalent mixing responsible for delocalization of the σ^*_{x2-y2} holes. The expansion of the c-axis in the superconducting phase is, according to this model, more a reflection of the presence of interstitial oxide ions in the La_2O_2 layers than of an increase in the c-axis Cu-O separation.

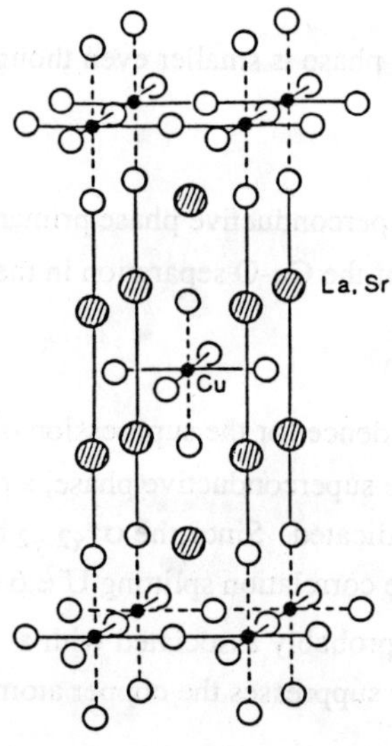

FIG.2. Tetragonal K_2NiF_4 structure.

BAND MODELS

Band-structure calculations[9-11)] for La_2CuO_4 are summarized schematically in the uncorrelated energy vs density-of-states diagram of Fig. 3. The Fermi

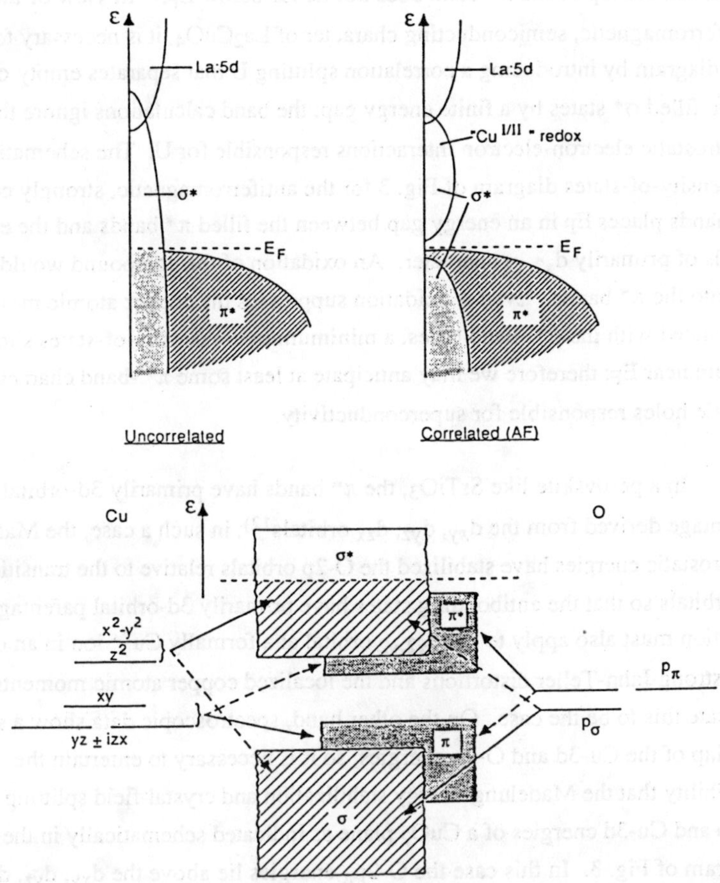

FIG. 3. Schematic energy versus density of electron states for uncorrelated and correlated σ* bands; <u>bottom</u>: location of Cu-3d and O-2p energies that would give π* bands a primarily O-2pπ parentage, but σ* bands of primarily Cu-3d parentage.

energy E_F lies in a σ^* band that is primarily σ^*_{x2-y2} in character at the top of the band; but the top of the π^* band does not lie far below E_F. In view of the antiferromagnetic, semiconducting character of La_2CuO_4, it is necessary to modify this diagram by introducing a correlation splitting U that separates empty σ^* states from filled σ^* states by a finite energy gap; the band calculations ignore the electrostatic electron-electron interactions responsible for U. The schematic energy vs density-of-states diagram of Fig. 3 for the antiferromagnetic, strongly correlated σ^* bands places E_F in an energy gap between the filled π^* bands and the empty σ^* bands of primarily d_{x2-y2} character. An oxidation of the compound would lower E_F into the π^* bands. Even if oxidation suppresses the copper atomic moments associated with the σ^* -band holes, a minimum in the density-of-states should remain near E_F; therefore we may anticipate at least some π^* -band character for the mobile holes responsible for superconductivity.

In a perovskite like $SrTiO_3$, the π^* bands have primarily 3d-orbital parentage derived from the d_{xy}, d_{yz}, d_{zx} orbitals[12)]; in such a case, the Madelung electrostatic energies have stabilized the O-2p orbitals relative to the transition-metal 3d orbitals so that the antibonding states have primarily 3d-orbital parentage. This situation must also apply to the d_{x2-y2} orbital at a formally Cu^{2+} ion in an oxide. The strong Jahn-Teller distortions and the localized copper atomic moments clearly indicate this to be the case. On the other hand, spectroscopic data show a strong overlap of the Cu-3d and O-2p energies, so it is necessary to entertain the possibility that the Madelung-energy stabilization and crystal-field splitting leave the O-2p and Cu-3d energies of a CuO_2 plane as indicated schematically in the lower diagram of Fig. 3. In this case the O-$2p_\pi$ energies lie above the d_{xy}, d_{yz}, d_{zx} energies so that covalent mixing produces a π^* band of primarily O-$2p_\pi$ parentage. In such a case, holes introduced into the π^* bands would be located more on the oxygen than the copper, and they could be more mobile than strongly correlated 3d

In order to obtain experimental evidence as to whether the π^* -band holes have primarily oxygen character, we[13,14)] have investigated whether compounds having the $Ba_2YCu_3O_{6+x}$ structure trap holes at oxygen clusters associated with a-axis oxygen.

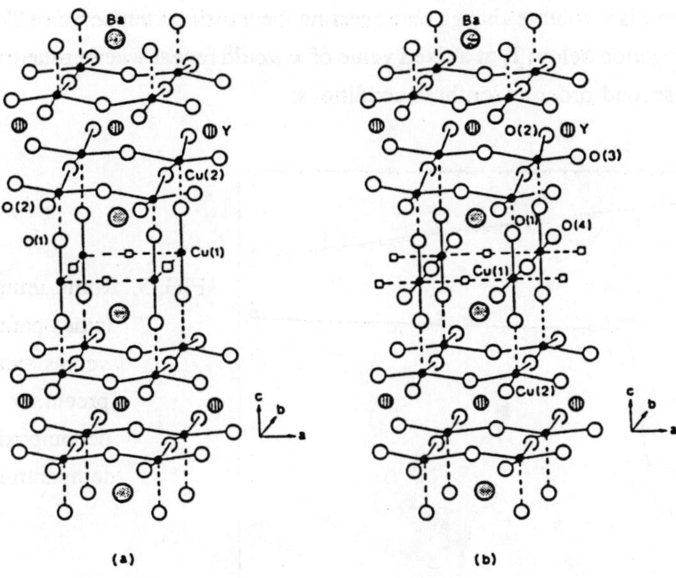

FIG. 4. Structures of (a) tetragonal $Ba_2YCu_3O_6$ and (b) orthorhombic $Ba_2YCu_3O_{6.94}$.

DISORDERED AND SUBSTITUTED $Ba_2YCu_3O_{6+x}$

Fig. 4 shows the structures of tetragonal $Ba_2YCu_3O_6$ and orthorhombic $Ba_2YCu_3O_{6.94}$. The tetragonal phase is antiferromagnetic and semiconducting; the orthorhombic phase is superconductive ($T_c \simeq 90K$). The oxygen stoichiometry can be varied, apparently continuously, over the entire compositional range $0 \leq x \leq 0.94 \pm 0.01$ for $Ba_2YCu_3O_{6+x}$. A smooth order-disorder orthorhombic-tetragonal transition occurs with continuous loss of oxygen on raising the temperature of an x = 0.94 sample in air above 400 °C[15]. However, on cooling tetragonal samples of composition $0.3 < x < 0.5$ in N_2, which holds the oxygen concentration to a fixed value of x, a pronounced increase in the $c/\frac{1}{2}(a+b)$ ratio on passing from the tetragonal to the orthorhombic phase is predicted. A careful investigation of whether there is a volume change on traversing the transition temperature T_t and a phase segregation below T_t at a fixed value of x would reveal whether the transition is first- or second-order under these conditions.

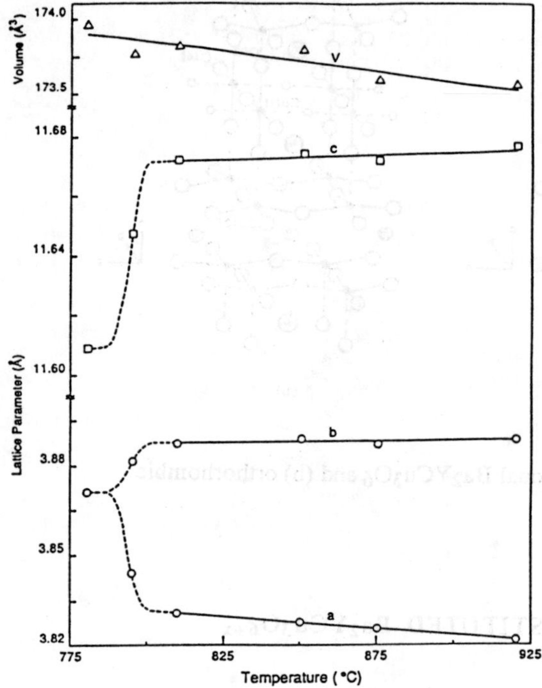

FIG. 5. Room-temperature lattice parameters versus oxalate-precursor decomposition temperature.

In the orthorhombic, superconductive phase of $Ba_2YCu_3O_{6+x}$, the oxygen are ordered on b-axis positions of the Cu(1) planes. Oxygen can be introduced into the a-axis positions by either quenched-in disorder or by introducing excess oxygen ($x > 7.0$). We[13] have prepared tetragonal $Ba_2YCu_3O_{6.70}$ by decomposing an oxalate precursor in air at 780 °C; orthorhombic $Ba_2YCu_3O_{6.70}$ prepared by rapid cooling from a N_2 anneal is superconductive ($T_c \approx 60$ K). Fig. 5 illustrates the sharp increases in the room-temperature $c/\frac{1}{2}(a+b)$ observed on changing from the tetragonal, semiconducting ($x = 0.67$) phase to the orthorhombic, superconductive phase on raising the temperature of oxalate decomposition from 780 °C to 800 °C.

All samples were subsequently annealed in air at 400 °C. We[14)] have also prepared compositions with x > 7.0 by substituting La for Ba. We were able to interpret the decrease in T_c with increasing x > 7.0, Fig. 6, in terms of trapping of two holes at each a-axis oxygen; the trap site was postulated to be a peroxide ion $(O_2)^{2-}$. Takita et al[16)] found similar behavior in the system $Ba_{2-x}Nd_xNdCu_3O_{6+x}$; they showed that T_c changed with the mobile-hole concentration measured by the Hall effect and were forced to postulate a trapping of holes in the Cu(1)-O chains; they did not specify the nature of the hole trap.

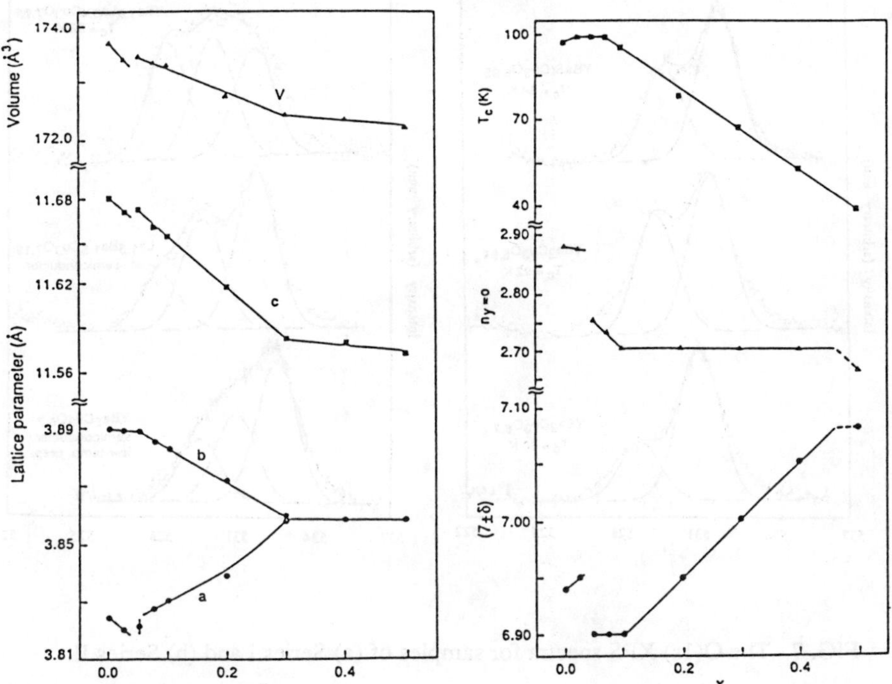

FIG. 6. Variation of lattice parameters, volume, T_c, formal Cu(1) oxidation state if Cu(2) assumed Cu^{2+}, and oxygen content versus composition x of $Ba_{2-x}La_xYCu_3O_{6+y}$.

In order to provide more direct evidence for the trapping of mobile holes at an oxygen-cluster species -- of which the peroxide $(O_2)^{2-}$ ion would be the simplest and most probable, we[17)] performed XPS measurements on two series of samples.

All samples of Series I were orthorhombic with all Cu(1) -plane oxygen only on b-axis positions; all samples of Series II contained oxygen on a-axis positions. In no sample was there any spectroscopic evidence for Cu(III) species; only Cu(II) and Cu(I) could be identified. On the other hand, the O(1s) XPS spectra showed a clear distinction between the samples of Series I and Series II, see Fig. 7.

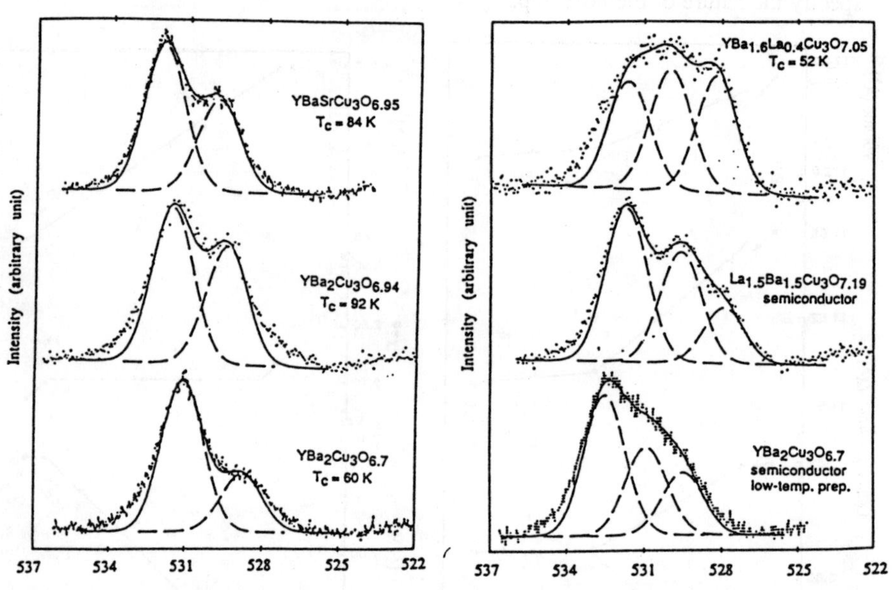

FIG. 7. The O(1s) XPS spectra for samples of (a) Series I and (b) Series II.

The highest binding-energy peak increases with aging in the spectrometer; it is clearly associated with surface oxygen present as O^-, O_2^-, OH^-, $\cdots$. The variability of the amplitude of this peak made it impossible to obtain quantitative information from relative peak heights; however, the data clearly show a splitting of the peak associated with the bulk oxygen in the case of samples of Series II. Just such a splitting was anticipated for a model in which the mobile holes introduced by oxidation of the CuO_2 layers beyond the formal valence $(CuO_2)^{2-}$ occupy π^* band states that are primarily of O-2p parentage. A formal valence $O^{(2-\delta)-}$ due to itinerant holes in the π^* bands would increase the O(1s) binding energy relative to that at an O^{2-} ion; trapping of holes at a peroxide ion $(O_2)^{2-}$ would increase the binding energy even more at the oxygen cluster, but not to the extent found at an isolated O^- ion such as is found on the surface.

ROLE OF c-AXIS OXYGEN

Finally, a possible role for the c-axis oxygen needs to be considered in view of the remarkable changes in axial ratio that occur in the antiferromagnet $\rightleftharpoons$ superconductor transitions. In $Ba_2YCu_3O_6$, the Cu(1) atoms of Fig. 4(a) have only two c-axis oxygen near neighbors; these copper have the formal valence Cu^+ and carry no localized atomic moment. The $(CuO_2)^{2-}$ planes become antiferromagnetic below a T_N; the planar Cu(2) have a formal valence Cu^{2+}. The Cu(1)-O distance is relatively short.

Fig. 8 shows the celebrated T_c vs x plot reported by Cava et al[18]. This figure invites consideration of the ordering of the oxygen in the Cu(1) planes in the orthorhombic phase. At the composition x = 0.5, two types of order can be imagined, the intrachain and the interchain ordering of Fig. 9. Electron microscopy has now established that the interchain ordering occurs[19]; it conserves square-coplanar coordination about the Cu(1) atoms that are oxidized beyond the formal valence state Cu^+ and it minimizes the O^{2-} - O^{2-} electrostatic interactions within the

Introduction of additional oxygen could be expected to have a lowest-energy configuration with intrachain ordering on the partially filled set of b-axis chains. No confirmed evidence for order other than the doubling along the a-axis has been observed by electron microscopy. Since this doubling is found over a wide range of compositions, we may assume that the chains filled at x = 0.5 remain filled at x > 0.5; the additional oxygen atoms occupy the alternate chains, but long-range intrachain order is difficult to establish.

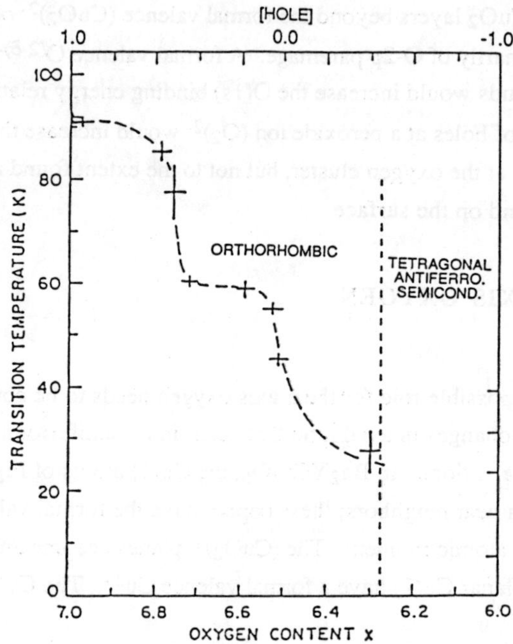

FIG. 8. Superconductor transition temperature T_c versus oxygen parameter x for $Ba_2YCu_3O_{6+x}$, after Cava et al. [15].

The step in Fig. 8 in the interval $0.6 \lesssim x < 0.75$ has suggested[21] a two-phase region between x = 0.5 and x = 0.75 with the intrachain ordering of Fig. 9(a) on the partially filled chains at x = 0.75. On the other hand, evidence for a completely flat plateau has not been established, so the model for the origin of the step needs to be modified. To this end, it may be useful to reconsider the oxidation process.

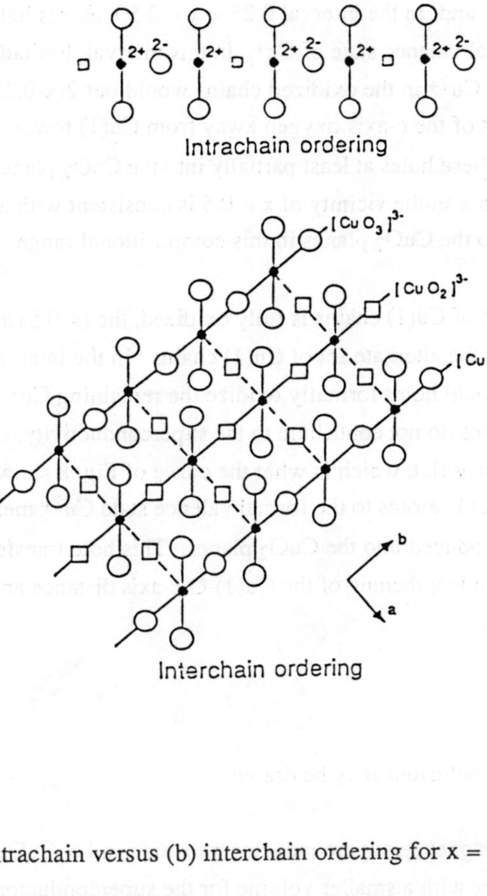

FIG. 9. (a) Intrachain versus (b) interchain ordering for x = 0.5.

In the $Ba_2YCu_3O_{6+x}$ system, location of the holes introduced by increasing x is important. Two possibilities exist: they may enter σ^* states at the Cu(1) atoms to oxidize them from Cu^+ to Cu^{2+} or they may enter π^* -band states. For any given chain, oxidation of the Cu^+ to the Cu^{2+} may be anticipated if the chain is half-filled or less. Oxidation of Cu(1) atoms from Cu^+ to Cu^{2+} would not oxidize the CuO_2 planes and would therefore have little influence on T_c. Therefore compositions in the interval $0 \leq x \leq 0.25$ remain tetragonal, antiferromagnetic, and semiconducting. On further oxidation, long-range interchain ordering establishes orthorhombic symmetry and, in the interval $0.25 < x < 0.50$, leaves half of the Cu(1) atoms in the formal valence state of Cu^+. In this interval, limitation of the Cu(1) formal valence to Cu^{2+} in the oxidized chains would put 2(x-0.25) holes into the π^* bands; and a shift of the c-axis oxygen away from Cu(1) toward Cu(2) atoms would introduce these holes at least partially into the CuO_2 planes. The sharp increase in T_c with x in the vicinity of x = 0.5 is consistent with a transfer of mobile π^* -band holes to the CuO_2 planes in this compositional range.

Once the first set of Cu(1) chains is fully oxidized, the (x-0.5) additional oxygen are inserted into the alternate set of Cu(1) chains. In the interval $0.50 < x < 0.75$, the 2(x-0.5) additional holes formally oxidize the remaining Cu^+ ions to Cu^{2+}; if the d-orbital holes do not contribute to the superconductivity, the T_c vs x curve would remain nearly flat, which is what the curve of Fig. 8 shows. For $x > 0.75$, limitation of the Cu(1) atoms to the formal valence state Cu^{2+} means that π^* -band holes are again introduced into the CuO_2 planes. This hole transfer would also be accompanied by a lengthening of the Cu(1)-O c-axis distance and a

CONCLUSIONS

The following conclusions may be drawn:

- The antiferromagnet $\rightleftharpoons$ superconductor transition in $La_2CuO_{4+\delta}$ appears to be first-order with a smaller volume for the superconductor phase and shorter Cu-O distances both within and perpendicular to the superconductor CuO_2 planes.

- The mobile holes active in superconductivity occupy π^* -band states that are primarily of O-$2p_\pi$ parentage.

- The localized atomic moments of the antiferromagnetic phase are probably suppressed in the superconductive phase; T_N may disappear before the copper atomic moments because of the two-dimensional character of the interatomic coupling, but a first-order transition occurring after the vanishing of T_N would imply a loss of atomic moment.

- The c-axis oxygen move toward the copper of the CuO_2 planes in the superconductive $Ba_2YCu_3O_{6+x}$ phase, and this motion not only transfers mobile π^* -band holes to the CuO_2 planes, but also increases the overall Cu-O covalent mixing so as to suppress any localized atomic moment on the copper.

- A maximum in T_c versus oxygen content at $x \simeq 1.0$ for the system $Ba_{2-y}La_yYCu_3O_{6+x}$ is due to a trapping of two holes per (x-1) a-axis oxygen introduced by an $y > 1$; hole trapping reduces the concentration of mobile π^* -band holes. Although interstitial oxygen in $La_2CuO_{4+\delta}$ do not trap holes, presumably because they enter the positively charged La_2O_2 layers we should expect that in the system $La_{2-x}Sr_xCuO_{4+\delta}$ any interstitial oxygen will trap two holes for x greater than some critical value x_c.

These conclusions are most compatible with a polarization-fluctuation mechanism for the electronic enhancement of the Cooper pair potential.

The experimental support given by several colleagues cited in the references to our work is gratefully acknowledged. The Robert A. Welch Foundation of Houston, Texas, and the Texas Advanced Research Program are thanked for their financial support.

REFERENCES

1. G. Shirane, Y. Endoh, R.J. Birgeneau, M. Kostner, Y. Hidaka, M. Oda, M. Suzuki, and T. Murakari, Phys. Rev. Lett. 14, 1613 (1987).

2. P.M. Grant, S.S.P. Perkin, V.Y. Lee, E.M. Engler, M.L. Ramirez, J.E. Vazquez, G. Lim, R.D. Jacowitz, and R.L. Greene, Phys. Rev. Lett. 58, 2482 (1987).

3. D.C. Johnston, J.P. Stokes, D.P. Goshorn, and J.T. Lewandowski, Phys. Rev. B36, 4007 (1987).

4. S.W. Hsu, S.Y. Tsauer, and H.C. Ku, Phys. Rev. B38, 856 (1988).

5. J.E. Schirber, B. Morosin, R.M. Merrill, P.F. Hlava, E.L. Venturini, J.F. Kwok, P.J. Nigrey, R.J. Baughman, and D.S. Ginley, Physica. C152, 121 (1988).

6. G. Demazeau, F. Tresse, Th. Plante, B. Chevalier, J. Etourneau, C. Michel, M. Hervin, B. Raveau, P. Lejay, A. Sulpice, and T. Tournier, Physica. C153-155, 824 (1988).

7. J. W. Rogers, Jr., N.D. Shinn, J.E. Schirber, E.L. Venturini, D.S. Ginley, and B. Morosin, Phys. Rev. B38, 5021 (1988).

8. Jianshi Zhou, Sanjai Sinha, and J.B. Goodenough (submitted).

9. L.F. Mattheiss, Phys. Rev. Lett. 58, 1028 (1989).

10. Jaejun Yu, A. J. Freeman, and J.H. Xu, Phys. Rev. Lett. 58, 1035 (1982).

11. J. Yu, S. Massidda, and A.J. Freeman, Phys. Lett. A122, 203 (1987).

12. J. B. Goodenough, Prog. Solid State Chem. 5, 145 (1972).

13. A. Manthiram and J.B. Goodenough, Nature 329, 701 (1987).

13. A. Manthiram, X.X. Tang, and J.B. Goodenough, Phys. Rev. B37, 3734 (1988).

15. M.O. Eatough, D.S. Ginley, B. Morosin, and E.L. Venturini, Appl. Phys. Lett. 51, 367 (1987).

16. K. Takita, H. Akinaga, H. Katoh, H. Asano, and K. Masuda, Jpn. J. Appl. Phys. 27, L67 (1988); W. Wong-Ng, L.P. Cook, C.K. Chiang, L.J. Swartzendruber, L.H. Bennett, J. Blendell, and D. Minor, J. Mater. Res. 3, 832 (1988) and ref. therein.

17. Y. Dai, A. Manthiram, A. Campion, and J.B. Goodenough, Phys. Rev. B38, 5091 (1988).

18. R.J. Cava, B. Batlogg, C.H. Chen, E.A. Rietman, S.M. Zahurak, D. Werder, Phys. Rev. B36, 5719 (1987).

19. C. Chaillout, M.A. Alario-Franco, J. J. Capponi, P. Strobel, and M. Marezio, Solid State Comm. 65, 283 (1988).

20.. L.T. Wille and D. deFontami, Phys. Rev. B37, 2227 (1988).

21. J.B. Goodenough, Mat. Res. Bull. 23, 401 (1988).

ELECTRON CORRELATIONS AND MAGNETIC INTERACTIONS IN HIGH T_C SUPERCONDUCTORS

C. S. WANG

Center for Superconductivity Research
Department of Physics, University of Maryland
College Park, Maryland 20742-7111

ABSTRACT

The general trends in the electronic structure of the high-T_c superconductors will be discussed based on energy band calculations. Superconductivity depends on (1) the hole concentration of the antibonding-pdσ-bands of the Cu-O_2 layers, which can be controlled by doping or oxygen stoichiometry,and (2) the metallic nature of the planes between Cu-O_2 layers, which determines the interlayer coupling between the Cu-O_2 layers. The failures of the band theory to stabilize the observed antiferromagnetic ground state in La_2CuO_4 will be analyzed based on our understanding of the Mott insulators, and of the heavy fermion superconductors. Recent theoretical efforts to determine the appropriate model Hamiltonian will be reviewed, and the results of our recent quantum Monte-Carlo simulations of an extended Hubbard model of Cu-O_2 lattice will be presented. In particular, we found the effects of doping is to suppress long range antiferromagnetic order, while the effects of oxygen vacancy is to promote ferromagnetic correlations between the neighboring Cu atoms of the vacancy site in an otherwise antiferromagnetic background.

I. INTRODUCTION

Following the discovery of superconductivity above 30 K by Bednorz and Müller in $La_{2-x}Ba_xCuO_4$[1], there has been an unprecedented effort to search for new superconductors with higher T_c's, stronger critical currents, and larger critical field. Much of the theoretical effort has been focussed on the origin of the high temperature superconductivity. The transition temperatures of the copper oxides are three to five times higher than those of intermetallics with comparable densities of states at the Fermi energy (E_F). The extremely small isotope effects[2] and many other unusual electronic properties have led to the widespread belief that the electron-phonon interaction by itself cannot be the dominant mechanism for the superconductivity. Instead, the occurrence of antiferromagnetic correlation in the copper oxides seems to support the notion that superconductivity in these compounds may be due to magnetic interactions[3,4]. The only exception is the $Ba_{1-x}K_xBiO_{3-y}$ which has an onset temperature of 30° K, which has the cubic perovskite structure, and most importantly, the $Cu\text{-}O_2$ layer is missing[5]. Instead of antiferromagnetism (spin density wave), the superconducting phase is near a planar breathing type displacement of the oxygen atoms away from one of the Bi neighbors (charge density wave), which seems to suggest alternative pairing mechanism due to charge fluctuations[6,7]. In addition, the movements of the oxygen atoms also suggests a stronger electron-phonon interaction strength. It remains interesting to see if this new material has the same mechanism to achieve higher T_c as the cuprate superconductors.

The band structures of $La_{2-x}(Sr,Ba)_xCuO_{4-y}$[8,9], $YBa_2Cu_3O_{7-x}$[10,11,12,13], and the new bismuth[14,15,16,17] and thallium[18,19] based superconductors have been calculated using a number of different band structure methods. The unusual superconducting transition phase diagram of $YBa_2Cu_3O_{7-x}$ as a function of oxygen vacancy concentration can be understood from (1) the hole count on the $Cu\text{-}O_2$ layer which modifies the electron-electron correlations, and (2) the metallic nature of the chain layers near E_F which control the interlayer coupling between the $Cu\text{-}O_2$ layers. Similar arguments can also be applied to the new bismuth and thallium based copper oxides where T_c increases with increasing number of $Cu\text{-}O_2$ layers. In this paper, we discuss the effects of these parameters on electron-electron correlations, magnetic interactions, and superconducting properties.

Fermi surfaces have been observed in some of these systems which seem to correlate well with the prediction of local density band theory and lend support to the Fermi liquid picture. However, there are still several important discrepancies between band theory and experiment. This is evident from comparisons with the photoemission experiments which show significant renormalization from the theoretical density of states. Furthermore, the observed antiferromagnetic ground state in La_2CuO_4 can not be stabilized in the standard spin-polarized band structure calculations. These results reflect an inadequacy of the local spin density approximation in describing the intraatomic Coulomb repulsion U.

The source of the error will be analyzed based on our understanding of two related compounds: Mott insulators and heavy fermion superconductors.

II. THE EFFECTS OF DOPING AND OXYGEN VACANCIES.

Fig. 1 shows the experimental phase diagram for $La_{2-x}(Sr,Ba)_xCuO_{4-y}$ and $YBa_2Cu_3O_{7-x}$. Superconductivity in these copper oxides occurs near a structural instability (orthorhombic ⇐⇒ tetragonal) as well as near a metal-insulator transition which can be controlled by doping or oxygen stoichiometry. At low temperature, the metallic phase becomes superconducting while the insulating phase shows long range antiferromagnetic order. The structure of most of the high T_c superconductors is **layered** perovskite with large anisotropy, and the most unique feature is the presence of the $Cu\text{-}O_2$ layers. This structure gives rise to a $Cu\text{-}O_2$ pdσ antibonding band which crosses the Fermi energy. It appears to be a general characteristic that this band is always less than half-filled for superconductivity to occur. The simplest example is La_2CuO_4, for which the pdσ band is exactly half-filled. This is an antiferromagnetic insulator, due to Fermi surface nesting. The compound becomes superconducting when the divalent La atoms are replaced by the trivalent Ba or Sr atoms. Since the La atoms contribute very little to the occupied conduction bands, substitutional alloying for trivalent La with either a divalent (Sr or Ba) or tetravalent element may be described

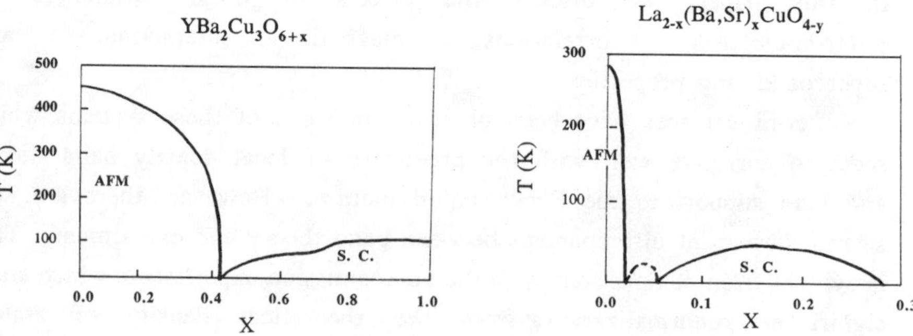

Fig. 1. Experimental phase diagrams for $YBa_2Cu_3O_{7-x}$ and $La_{2-x}(Ba,Sr)_2CuO_{4-y}$.

by the simple rigid-band model. It is assumed, that the changes in the density of states (*DOS*) are negligible and E_F can be shifted rigidly to accommodate the extra charge. In the range of doping for superconductivity to occur, the rigid band picture has been confirmed by tight-binding CPA calculations[20]. Thus, the effects of doping can be characterized by shifting the number of electrons in the antibonding pdσ band away from the half-filled limit.

Unlike the La atom which does not contribute much to the occupied conduction band, the oxygen atoms actively participate in the bonding and antibonding states. Therefore, the effects of oxygen vacancies cannot be understood purely on the basis of the rigid band model. Recently, Sterne and Wang[21] have studied the effects of oxygen vacancies on the electronic properties of La_2CuO_4 using a supercell geometry. The supercell is constructed by doubling the tetragonal unit cell in the two planar directions with 28 atoms in the unit cell. One out of the 16 oxygen atoms in the supercell was removed from the two dimensional Cu-O_2 network which formally corresponds to $La_2CuO_{3.75}$.

The effects of the oxygen vacancy are clearly indicted in Fig. 2 where the band structure of La_2CuO_4 in this 28 atom supercell (Fig. 2a) is compared with that of the $La_2CuO_{3.75}$ (Fig. 2b). Oxygen vacancies alter the band structure significantly near E_F with the fourfold degeneracy at **a** and **b** completely lifted. Removing an oxygen atom reduces the coordination numbers of the neighboring copper atoms which leads to a narrower conduction band width. It is interesting to note that the system is very close to being a semiconductor. The electron pocket around **b** is compensated by a hole pocket elsewhere in the Brilliouin zone but the overlap is clearly very small.

Another consequence of the oxygen vacancy is to produce an upward shift of E_F which can be seen by comparing the fourfold degenerate levels at **a** and **b** that lie **above** E_F in Fig 2a with the corresponding gaps **at** E_F in Fig. 2b. A similar conclusion can be drawn by comparing the band structures of $YBa_2Cu_3O_7$ and that of $YBa_2Cu_3O_6$[22]. This can be understood from the ionicity of the oxygen atom in a manner similar to the case of the Ba or Sr doping. If we assume that the oxygen atom is present as O^{2-}, then removing an oxygen atom will leave two extra electrons which

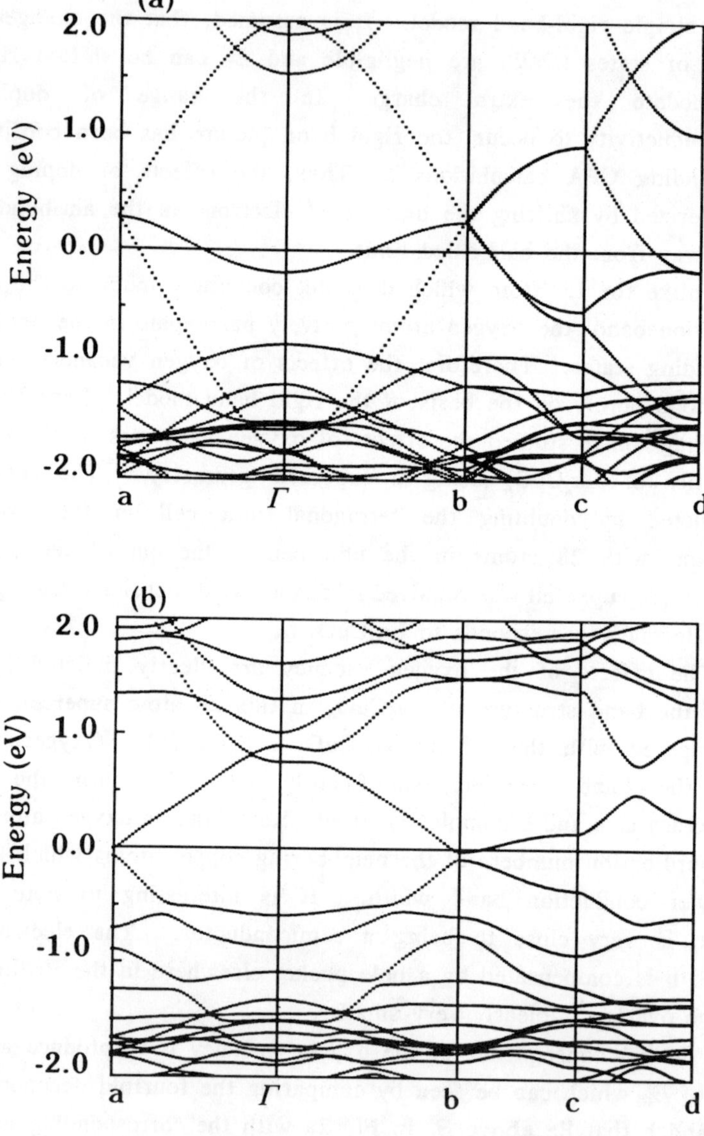

Fig. 2. Energy bands of (a) La_2CuO_4 folded back to the reduced zone of the 28 atom supercell, and (b) $La_2CuO_{3.75}$ with one out of sixteen oxygen atoms removed from the two-dimensional Cu-O plane in the supercell, from Sterne and Wang[21].

are accommodated in part by an upward shift of E_F. It is assumed that removing a neutral oxygen atom with four valence p-electrons will at the same time remove six valence p-states, thereby pushing E_F upward. We found 0.3 electrons remain in a sphere of radius 1.9 a.u. around the vacancy site, which is stabilized by the Madelung potential somewhat similar to the F-center in ionic oxides. These results indicate the importance of self-consistency to model oxygen vacancy. Recently, Papaconstantopoulos et al.[20] have used the tight-binding coherent potential approximation to understand the disorder phase of oxygen vacancies and found that deoxygenation does not raise E_F contrary to simple doping arguments. However, the vacancy onsite energy is taken as infinite in their calculation so that all charge is expelled from the vacancy sites. It remains interesting to see if the discrepancy reflects errors of non-self-consistent charge distribution in the CPA calculations or effects of disorder not included in the supercell model.

III. INTERLAYER COUPLING

The recent discovery of high temperature superconductivity in bismuth[23] and thallium[24] based copper oxides has provided valuable insight to the mechanism of superconductivity. The structures of these materials have been determined[25,26,27], and T_c depends on the number of Cu-O_2 layers in a very simple way: The more Cu-O_2 layer, the higher the T_c. For example, $Bi_2Sr_2CuO_6$ which has only one Cu-O_2 layer has a comparatively low T_c of about 6-22K[28], while the two-layer compound $Bi_2Sr_2CaCu_2O_8$ superconducts at 84K.

Sterne and Wang[17] have calculated the electronic structure for the one- and two-layer bismuth compounds and compared them to see what may be responsible for the large difference in their transition temperatures. The resulting band structures are shown in Fig. 3. We found the high T_c compound has Bi-O bands crossing the Fermi energy so that the Bi-O planes which lie between the CuO_2 layers are metallic while in the low T_c compound these Bi-O bands are almost unoccupied so that the Bi-O layers are nearly insulating. Since the metallic nature of the Cu-O chain layer bands in $YBa_2Cu_3O_{7-x}$ also decreases with decreasing T_c as oxygen vacancies are introduced, we proposed that the metallic nature of the layers between the Cu-O_2 layers could enhance T_c significantly. In this

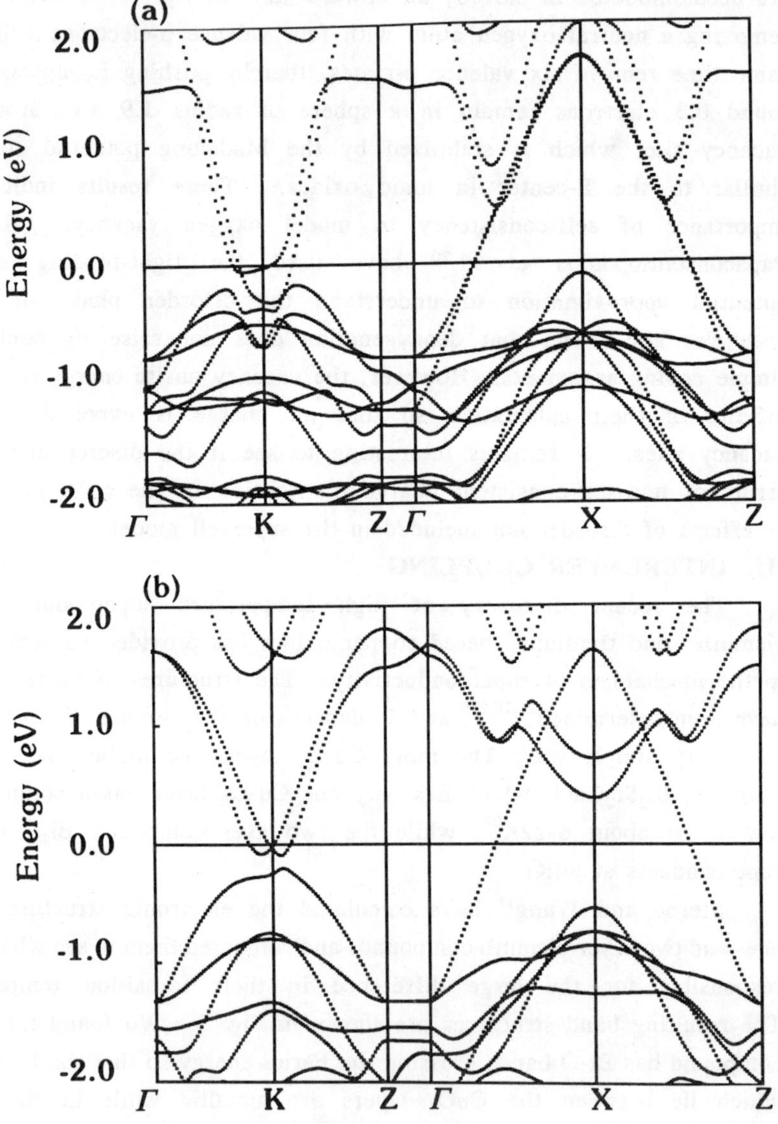

Fig. 3. Energy band structures around the Fermi energy for (a) the high temperature superconductor $Bi_2Sr_2CaCu_2O_8$ and (b) the low temperature superconductor $Bi_2Sr_2CuO_6$, from Sterne and Wang[17]. Note the presence of the Cu-O antibonding band in both systems and the difference in the bismuth bands around the K point.

picture, we assume that the superconducting transition occurs through Bose condensation of quasi-particles in the two-dimensional Cu-O_2 layers. According to the Mermin-Wagner theorem, this transition cannot be entirely two-dimensional, since fluctuations prevent the transition from occurring at a finite temperature. The intervening metallic layers enhance the coupling between neighboring Cu-O_2 layers to make the system weakly three-dimensional.

Table 1. Interlayer separations and coherence lengths (in Å) for a number of high temperature superconductors [17]. $d_{cu\text{-}cu}$ is the perpendicular distance between Cu-O_2 planes. $d_{Cu\text{-}M}$ is the perpendicular distance from the Cu-O_2 layer to the intervening metallic layer, Cu-O in $YBa_2Cu_3O_7$ and Bi-O in the bismuth compounds. ξ_c is the coherence length perpendicular to the Cu-O_2 plane.

	$d_{Cu\text{-}Cu}$	$d_{Cu\text{-}M}$	ξ_c
$La_{2-x}Sr_xCuO_4$	6.6[a]	-	7 - 13[e]
$YBa_2Cu_3O_7$	8.2[b]	4.1[b]	4.8-7.0[f]
$Bi_2Sr_2CaCu_2O_8$	12.1[c]	4.4[c]	4.0[g]
$Bi_2Sr_2CuO_6$	12.3[d]	4.5[d]	-

[a]Reference [29] [b]Reference [30] [c]Reference [31]
[d]Reference [32] [e]Reference [33] [f]Reference [34]
[g]Reference [35]

The metallic nature of the Bi-O_2 layers in $Bi_2Sr_2CaCu_2O_8$ or the Cu-O chain layers in $YBa_2Cu_3O_{7-x}$ are directly tied to the hole counts in the Cu-O_2 layers since they arise from charge transfer from one layer to another. The metallic interlayers enhance coupling between the Cu-O_2 planes, while the hole concentration in the Cu-O_2 layers is an important parameter for the strength of electron correlation. To emphasize the effects of interlayer coupling, we show in Table 1 the experimentally measured coherence lengths normal to the Cu-O_2 planes, ξ_c, and the Cu-O_2 interlayer spacing, $d_{Cu\text{-}Cu}$ for the $La_{2-x}Ba_xCuO_4$, $YBa_2Cu_3O_7$ and $Bi_2Sr_2CaCu_2O_8$ systems. In $La_{2-x}Sr_xCuO_4$, where there is no metallic layer between the Cu-O_2 planes, the coherence length is long enough to allow supercurrent to flow directly between the layers, but in $YBa_2Cu_3O_7$ and $Bi_2Sr_2CaCu_2O_8$, ξ_c is smaller than the distance between the two Cu-O_2 planes. In both cases, however, ξ_c is comparable to the distance $d_{Cu\text{-}M}$ from the Cu-O_2 layer to the metallic chain layer or Bi-O plane, so the supercurrent could then flow between Cu-O_2 layers by taking advantage of the metallic states on the intervening layers, essentially hopping from copper-oxygen layer to copper-oxygen layer by tunneling through the metallic interlayer. Since the coherence length can be defined within the phenomenological Ginzburg-Landau theory, the above argument is in principle applicable to most pairing mechanisms that have been proposed for the high-T_c superconductors. However, the unusually low coherence length could lead to an unusual flux-pinning picture and thus the critical field measurements may need to be reevaluated in this light[36]. Apparently, there is a strong frequency dependence of the critical fields and the coherence lengths may be even shorter than those shown in Table 1.

Our model is also applicable to $Tl_2Ba_2CuO_6$, $Tl_2Ba_2CaCu_2O_8$, and $Tl_2Ba_2Ca_2Cu_3O_{10}$ which have T_c of 80K, 110K, and 125K, respectively. The band structures of the two- and three- layer thallium compounds have been reported by Yu et al.[18] and that of the one- layer compounds by Hamann and Mattheiss[19]. Once again, the intervening metallic Tl-O layers play the dual role of modifying the hole concentration of the $Cu\text{-}O_2$ layers and enhancing the interlayer coupling between the $Cu\text{-}O_2$ layers. These effects increase with increasing number of $Ca{:}Cu\text{-}O_2$ layers and therefore increasing T_c. However, the degree of charge transfer among the Tl-O and $Cu\text{-}O_2$ layers appears to be much weaker than that of the bismuth compounds considering how high the superconducting temperatures are. This apparent discrepancy can be understood from recent neutron diffraction measurements on many different samples of $Tl_2Ba_3CuO_6$ which have shown that T_c varies from 4K to 70K depending on the precise structure arrangement[37]. In particular, there is an orthorhombic phase with well-ordered superstructure which is non-superconducting, while the superconducting material is pseudo-tetragonal with **disordered** oxygen within the Tl-O plane.

Recently, Lee et al.[38] have observed significant frequency shifts of thallium nuclear magnetic resonance in the superconducting state of the high temperature superconductor $Tl_2Ba_3Ca_3Cu_4O_{10+x}$, which suggests that the Tl-O layers participate directly in the superconductivity. Within our model, the superconductivity occurs in the two-dimensional $Cu\text{-}O_2$ layers, and interlayer tunneling is essential because the coherence length normal to the Cu-O_2 layer is so short. Thus, the observed superconductivity in the metallic Tl-O layer may be interpreted as arising from proximity effects.

IV. LIMITATIONS OF THE LOCAL DENSITY BAND THEORY.

The observed antiferromagnetic phase of La_2CuO_{4-y} is not the ground state of the local-density band theory. Sterne and Wang have carried out spin-polarized band structure calculations on stoichiometric La_2CuO_4 and two ordered oxygen-vacancy supercells corresponding to $La_2CuO_{3.75}$.[21] Both ferromagnetic and antiferromagnetic band structures were calculated for La_2CuO_4 and found to be unstable. These conclusions are similar to an earlier calculation by Leung et al for Sc_2CuO_4[39].

The failure of the band theory to stabilize the observed antiferromagnetic ground state indicates an inadequacy of the local-spin-density approximation in describing the intra atomic Coulomb repulsion U. This is also evident from the comparison with photoemission experiments which indicates that the theoretical Cu_{3d} density of states are much too narrow and much too close to the Fermi energy (E_F)[40]. Furthermore, the *XPS* and *BIS* intensities at E_F are weaker than the theoretical estimates. Large Coulomb correlation energies have also been inferred from the satellite feature at about 12.4 eV below E_F in the valence band photoemission spectra, which corresponds to Cu-$3d^8$ final states pushed out of the valence band by U_d. The fact that the $3d^8$ state occurs at such high energy indicates that charge fluctuation to $3d^8$ must be strongly suppressed in the ground state. Nücker et al.[41] studied excitations of the O_{1s} electron into the local unoccupied part of the *DOS* at the O atom by electron energy-loss spectroscopy. They proposed that the semiconducting compound is a charge-transfer semiconductor with an almost completely filled O_{2p} valence band; while for the superconducting compound, holes are created in the O_{2p} band which dominates the *DOS* at E_F. Since the cross section for the O_{2p} states are much weaker than that of the Cu_{3d} states, this model naturally explains the low *BIS* and *XPS* intensity at E_F. It also explains why very little change was found in the Cu_{2p} *XAS* spectra[42], which measures holes on the Cu site, upon variation of x and y. The effects of the strong electron correlations in the high temperature superconductors are not fully understood, but some insight can be gained from our understanding of two related systems: (1) Mott insulators, and (2) heavy fermion superconductors.

V. MOTT INSULATORS.

The mott insulators represent one of the most dramatic failures of the local density band theory. It has been shown by Terakura et al.[43] that the *AFM* ground states are obtained from band theory; but in the case of FeO or CoO the observed insulating gaps were not obtained, and in the case of NiO they are an order of magnitude too small. The mechanism for the breakdown of the local-density band theory in Mott insulators has recently been reviewed by Brandow[44]. Because of crystal-field splitting of the 3d states, the different-orbital exchange parameter J can be much smaller than the self-exchange parameter U. Orbital dependence of the exchange and correlation potential is neglected in the local-spin-density approximation, which amounts to averaging over these two very different quantities. It thereby misses the main point of the Mott picture, namely, that the unoccupied orbital lies above that of an occupied orbital by an amount of order U. The problem is greatly reduced in wide-band materials where U is strongly screened so that the mechanism for the Mott splitting is weakened. In the case of the monoxides, the O_p states lie significantly lower in energy than the transition metal 3d states so that the overall band width W is smaller than U. In high T_c superconductors, $E_d \cong E_p$ which leads to a rather large band width (W ≥ 9eV). Nevertheless, the system is near a metal-insulator transition so that the screening is not very effective.

Recently, Savane and Gunnarsson[45] have included self-interaction correction (SIC) to the local-spin-density (LSD) approximation of a two-dimensional Hubbard model of a $Cu\text{-}O_2$ lattice. The SIC approximation removes the unphysical Coulomb interaction of an electron with itself as well as the corresponding LSD exchange-correlation potential. Unlike LSD approximation, this potential is orbital-dependent. They used a set of realistic values for the screened Coulomb repulsion (U_d=8 eV and U_p=5 eV) in the LSD calculation, and subtracted "bare" interactions U_d=25 eV and U_p=20 eV in the SIC calculation since the other electrons do not feel the SIC potential. They found the SIC-LSD approximation gives a somewhat too large moment for the system when compared with experiments, while the

LSD approximation greatly underestimates the tendency to antiferromagnetism. These results suggest the possibility of studying the antiferromagnetic phase by including SIC in the band structure calculation. However, the superconducting phase may be dominated by short-range spin or charge fluctuations which can not be understood from a single particle picture.

The Mott-Hubbard theory has been generalized by Zannen et al.[46] to describe a wide range of different transition metal compounds. They solved the Anderson impurity Hamiltonian which includes quantum charge fluctuations beyond the mean field approximation. They showed that the correlation gap can be either *d-d* or charge transfer character depending on the relative magnitude of the parameters U_d and U_{pd}, and that the gap can go to zero even if U_d is large. There is a metallic ground state in which the 3d electrons are delocalized even for U_d large, provided U_{pd} is less than half of the ligand valence band width. This criteria appears to be satisfied by the high T_c superconductors although the overall band width of the superconductors is significant larger. In addition, there are RKKY interactions between localized moments in a solid, which are neglected in the Anderson impurity model. Spin fluctuations derived from the RKKY interaction may be the pairing mechanism for heavy-fermion superconductors.

VI. Heavy-Fermion Superconductors

The heavy fermion superconductors are characterized by extremely large values of the linear coefficient of the specific heat and of the magnetic susceptibility, which is interpreted as implying fermionic quasi-particles bearing a very large effective mass m* at low temperature[47]. By contrast, values of m* deduced from transport measurements are not significantly enhanced, which suggested that the energy dependence of the self-energy may dominate its momentum dependence[48]. As the temperature is raised, m* rapidly decreases to normal values. This suggests that the self-energy corrections diminish rapidly away from E_F and explains the good agreement between the band structure and angle resolved photo emission for UPt_3 even for f-states that lie within 0.1 eV from E_F (m* is a factor of 21 larger than the already heavy band mass). The most striking result is that the complicated Fermi surfaces predicted by band theory[49] for UPt_3 was layer confirmed by experiment[50]. The picture that emerges is that the f-electrons in UPt_3 are itinerant with quasi-particle energies $E(k)$ (solid line in Fig. 4) that are well described by the local density band structure $\varepsilon(k)$ (dashed line in Fig. 4), except near the chemical potential μ where there is a strongly frequency dependent self-energy correction $\Sigma(k,\omega)$ due to spin fluctuations:

$$E(k) - \mu = \varepsilon(k) - \mu + \Sigma(k\ ,\ E(k)-\mu) \tag{1}$$

The magnetic fluctuations responsible for the mass enhancement in the thermodynamic measurements may also be the pairing mechanism for its superconductivity. To study the effects of quantum fluctuations, McQueen and Wang[51] have performed finite-temperature Monte-Carlo simulations of a two-dimensional 4x4 Anderson lattice. We find strong *AFM* correlation above the single-impurity Kondo temperature (T_K). Below T_K, both the local moments and the *AFM* correlations are suppressed somewhat, and *FM* correlations start to develop. While *AFM* correlations suggest nearest neighbor anisotropic singlet pairing, the *FM* correlations could support $E(k)$ p-wave triplet-pairing superconductivity. Within Stoner theory, the enhancement of the *FM* correlations at lower temperature can be understood from the mass enhancement factor. As can be seen from Fig. 4, only

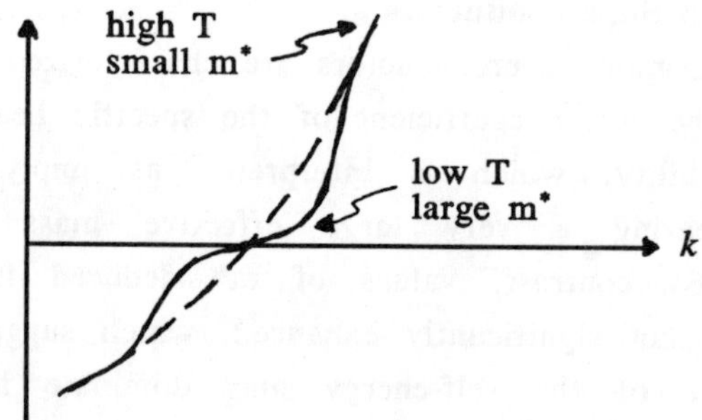

Fig. 4. Quasi-particle energy $E(k)$ (solid line) vs. the local density band energy $\varepsilon(k)$ dashed line). At low temperature, only states of heavy m^* are occupied. At sufficiently high temperature, states of normal m^* will be populated.

states of heavy mass are occupied at low temperature, but m* decrease rapidly with increasing temperature as the low lying excited states of normal m* start to be populated.

Experimentally, there is a T^2 dependence in the resistivity of UPt_3 just above its superconducting transition temperature T_c, as well as a $T^3 \log T$ term in its low-temperature normal-state specific heat[52]. Both of these behaviors have been attributed in systems like UAl_2 and $TiBe_2$ to long-range **ferromagnetic** spin fluctuations (paramagnons). This is in contrast to neutron scattering measurements[53] that show strong **antiferromagnetic** correlations in UPt_3 which increase linearly with decreasing temperature but appear to level off below T_c. These results may be consistent with our picture that the energy dependence of the self-energy may dominate its momentum dependence, and therefore the momentum dependence of the magnetic structure function is very weak.

VIII. THE IMPLICATIONS OF THE OBSERVED FERMI SURFACES.

Recently, Fermi surfaces have been measured by two-dimensional angular correlation of the positron annihilation radiation (2D-ACPAR) technique on a single crystal $YBa_2Cu_3O_{7-x}$ [54]. The momentum distribution shows structures associated with the higher zones, which indicates positron annihilation with extended rather than localized electrons. Three nearly cylindrical Fermi surface sheets have been observed. There is a square piece centered at the *S* point, which corresponds to the two bands of predominantly $Cu\text{-}O_2$ layer character which are too close to one another to be resolved. Qualitatively, the agreement for the other two bands of predominantly chain character appears to be reasonable considering the facts that (1) they are much more sensitive to oxygen vacancies, and (2) there are some discrepancies in different theoretical predictions for one of the chain bands with very large effective mass.

Qualitative agreement between theory and experiments for the topology of the Fermi surfaces raises serious doubt about the resonating valence bound theory, which is probably the most innovative model that has been proposed for superconductivity in these materials. In this picture, the normal state consists of highly degenerate, short range, singlet pairs of electrons, and superconductivity arises from Bose condensation of topological disorders. In the half-filled limit, the system is an insulator. Away from the half-filled limit, there may be a "pseudo Fermi surface" in the normal state, if the band gap vanishes along certain lines or points in the Brillioin zone, but it will be very different from that of a paramagnetic band structure.

The agreement between band theory and experiments for the Fermi surfaces is consistent with strong electron correlations. The best example is UPt_3 discussed in section VI, which is characterized by a strongly frequency dependent self-energy correction $\Sigma(k,\omega)$ {see Eq. (1)}. For low energy, $\Sigma(k,\omega)$, which is dominated by its frequency dependence, can be expanded linearly to yield:

$$E(k) - \mu = z(k) \left[\varepsilon(k) - \mu + \Sigma(k, 0) \right] \tag{2}$$

where

$$z(k) = \frac{1}{\left| 1 - \frac{\partial \Sigma(k.\omega)}{\partial \omega} \right|_{\omega=0}} \tag{3}$$

is the renormalization factor. The quasi-particle Fermi surface, which is determined by $E(k)=\mu$, will not be renormalized from the band Fermi surface $(\varepsilon(k)=\mu)$ if $\Sigma(k, 0) \cong 0$, no matter how large $z(k)$ is. For an interacting electron gas, the energy dependence of the self-energy vanishes at μ $(\Sigma(k,0)=0)$[55]. Moreover, according to the Luttinger's theorem[56], the volume enclosed by the Fermi surface should be identical to the non-interacting case. These are two of the reasons for the success of the band theory in spite of the enormous mass enhancement factor in UPt_3.

The situation is quite different from some other heavy fermion compounds. For example, the experimental Fermi surfaces of UPd_3 can not be reproduced by band theory until the f-electrons are treated as localized core states that do not hybridize with the conduction bands[57]. In this case the f-electrons are more localized, with extremely narrow band width, and the system is not a superconductor.

To illustrate these ideas, we show in Fig. 5, a recent variational Monte-Carlo simulations of McQueen and Wang[58] for the momentum density $<n_k^+ n_k>$ of the one-band Hubbard model ($U=16t$) on a 10x10 square lattice with three different filling factor (a) ν = 1.000, (b) ν = 0.9375, and (c) ν = 0.800. The momentum density can be measured in

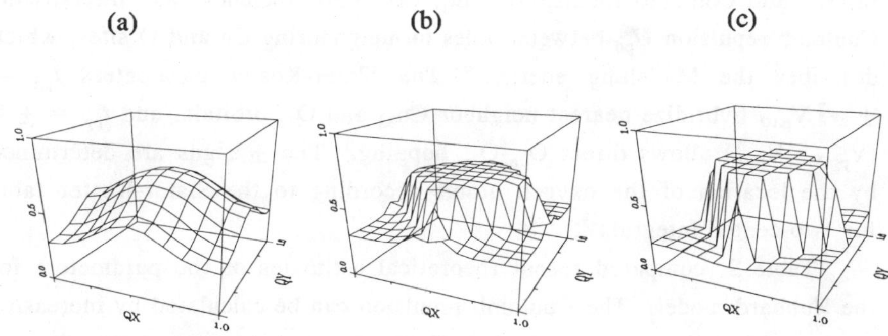

Fig. 5. The momentum distribution function $n(k)$ for a Hubbard model on a 10x10 lattice for U=16t, where t is the hopping integral, from McQueen and Wang[58]. The particle numbers are: (a) $\nu=1$, (b) $\nu=0.9375$, and (c) $\nu=0.8$.

2D-ACPAR experiments. In the half-filled limit (a), it is an antiferromagnetic insulator so that there is no Fermi surface effect. The system gradually loses long-range antiferromagnetic order as the filling factor is reduced. As can be seen from Fig. 5b and 5c, the volume enclosed by the Fermi surface is reproduced exactly in accordance with Luttinger's theory[56], while some of the spectral weight is shifted to higher momentum due to electron correlation. It is interesting to note in case (c), that there is a depletion of $n(k)$ for $k > k_F$ near the Fermi surface. The extra depletion leads to a jump at the Fermi surface which is the renormalization factor $z(k)$.

VIII. THE HUBBARD MODEL

Recently, much theoretical effort has been applied to estimate the effective Hubbard Hamiltonian for the antibonding pdσ band:

$$H = E_d \sum_i n_{d,i} + U_d \sum_i n_{d,i\uparrow} n_{d,i\downarrow} + E_p \sum_i n_{p,i} + U_p \sum_i n_{p,i\uparrow} n_{p,i\downarrow}$$
$$+ t_{pd} \sum_{<ij>} (d^+_{i\sigma} p_{j\sigma} + p^+_{i\sigma} d_{j\sigma}) + t_{pp} \sum_{<ij>} (p^+_{i\sigma} p_{j\sigma} + p^+_{i\sigma} p_{j\sigma})$$
$$+ U_{pd} \sum_{<ij>} n_{p,i} n_{d,j}, \qquad (4)$$

where i, j denotes site indices and σ (=$\uparrow,\downarrow$) denote spins. The d^+ (d) and p^+ (p) create (destroy) a **hole** on the Cu_{3d} and O_{2p} sites respectively; E is the on-site energy, and U_p, U_d are the corresponding intraatomic Coulomb repulsions. Eq. (4) also includes an interatomic Coulomb repulsion U_{pd} between holes on neighboring Cu and O sites, which describes the Madelung energy. The Slater-Koster parameters t_{pd} = $\pm\frac{1}{2}\sqrt{3}\, V_{pd\sigma}$ hybridize nearest neighbor Cu_{3d} and O_{2p} orbitals, and $t_{pp} = \pm \frac{1}{2}$ $(V_{pp\sigma} - V_{pp\pi})$ allows direct O_{2p}-O_{2p} hopping. The $\pm$ signs are determined by the location of the oxygen atoms according to the Slater-Koster table for two-center integrals[59].

Table 2. compared recent theoretical estimates of the parameters for the Hubbard model. The Coulomb repulsion can be calculated by increasing or decreasing the occupancy of a single impurity site in a superlattice, and then comparing the resulting total energies. This is more difficult for the itinerant O_{2p} bands than the localized Cu_{3d} state because the Wannier functions for itinerant states are not well defined. This explains the relatively large discrepancy in $U_p \cong$ 4-14 eV compared with $U_d \cong$ 8-10 eV, and $U_{pd} \cong$ 0-2 eV, among different calculations shown in Table 2.[60,61,62,63] Making the Anderson impurity approximation and using the methods of Gunnarsson and Schönhammer[64], McMahan et al.[60] found reasonably good agreement with experimental photoemission spectra. The impurity model calculations lead to a magnetic ground state with a correlation gap E_{BIS}-E_{XPS} = 1.3 eV, which should be added to crystal field splittings. In this picture, the ground state of stoichiometric La_2CuO_4 is an insulator if the lattice broadening of the two levels is not sufficient to fill in the gap.

Table 2. Parameters for the two-dimensional extended Hubbard model {Eq. (4)} of La_2CuO_4. Energies are given in eV.

	McMahan et al.[60]	Chen et al.[61]	Zannen et al.[62]	Schluter et al.[63]
E_d	-4.1			
E_p	-2.9			
t_{pd}	-1.6			
t_{pp}	0.65			
U_d	8.0	9	8.0	10±1
U_p	4.1 - 7.3	14.3		7±1
U_{pd}	≥0 6	1.6	0-1	2±1

Recently, Bhattacharya et al.[65] have performed finite temperature quantum Monte Carlo simulation of a two dimensional Hubbard model on a

4x4 Cu-O_2 lattice. Specifically, we studied the effects of doping within the rigid band model and the effects of oxygen vacancies as a function of the onsite vacancy levels. Among our major results, we found doping suppresses the long-range AFM correlations, while an oxygen vacancy promotes ferromagnetic (FM) coupling between the neighboring copper atoms in an otherwise AFM background. Our results are consistent with recent neutron scattering experiments that were carried out on a series of $La_2Sr_{2-x}CuO_4$ single crystals with a Meissner fraction less than 20%[66]. Apparently, doping with Sr does not change the magnetic moments but shortens the correlation length. Otherwise, there is no important difference in magnetic scattering in the normal and superconducting states.

The FM correlation surrounding an oxygen vacancy introduces magnetic frustration which can alter the distribution of the oxygen vacancies. This is an important issue in understanding the structural properties of high T_c superconductors. The transformation from AFM to FM can be understood as follows: Near the half-filled limit, antiferromagnetism is more favorable because it allows electrons to hop to their neighboring sites thereby reducing its kinetic energy. A vacancy disrupts the electron hopping and therefore reduces the gain in kinetic energy in antiferromagnetism. In addition, vacancies tend to reduce band width, making electrons more localized, sometimes opening band gaps; they also introduce defect levels in the correlation gap. These effects are smeared out a little by disorder, but ferromagnetism may become more favorable if the cost in kinetic energy of occupying only one spin band is reduced.

In many ways, our results are similar to those of Aharony et al.[67], who use a **classical** Heisenberg model to describe the dopant dependent phase diagram of $La_{2-x}(Sr,Ba)_xCuO_{4-y}$. They found the classical ground state for the coupling of the neighboring Cu atoms with a localized O^- hole (introduced by doping) is $|+ - +\rangle$. In our picture, the dopant induced oxygen holes are itinerant but the localized oxygen vacancies are somewhat similar to spins of a classical model.

ACKNOWLEDGEMENT

This review includes work done in collaboration with her post doc P. A. Sterne and her students P. G. McQueen and A. Bhattacharya. This work is supported by the National Science Foundation (Grant No. DMR-86-01708) and the Office of Naval Research (Contract No. N00014-86-K-0266). Computations were carried out under the auspices of the National Science Foundation at the Pittsburgh Supercomputer Center.

REFERENCE

1 Bednorz, J. G. and Müller, K. A., Z. Phys. B64, 18 (1986).

2 Batlogg, B., Cava, R. J., Jayaraman, A., van Dover, R. B., Kourouklis, G. A., Sunshine, S., Murphy, D. W., Rupp, L. W., Chen, H. S., White, A., Short, K. T., Mujsce, A. M., and Rietman, E. A., Phys. Rev. Lett. 58, 2333 (1987). Bourne, L. C., Crommie, M. F., Zettl, A., zur Loye, H-C., Keller, S. W., Leary, K. L., Stacy, A. M., Chang K. J., Cohen, M. L. and Morris, D. E., Phys. Rev. Lett. 58, 2337 (1987). Leary, K. J., zur Loye, H-C., Keller, S. W., Faltens, T. A., Ham, W. K., Michaels, J. N., and Stacy, A. M., Phys. Rev. Lett. 59, 1236 (1987). Faltens, T. A., Ham, W. K., Keller, S. W., Leary, K. J., Michaels, J. N., Stacy, A. M., zur Loye, H-C., Morris, D. E., Barbee, T. W., Bourne, L. C., Cohen, M. L., Hoen, S., and Zettl, A., Phys. Rev. Lett. 59, 915 (1987). Batlogg, B., Kourouklis, G., Weber, W., Cava, R. J., Jayaraman, A., White, A. E., Short, K. T., Rupp, L. W., and Rietman, E. A., Phys. Rev. Lett. 59, 912 (1987). Bourne, L. C., Zettl, A., Barbee, T. W. and Cohen, M. L., Phys. Rev. B36, 3990 (*RC*), (1987).

3 Anderson, P. W., Baskaran, G., Zou, Z. and Hsu, T., Phys. Rev. Lett. 58, 2790 (1987); Anderson, P. W. and Zou, Z., *ibid.* 60, 132 (1988); Kivelson, S. A., Rokhsar, D. S. and Sethna, J. P., Phys. Rev. B 35, 8865 (1987)

4 V. J. Emery, V. J., Phys. Rev. Lett., 58, 2794 (1987); Hirsch, J. E., *ibid*, 59, 228 (1987); Parmenter, R. H., *ibid.* 59, 923 (1987); Schrieffer, J. R., Wang, X.-G., Zhang, S.-C., *ibid*, 60, 944 (1988); Aharony, A., Birgeneau, R. J., Coniglio, A., Kastner, M. A., and Stanley, H. E., *ibid*, 60, 1330 (1988).

5 Cava, R. J., et al., Nature (London) 332, 814 (1988).

6 Varma, C. M., Schmitt-Rink, S. and Abrahams, E., Solid State Commun. 62, 681 (1987)

7 Ruvalds, J., Phys. Rev. B 35, 8869 (1987)

8 Mattheiss, L. F., *Phys. Rev. Lett.* 58, 1028 (1987).

9 Yu, J., Freeman, A. J., and Xu, J.-H., *Phys. Rev. Lett.* 58 1035 (1987).

10 Mattheiss, L. F. and Hamann, D. R., *Solid State Commun.* 63, 395 (1987).

[11] Massidda, S., Yu, J., Freeman, A. J. and Koelling, D. D., *Phys. Lett.* A122, 198 (1987); Yu, Massidda, S., Freeman, A. J. and Koelling, D. D., *Phys. Lett.* A122, 203 (1987).

[12] Krakauer, H. and Pickett, W. E., in *Novel Superconductivity*, ed. S. A. Wolf and V. Z. Kresin (Plenum, New York, 1987) p. 501.

[13] Krakauer, H. and Pickett, W. E., in *Novel Superconductivity*, ed. S. A. Wolf and V. Z. Kresin (Plenum, New York, 1987) p. 501.

[14] Hybertsen, M. S. and Mattheiss, L. F., *Phys. Rev. Lett.* 60, 1661 (1988).

[15] Krakauer, H. and Pickett, W. E., *Phys. Rev. Lett.* 60, 1665 (1988).

[16] Massidda, S., Yu, J., and Freeman, A. J., to be published in *Physica* C Superconductivity, (1988).

[17] Sterne, P. A. and Wang, C. S., *J. Phys. C* 21, L949 (1988)

[18] Yu, J., Massidda, S., and Freeman, A. J., *Physica* C 152, 273 (1988) (1987).

[19] Hamann, D. R. and Mattheiss, L. F., *Phys. Rev.* B38, 5138 (1988).

[20] Papaconstantopoulos, D. A., Pickett, W. E. and DeWeert, M. J., *Phys. Rev. Lett.* 61, 211 (1988).

[21] Stern, P. A. and Wang, C. S., Phys. Rev. B37 7472 (1988).

[22] Yu, J., Freeman, A. J. and Massidda, S., in *Novel Superconductivity*, ed S. Wolf, and V. Z. Kresin (Plenum, New York 1987) p. 367.

[23] Maeda, H., Tanaka, Y., Fukutomi, M. and Asano, T., Jpn. J. Appl. Phys. Lett. 27,L209 (1988); Chu, C. W., et al, Phys. Rev. Lett. 60, 941 (1988); Michel, C., Hervieu, M., Borel, M. M., Grandin, A., Deslandes, F., Provost, J. and Rave, B., Z. Phys. B 68, 421 (1987)

[24] Sheng, Z. Z., Hermann, A. M., El Ali, A., Almasan, C., Estrada, J., Datta, T., and Matson, R. J., Phys. Rev. Lett, 60, 937 (1988); Parkin, S. S. P., Lee, V. Y., Engler, E. M., Nazzal, A. I., Huang, T. C., Gorman, G., Savoy, R. and Beyers, R. (preprint 1988)

[25] Sunshine, S. A., Siegrist. T., Schneemeyer. L. F., Murphy, D W., Cava, R. J., Batlogg, B., van Dover, R. B., Fleming, R. M., Glarum, S. H., Nakahara, S., Farrow, R., Krajewski, J. J., Zahurak. S. M., Waszczak, J. V., Marshall, J. H., Marsh, P., Rupp, L. W., and Peck, W. F., preprint, AT&T Bell Labs., 1988 (to be published).

26 Hazen, R. M., Prewitt, T. C., Angel, R. J., Ross, N. L., Finger, L. W., Hadidiacos, C. G., Veblen, D. R., Heaney, P. J., Hor, P. H., Meng, R. L., Sun, Y. Y., Wang, Y. Q., Xue, Y. Y., Hung, H. J., Gao, L., Bechtold, J., and Chu, C. W., Phys. Rev. Lett. 60, 1174 (1988); Tarascon, J. M., Le Page, Y., Barboux, P., Bagley, B. G., Greene, L. H., McKinnon, W. R., Hull, G. W., Giroud, M. and Hwang, D. M., preprint, Bellcore, 1988 (to be published); Subramanian, M. A., Torardi, C. C., Calabrese, J. C., Gopalakrishnan, J., Morrissey, K. J., Askew, T. R., Flippen, R. B., Chowdhry, U., Sleight, A. W., Science 239, 1015 (1988).

27 Hazen, R. M., Finger, R. W., Angel, R. J., Prewitt, C. T., Ross, N. L., Hadidiacos, C. G., Heaney, P. J., Veblen, D. R., Sheng, Z. Z., El Ali, E., and Hermann, A. M., Phys. Rev. Lett. 60, 1657 (1988).

28 Torrance, J. B., Tokora, Y., LaPlaca, S. J., Huang, T. C., Savoy, R. J. and Nazzal, A. I., Solid State Commun. (in press).

29 Jorgensen, J. D., Schüttler, H.-B., Hinks, D. G., Capone, D. W., Zhang, H. K. and Brodsky, M. B., Phys. Rev. Lett 58, 1024 (1987).

30 Beech, F., Miraglia, S., Santoro, A. and Roth, R. S., Phys. Rev. B 35, 8778 (1987).

31 Sunshine, S. A., et al, preprint, AT&T Bell Labs., 1988 (to be published).

32 Torrance, J. B., Tokora, Y., LaPlaca, S. J., Huang, T. C., Savoy, R. J. and Nazzal, A. I., Solid State Commun. (in press).

33 Shamoto, S., Onoda, M., Sato, M. and Hosoya, S., Solid State Commun. 62, 479 (1987).

34 Worthington, T. K., Gallagher, W. J. and Dinger, T. R., Phys. Rev. Lett 59, 1160 (1987); Umezawa, A., et al, preprint (1987).

35 Batlogg, B., Palstra, T. T. M., Schneemeyer, F. W., vanDover, R. B. and Cava, R. J., Proceedings of the Interlaken Meeting, March 1987 (to be published in Physica).

36 Yeshurun, Y. and Malozemoff, A. P., Phys. Rev. Lett. 60, 2202 (1988)

37 Hewat, A. W., Bordet, P., Capponi, J. J., Chaillout, C., Chenavas, J., Godinho, M., Hewat, E. A., Hodeau, J. L., Marezio, M., submitted to Physica C, (1988).

38 Lee, M., Song, Y.-Q., Halperin, W. P., Tonge, L. M., Marks, T. J., Marcy, H. O., and Kannewurr, C. R., to be published.

39 Leung, T. C., Wang, X. W., and Harmon, B. M., Phys. Rev. B37, 384 (1988)

40 Fuggle, J. C., Weijs, P. J. W., Schoorl, R., Sawatzky, G. A., Fink, J., Nücker, N., Durham, P. J. and Temmerman, W. M., Phys. Re. 37, 123 (1988).

41 Nücker, N., Fink, J., Fuggle, J. C., Durham, P. J. and Temmerman, W. M., *Phys. Rev.* B37, 5158 (1988).

42 Bianconi, A., Congiu Castellano, A., De Santis, M., Rudolf, P., Lagarde, P., Flank, A. M. and Marcelli, A., *Solid State Commun.* 63, 1009 (1987)

43 Terakura, K., Oguchi, Williams, A. R., and Kubler, J., Phys. Rev. B30 4734, (1984).

44 B. H. Brandow, B. H.,to be published in the proceedings of the NATO Advanced Workshop, "Narrow Band Phenomena", Steverden, Netherlands, June 1-5 (1987).

45 Svane, A. and Gunnarsson, O., Phys. Rev. B37, 9919 (1988); Svane, A. and Gunnarsson, O., to be published.

46 Zaanen, J., Sawatzky, G. A., and Allen, J. W., Phys. Rev. Lett. 55, 428 (1985).

47 Lee, P. A., Rice, T. M., Serene, J. W., Sham, L. J., and Wilkens, J. W., Comments on Condensed Matter Physics 12, 99 (1986).

48 Varma, C. M., Phys. Rev. Lett. 55, 2723 (1985)

49 Wang, C. S., Norman, M. R., Albers, R. C., Boring, A. M., Pickett, W. E.,Krakauer, H., and N. E., Phys. Rev. B35 7260 (1987).

50 Taillefer, L., and Lonzarich, G. G., Phys. Rev. Lett., 60, 1570 (1988)

51 McQueen, P. G., and Wang, C. S., to be published.

52 Stewart, G. R., Reviews of Modern Physics 56, 755 (1984).

53 Aeppli, G., Bucher, E., Broholm, C., Kjems, J.K., Baumann, J. and Hufnagl, J., Phys. Rev. Lett. 60, 615 (1988)

54 Semdskjaer, L. C., Liu, J. Z., Benedek, R., Legnini, D. G., Lam, D. J., Stahulak, M. D., and Claus, H., to be published.

55 Hedin, L. and Lundqvist, S., Solid State Phys. 23, 1 (1969).

56 Luttinger, J. M., Phys. Rev. 119, 1153 (1960).

57 Norman, M. R., Oguchi, T., and Freeman, A. J., (unpublished).

58 McQueen, P. G., and Wang, C. S., to be published.

59 Slater, J. C. and Koster, G. F., *Phys. Rev.* 94, 1498 (1954).

60 McMahn, A. K., Martin, R. M. and Satpathy, S., to be published.

61 Chen, C. F., Wang, X. W., Leung, T. C., and Harmon, B. N., to be published.

62 Zaanen, J., Jepsen, O., Gunnarsson, O., Paxton, A. T., Andersen, O. K., and Svane, A., to be published. (Interlaken).

63 Schluter, M., Hybertsen, M. S., and Christensen, N. E., to be published (proc. Interlakken).

64 Gunnarsson, O. and Schönhammer, K., *Phys. Rev.* B28, 4315 (1983).

65 Bhattacharya, A., McQueen, P. G., Wang, C. S., and Einstein, T., to be published.

66 Shirane, G., Int. Conf. on Mag. (1988).

67 Aharony, A., Birgeneau, R. J., Coniglio, A., Kastner, M. A., and Stanley, H. E., Phys. Rev. Lett. 60, 1330 (1988).

PLASMONS IN THE CUPRATES AND THE PHONON-PLASMON MECHANISM OF HIGH T_c

Vladimir Z. Kresin
Materials and Chemical Sciences Division
Lawrence Berkeley Laboratory
1 Cyclotron Road
Berkeley, CA 94720

Hans Morawitz
IBM Research Division
Almaden Research Center
650 Harry Road
San Jose, CA 95120-6099

ABSTRACT

The collective excitations in the high T_c oxides are studied. The dependence of the electron-loss spectroscopy on temperature is predicted. The phonon-plasmon mechanism of high T_c is analyzed. The non-monotonic dependence of T_c on the carrier concentration can be explained in the framework of the proposed mechanism.

INTRODUCTION

In this paper we will study the collective excitations in highly anisotropic crystals such as high T_c oxides. The anisotropy leads to a strong modification of the plasmon spectrum relative to the isotropic 3D case. This modification along with peculiar values of normal parameters makes the plasmon contribution to high T_c favorable.

A great number of experimental data indicate that the new high T_c materials display the BCS type of superconductivity (pairing, presence of the energy gap), but we are dealing with its "exotic" version. This is due to unusual normal properties of the materials. The analysis of the normal properties carried out in [1-3] has shown that the cuprates are characterized by a very small value of the Fermi energy E_F (~0.1 eV) and a large value of the effective mass m* (e.g., for La-Sr-Cu-O $m^* \simeq 5\ m_e$). We think that such unusual values of the normal parameters along with a high anisotropy are key features of the materials. An appearance of the fluctuating pairing, a small value of the coherence length, unusual transport properties, etc., (see [3]) are the consequences of these features.

The structure of the paper is as follows: Sec. I describes the structure of the plasmon spectrum in layered crystals. In this section we also study the electron-loss spectroscopy (ELS) in such crystals and predict its temperature dependence. The phonon-plasmon mechanism of superconductivity is discussed in Sec. II. Section III addresses the problem of the dependence of T_c on the carrier concentration.

I. Plasmons in Layered Structures

Plasmon Branches. The presence of the layered structure leads to a strong modification of the plasmon spectrum relative to the usual isotropic 3D case. The modification along with a large value of the effective mass makes a plasmon contribution to the superconducting state very important. The phonon-plasmon mechanism of high T_c superconductivity has been discussed in our papers [4,5]. Afterwards different aspects of the plasmon mechanism were studied in [6].

The plasmons describe the collective motion of the carriers. First of all, we want to draw a clear distinction between the plasmon contribution to superconductivity and the excitonic mechanism. The later assumes spatial separation between the superconducting system and the system providing the pairing. In contrast, plasmons, like phonons, are located in the same spatial region as the paired carriers.

Contrary to isotropic 3D case the plasmon frequency $\omega_{p\ell}(\vec{q})$ does not have an energy gap at q = 0 (q is a momentum of the plasmon). They form the plasmon band $\omega(\vec{q}_{\parallel}, q_z)$; the axis Z has chosen to be perpendicular to the layers. The values of ω are restricted between the upper ($q_z = 0$) and lower ($q_z = \pi/c$; c is the interlayer distance) branches. It has been shown in our paper [5] that the plasmon density of states is peaked at the boundaries. Hence, the plasmon spectrum can be approximated as a set of two branches: upper (U) and lower (L), that is we are dealing with a combination of 2D [7] and 3D dispersion relations.

The electronic wave function $\psi_{\vec{k}}(\vec{r})$ in the layered crystal can be written in the form:

$$\psi_{\vec{k}}(\vec{r}) = e^{i\vec{\kappa}\vec{\rho}} \chi_{k_z}(z) \tag{1}$$

Here $\vec{k} = \{\vec{\kappa}, k_z\}$ is a quasi-momentum and

$$\chi_{k_z}(z) = \sum_n \phi(z-n)\, e^{ik_z n} \tag{2}$$

Summation is taken over the layers, $\phi(z)$ is a wave function of an isolated layer or a set of close layers in the unit cell. The wave function (1) is a combination of the free electron model (in the plane) and the tight-binding approximation for z direction. The wave function (1) reflects a high anisotropy of the crystal.

Note that the main contribution of the pairing comes from L branch. On the other hand, it is difficult to observe this branch experimentally (infrared absorption provides information about the U branch). In the next section we will propose the novel method of the study of plasmon spectrum in layer materials, based on the presence of the temperature dependent loss features.

Electron-Loss Spectroscopy (ELS). Consider the particle passing through the layer crystal. The energy lost is described by the equation

$$\gamma \equiv \frac{\partial E}{\partial t} = (2\pi)^{-3} \int (\epsilon_{\overline{p}} - \epsilon_{\overline{p-q}})\ W_{\vec{q}}\ d\vec{q} \tag{3}$$

where $W_{\vec{q}}$ is the total probability of the transition $\vec{p} \rightarrow \vec{p} - \vec{q}$ ($\vec{p}$ is the momentum of the particle). The quantity $W_{\vec{q}}$ can be evaluated with the use of the method of the thermodynamic Green's function, in analogy with [8]. We obtain

$$\gamma \equiv \frac{dE}{dt} = \frac{1}{4\pi^3} \int \omega V^2_{\vec{q},\vec{q}_z} (N+1) \operatorname{Im}[\Pi^{-1}(\vec{q}_{\parallel},\omega) - I]^{-1} d\vec{q} \tag{4}$$

Here ω is the transferred energy, I described the Coulomb interaction of the electron's in the Bloch states, $N = (e^{\omega\beta}-1)^{-1}$. The main contribution to the integral (18) comes from the poles of the total plasmon vertex, that is the zeroes of $S = \Pi^{-1}(q_{\parallel},\omega) - I$. The poles correspond to the plasmon band $\omega(q_{\parallel},q_z)$ (see above). The plasmon spectrum consists of two branches (see above). For U-branch $\omega >> T$ and the corresponding term does not depend on temperature. The contribution of U-branch is similar to that in the usual 3D case; in the last case dE/dT does not depend on T.

The contribution of the L branch appears to the temperature dependent. As a result we obtain

$$\gamma = \gamma_u + \gamma_L(T) \tag{5}$$

There is a total similarity between the independence of γ_u on T and the absence of the temperature dependence (because $\omega_{p\ell} >> T$) in the usual isotropic 3D case.

The term γ_L depends on temperature. One can show that up to $T \lesssim 5 \times 10^2$ K this dependence has a form: $\gamma_L \sim (T/\epsilon_F)^3$. A small value of E_F makes this contribution noticeable. Note that the "demon" branch also displays the temperature dependence, but a large value of the group velocity leads to a smaller value of γ_D.

The dependence of the electron (positron) loss on T is a novel feature of ELS. It would be interesting to carry out the experiments aimed at the observation of this dependence.

II. Phonon-Plasmon Mechanism

High T_c is caused by the coexistence of the phonon and plasmon mechanisms. At present, there is a number of experimental data on thermal conductivity, sound attenuation, Raman scattering, etc., which show the importance of the electron-phonon coupling in the cuprates.

For example, the increase of the thermal conductivity k(T) at temperatures $T < T_c$ observed in [9] means (see, e.g., [10]) that the phonons make a major contribution to the total thermal flow, and the electron-phonon interaction (EPI) is a main relaxation mechanism.

The presence of the isotope shift is also a manifestation of the electron-phonon coupling. In connection with it, one should note (a more detailed discussion see [5]) that it is incorrect to correlate directly the numerical value of α with the strength of the coupling.

The analysis of the experimental data on tunneling leads to the conclusion that for La-Sr-Cu-O, $\lambda \simeq \lambda_{e\text{-}ph} \simeq 1.5\text{-}2$. According to [11,12], for Y-Ba-Cu-O, $\lambda \simeq 2$. The last value is obtained with the the use of data on heat capacity [13].

Hence, the electron-phonon coupling plays an important role, but, nevertheless, it is not strong enough in order to provide high T_c. There is a need for an additional mechanism and it is provided by the electron-plasmon coupling.

The generalized Eliashberg equation in the presence of both, phonon and plasmon mechanisms, can be written in the form:

$$\Delta(p_z,\omega_n)Z = T\sum_{\omega_n'}\int dp_z'd\Omega \left\{ \left[\lambda_{ph}(\Omega,p_z,p_z') \frac{\Omega^2}{\Omega^2+(\omega_n-\omega_n')^2} - V_c\theta(|\omega_n|-\omega_o) \right] + \lambda_{p\ell}(\Omega,p_z,p_z') \frac{\Omega^2}{\Omega^2+(\omega_n-\omega_n')^2} \right\} \frac{\Delta(\omega_n',p_z')}{|\omega_n'|}\Bigg|_{T=T_c} . \qquad (6)$$

The concept of coexistence means that the electron-phonon interaction plays an important role. The Coulomb repulsion is overcome mainly by the the electron-phonon interaction. As for the plasmon contribution, one can see directly from Eq. (3) that the electron-plasmon interaction provides an additional mechanism of electron-electron attraction, and in the presence of electron-phonon interaction it leads to an additional increase in T_c.

III. The Dependence of T_c on the Carrier's Concentration

The new high T_c oxides display a strong dependence T_c on the carrier concentration n. This dependence appears to be non-monotonic, and there is a maximum T_c at some value n_m.

Speaking of the origin of such dependence, one should note that a similar phenomenon has been observed in superconducting semiconductors (see [14]).

The problem of the dependence $T_c(n)$ has been studied by one of the authors in [15]. We think that a similar explanation is applicable for the cuprates. Indeed, the electron-phonon coupling constant λ_{e-ph} can be written in the form (cf., [15]): $\lambda_{e-ph} = \int \zeta\, dk_z\, d\phi\, (u_i^2q^2/\omega_i^2(q))\, \gamma_i(q)$.

Here ζ is the Fröhlich parameter: $\xi = \tilde{\xi}k_F$, q is the phonon transfer momentum, u_j is the sound velocity, $\gamma_j(q)$ slowly depends on q, and ϕ is the polar angle between the electron momenta $\vec{k}_{\parallel}$ and $\vec{k}_{\parallel}'$; one should take summation over phonon branches j. Making transformation $\phi \rightarrow q$, we obtain

$$\lambda_{e-ph} \simeq \frac{\tilde{\zeta}(\pi/c)}{2k_F}\int_0^{k_1} dq.q(\sin\phi)^{-1}\frac{u_j^2q^2}{\omega_j^2(q)}\gamma_j(q) , \qquad (7)$$

where

$$\sin\phi \simeq \left[1-(1-(q/k_F\sqrt{2}\,))^2\right]^{1/2} \quad ,$$

and

$$k_1 = \min\,\{2k_F, q_D\} \quad .$$

The value of k_F depends on the carrier concentration: $k_F \sim n^{1/2}$. Consider the case, when n is small, so $2k_F < q_D$. Then $k_1 = 2k_F$. In this case λ_{e-p} increases with increasing n. But eventually we came to the situation when $2k_F > q_D$. Then $k_1 = q_D$ and $\lambda_{e-p} \sim k_F^{-1}$; hence, λ_{e-ph} became the decreasing function of n. Therefore, in the region $2k_F \sim q_D$ one should observe the maximum of T_c.

Hence, the analysis of the electron-phonon interaction leads to an appearance of the maximum $T_c(n)$ [16]. The total $\lambda = \lambda_{ph} + \lambda_{p\ell}$. The behavior of $\lambda_{p\ell}(n)$ will be studied in detail elsewhere, but it is clear that the observed dependence $T_c(n)$ can be explained in the framework of the proposed phonon-plasmon mechanism.

SUMMARY

In this paper we describe the properties of collective excitations in layered crystal and the phonon-plasmon mechanism of high T_c. The main results can be summarized as follows:

1. The plasmon spectrum can be presented as a set of U and L branches; the main contribution to the superconductivity comes from L-branch.
2. The presence of the L-branch leads to the temperature dependence of the electron (positron)-loss spectroscopy. It would be interesting to verify this prediction experimentally.
3. High T_c is caused by a coexistence of a strong electron-phonon coupling and the plasmon mechanism.
4. The electron-phonon coupling constant depends on the carrier concentration n and it leads to an appearance of the maximum of the dependence $T_c(n)$.

ACKNOWLEDGMENT

Work of VZK was supported in part the Office of U. S. Naval Research under Contract No. N00014-86-F-0015 and carried out at the Lawrence Berkeley Laboratory under Contract No. DE-AC03-76SF00098.

REFERENCES

1. Kresin, V. Z. and Wolf, S. A., Solid State Comm. 63, 1141 (1987); J. Superconductivity 1, 143 (1988).
2. Kresin, V. Z. and Wolf, S. A., Novel Superconductivity, S. Wolf and V. Kresin, eds. (Plenum Press, New York, 1987), p. 237.
3. Kresin, V. Z., Deutcher, G., and Wolf, S. A., Proc. Latin Amer. Conf. on High T_c, Rio de Janeiro, 1988, World (in press); J. Superconductivity (in press).
4. Kresin, V. Z., Proc. Materials Res. Soc. Mtg., Anaheim, April 1987, p. 19, Phys. Rev. B 35, 8716 (1987); Ref. 2, p. 309.
5. Kresin, V. Z. and Morawitz, H., Ref. 2, p. 445; Phys. Rev. B 37, 7854 (1988); J. Superconductivity 1, 89 (1988); Physica C 153, 1327 (1988), Proc. Latin Amer. Conf. on High T_c, Rio de Janeiro, 1988, World (in press).
6. Ruvalds, J., Phys. Rev. B 35, 8869 (1987); Ihm, J. and Lee, D., Ref. 2, p. 451; Ashkenazi, J., Kuper, C., and Tuk, R., Solid State Comm. 63, 1144 (1987); Griffin, A., Phys. Rev. B 37, 5943 (1988); Pashitskii, E. (preprint).
7. Stern, F., Phys. Rev. Lett. 18, 546 (1967).
8. Larkin, A., Sov. Phys.-JETP 37, 186 (1960).
9. Jerowski, A., et al., Phys. Lett. A 122, 431 (1987); Morelli, D., Heremans, J., and Swets, D., Phys. Rev. B. 36, 3917 (1987); Uher, C., et al., Phys. Rev. B 36, 5676, 5680 (1987).
10. Geilikman, B. and Kresin, V., Non-Steady and Kinetic Effects in Superconductors (Wiley, New York, 1974).
11. Panova, et al., Sov. Phys.-JETP Lett. 46, 79 (1987).
12. Kresin, V. Z. and Wolf, S. A., preprint.
13. Phillips, N., et al., Ref. 2, p. 739.
14. Hulm, J., et al., Proc. LT-X, Moscow, 1965, p. 86.
15. Kresin, V. Z., J. Low Temp. Phys. 5, 565 (1971).
16. Torrance, G. B., et al., Phys. Rev. Lett. 61, 1127 (1988).

APPLICABILITY OF CONVENTIONAL BAND THEORY AND THE ROLE OF PHONONS IN HIGH T_C SUPERCONDUCTORS

R.E. Cohen[a], W.E. Pickett[a], H. Krakauer[b], D. Singh[a*], D.A. Papaconstantopoulos[a], and P.B. Allen[c]

[a] Complex Systems Theory Branch, Naval Research Laboratory, Washington, D.C., 20375-5000

[b] Department of Physics, College of William and Mary, Williamsburg, VA 23185

[c] Department of Physics, State University of New York, Stony Brook, NY 11794

* National Research Council--NRL Associate

ABSTRACT

State-of-the-art electronic structure and total energy calculations within the local density approximation (LDA) describe many of the features of the high T_c oxides, including transport properties, structures, and phonon frequencies. LDA, however, does not give sufficient exchange splitting to give a stable antiferromagnetic solution in La_2CuO_4, even when the symmetry is relaxed and the in-plane oxygens are allowed to spin polarize. Nevertheless, the accuracy of the predictions of paramagnetic calculations suggests that the spin fluctuations play only a minor role in determining transport and lattice dynamics in the superconducting samples.

One of the most important basic questions about the high Tc superconductors is whether traditional band structure approaches are applicable. If they are applicable, we can in principle calculate superconducting properties, including T_c. However, the applicability of band theory to the high T_c superconductors has been questioned; these materials have often been called "highly correlated" and much has been made of a large Hubbard U for the copper ions. The burden of proof has laid heavily on band theory, since models that attempt to include highly localized correlations have not developed to the point of actually being able to quantitatively calculate properties, primarily because of impossibly huge computational requirements. Three observations have been used to discredit the band picture for these materials. Firstly, La_2CuO_4 is an insulator, whereas a simple band picture makes it a metal. Secondly, the linear resistivity with respect to temperature was said to be "too linear" to be explained by Fermi liquid theory, such as would arise from band quasiparticles.

Thirdly, early calculations found an unstable planar breathing mode, contrary to experiment, in La_2CuO_4. Each of these points will be addressed below.

MAGNETISM

Undoped La_2CuO_4 and $YBa_2Cu_3O_6$ are semiconductors, rather than metals. The former has an odd number of electrons per unit cell, so that paramagnetic band theory must give a metallic, rather than insulating ground state. However, if the unit cell is doubled by formation of a charge or spin density wave, an insulator can be formed since there is then an even number of electrons per unit cell. Experimentally, it is known that antiferromagnetic spin density waves condense, and this is consistent with the band picture. The problem arises in quantitative calculations based on the local spin density approximation (LSDA). Numerous calculations have been published in the literature, of which the best calculations find either too small a magnetic moment, or no moment at all, and no gap is formed as would be appropriate for an insulating ground state.[1)2)] All of these calculations assume symmetry such that no net moment can form on the in-plane oxygen atoms. Therefore, we performed a Linear Augmented Plane Wave study of La_2CuO_4 in the observed orthorhombic structure, but with symmetry lowered to triclinic, retaining only inversion symmetry. We used several different (small) k-point sets, including sets with points on or very close to the Fermi surface, and we were unable to find a stable antiferromagnetic solution. In fact, even when a staggered field was put on the oxygens, only a small moment was generated on the oxygens. This adds support to the generally accepted conclusion that LSDA simply does not properly describe the ground state in the undoped semiconductors represented by La_2CuO_4.

On the other hand, all this necessarily demonstrates is that LSDA is not sufficiently accurate to stabilize a moment and open a gap in the density of states. Calculations which artificially "widen the gap" do lead to an antiferromagnetic solution. Fig. 1 shows the band structure of undistorted tetragonal La2CuO4 folded back into the orthorhombic Brillouin zone. The band that crosses the Fermi level is doubly degenerate, and exchange splitting will split this band; if the splitting is greater than ~0.4 eV a gap will be opened. Local correlations not included in LSDA could contribute to opening this gap, but LDA sometimes has similar problems in other materials which are not considered "highly correlated." For example, Ge is calculated to have zero gap and to be a semimetal, instead of an insulator.[3)]

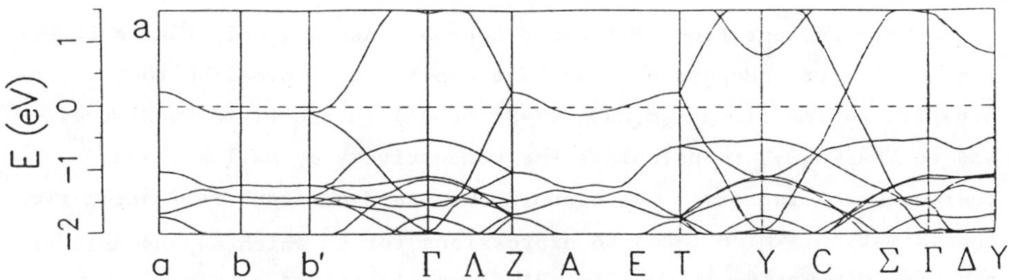

Figure 1. Band structures for the undistorted tetragonal structure, with bands folded back into the orthorhombic Abma Brillouin zone.

Fortunately, many of the exciting properties of the metallic oxides occur after the materials are doped, which destabilizes the spin density wave. In this regime, it appears that LSDA gives results as accurate as are obtained in any other material, as is discussed below.

LDA-BASED STUDIES OF ELECTRICAL TRANSPORT

An understanding of the electrical transport properties in the normal state is essential in developing a correct picture of these new high T_c superconductors. First, the transport arises (in most pictures) from the same carriers which become paired in the superconducting state, so transport studies provide a fertile field for identifying important characteristics of the carriers themselves. Second, it is likely that the interactions which limit the resistivity will be related to the interactions which lead to the superconducting pairing.

We have carried out detailed calculations to test the hypothesis that (1) the resistivity is due to quasiparticle behavior within a Fermi liquid picture, and (2) these quasiparticles are described approximately by the band electrons and holes resulting from the LDA calculations.[4] Such an approach can have some success even if correlations beyond LDA are important for other properties (magnetism, high energy spectra) if the effect of such correlations near E_F results only in a modest correction to the band mass and negligible rearrangement of the Fermi surface.

Otherwise, our Bloch-Boltzmann approach assumes only that scattering events are independent. We have applied the formalism to the metallic regime of $La_{2-x}M_xCuO_4$ (M=Sr or Ba) in the rigid band model, and to $YBa_2Cu_3O_7$, to calculate the conductivity σ, Hall R^H, and thermopower S tensors. For σ and R^H we use a constant relaxation time approximation, which leads to expressions for R^H which depend only on the band dispersion at E_F. In this approximation S vanishes, so we have investigated three models of the energy dependence of the scattering. Due to the lack of agreement in the published single crystal data for S and the dependence on the scattering model, nothing of a conclusive nature has been achieved; the details have been discussed elsewhere.[4]

An important result which is related only to the question of quasiparticle transport (versus some more exotic picture) is the question of the linearity of the resistivity $\rho(T)$, which has been suggested to be too linear to arise from normal quasiparticle transport. We have calculated the standard expression for $\rho(T)$, assuming the scattering is due to phonons and that the electron-phonon spectral function is proportional to the measured phonon density of states. The resulting curves for both systems, shown in Fig. 2, demonstrate that the expected resistivity from phonon scattering is just as linear as is the data, so the resistivity data is entirely consistent with quasiparticle behavior of the carriers.

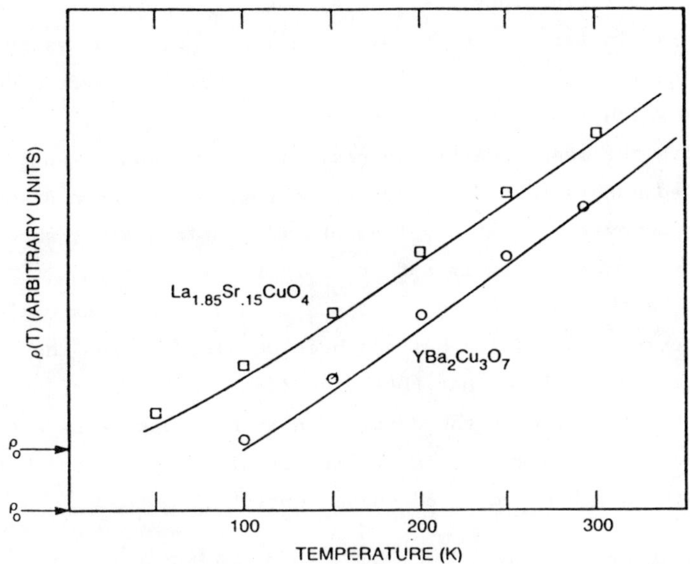

Figure 2. Expected shape of the electron-phonon part of the resistivity according to Boltzmann theory. The residual resistivity and the magnitude have been adjusted to allow a clear comparison of the shape.

Calculations of the Hall tensor have led to surprising results, which are in substantial agreement with data. Calculations of the concentration x dependence for $La_{2-x}M_xCuO_4$ correctly indicate a hole-like sign for carrier motion in the Cu-O layer, and a magnitude for x=0.06 (T_c=20 K sample) within about a factor of two of the experimental data of Suzuki and Murakami[5)]. Two features of this calculation were surprising: the sign for carrier motion perpendicular to the layers (the field in the layers) was predicted to be electron-like, and for larger doping concentrations, R^H was predicted to decrease in magnitude and change sign around x=0.25. Such a change of sign of R^H with respect to field orientation or from doping can occur naturally in a band picture, but is difficult to explain in other models.

Calculations for $YBa_2Cu_3O_7$ led to similar unexpected results, that is, differing "sign of the carriers" for fields perpendicular to and parallel to the Cu-O layers. This prediction was subsequently borne out by measurements by Tozer et al.[6)] on measurements on twinned crystals. For fields in the plane, their data is T-independent, is in quantitative agreement with our predictions, and has <u>electron-like</u>

sign. For the field along the c axis, the sign is hole-like and the magnitude is similar to the calculated value. For both systems, R^H for in-plane motion of the carriers seems to have an inverse dependence on T which is not understood.

Finally, the magnitude of the resistivity, or equivalently $d\rho/dT$, has received much attention. Several scenarios have been discussed in Ref. [4]. However, recent developments in sample quality show that the intrinsic resistivity is a factor of two or three less than was thought to be the case several months ago, and it remains unclear whether the intrinsic value has yet been obtained. From the above comparisons, we conclude that LDA-based Bloch-Boltzmann transport theory is consistent with the data which is most appropriate for comparison, and has even made startling predictions for the Hall tensor which have been verified subsequently.

LATTICE DYNAMICS

One of the earliest results that suggested that conventional band theory was not applicable to the high Tc copper oxides was Weber's result of an unstable oxygen breathing mode in La_2CuO_4,[7] whereas experimentally the breathing mode is a high frequency mode, and a tilt mode goes unstable. Though Weber's calculations gave an unstable breathing mode, ionic calculations using an ab initio model gave the breathing mode to be the highest frequency mode at the X-point, suggesting that ionic forces, which Weber neglected, are very important [8][9] and that his tight-binding Hamiltonian was not sufficiently accurate.

We have now performed a series of general potential Linearized Augmented Plane Wave (LAPW) calculations[10] in order to test this result, as well as to test the accuracy of LDA for other phonon frequencies.[11] These calculations make no assumptions about the shape of the potential or charge density, and are fully self-consistent. Fig. 3 shows the energy versus displacement for the symmetry modes calculated. LDA gives the observed tilt mode to be unstable, with a minimum at about the experimental distortion. Thus the tilt is not due to many-body interactions not present in LDA, as had been suggested earlier.[12]

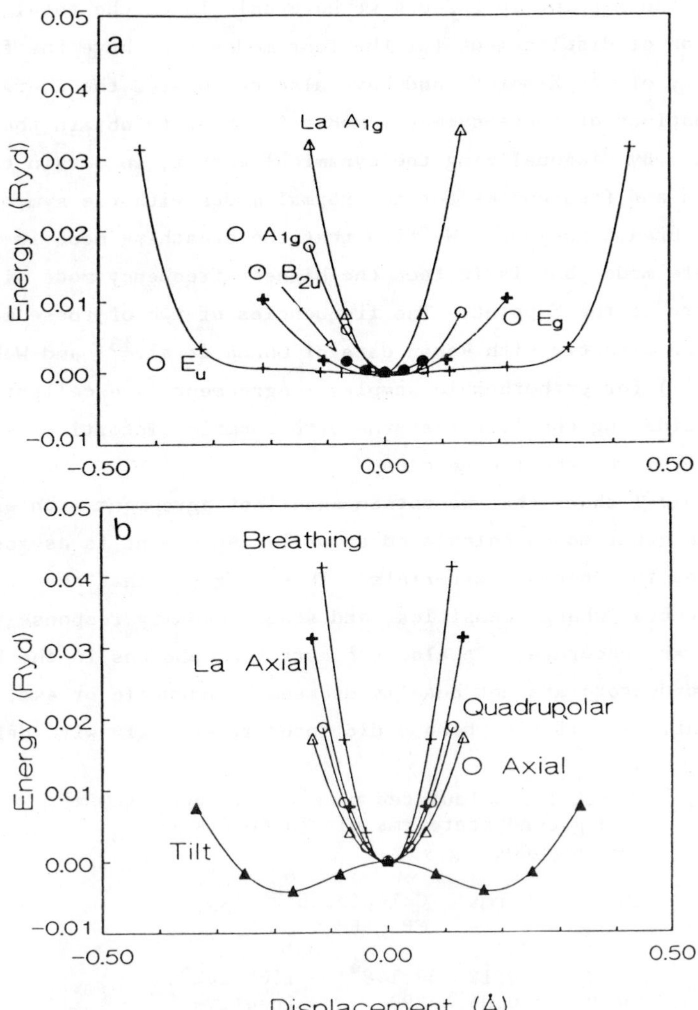

Figure 3. Energy versus displacement for (a) zone center (Γ-point) and (b) zone boundary (X-point) modes. The energies are per 7 atom unit cell in (a) and 14 atom supercell in (b). The solid lines are quadratic fits to the calculations, except for the E_u and tilt modes, where sixth order polynomials were used. For the tilt mode, the x-coordinate is the displacement of the O_z ion in the xy plane.

At the X-point in La2CuO4 we have calculated the total energy as a function of displacement for the four modes that have the full symmetry of the X-point, and have also calculated the energy for combinations of these symmetry modes in order to obtain the dynamical matrix. By diagonalizing the dynamical matrix, we obtain the eigenvectors and frequencies for the normal modes with the symmetry of the planar breathing mode. We find that the breathing mode is not an unstable mode, but is in fact the highest frequency mode with Ag symmetry at the X-point. The frequencies of two of these modes can be compared directly with Raman data of Ohana et al.[13] and Weber et al.[14] (Table 1) for orthorhombic samples. Agreement is excellent, especially considering the fact that the orthorhombic distortion is likely to slightly lower the frequencies.

Table 1 shows that we obtain excellent agreement with experiment for the other modes calculated as well. Agreement is as good as is obtained for "normal" materials. This suggests that the total energy differences, charge densities, and static density response predicted by LDA are accurate. It also suggests that phonons in the high Tc superconductors are not heavily dressed by magnetic or excitonic fluctuations. If the phonons did interact strongly with degrees of

TABLE I. Calculated phonon frequencies (cm^{-1}) and ground state rms displacements (Å) for La_2CuO_4.

Mode	rms	Calculated FP	Calculated EM	Exp.
Γ:				
O E_u	0.12	39,148*		110[a] 162[b]
O B_{2u}	0.06	293		inactive
O E_g	0.07	233		228[c]
O A_{1g}	0.05	415	417	429[c] 424[d]
La A_{1g}	0.02	224	220	228[c] 231[d]
X:				
La axial	0.03	155	156	137[e]
O axial	0.06	339	299	
Quad.	0.05	404	475	
Breath.	0.04	609	731	710[c] 680[d]

FP: frozen phonon (symmetry mode)
EM: eigenmode
* The second number is for an excitation from the ground state to the second odd symmetry state.

References: a. Sugai et al.; b. Gervais et al.; c. Weber et al.; d. Ohana et al., Boni et al. Weber et al.

freedom that are not included in our calculations, this would effect the observed frequencies, and we would expect significant differences between our calculations and experiment. Thus it appears unlikely that the large isotope effect observed in the oxide superconductors is due to "parasitic involvement of phonons" in the superconducting mechanism.[16)]

ELECTRON-PHONON MECHANISM

The observation of an isotope effect in all of the high Tc superconductors indicates phonon involvement in the superconductivity mechanism. The absence of magnetism in (BaPb)BiO_3 (T_c=13 K) and (BaK)BiO_3 (T_c=30 K) suggests a non-magnetic mechanism, since a reasonable assumption is that the mechanism is the same in all of the oxide superconductors. The small isotope effect in $YBa_2Cu_3O_7$ suggests either another mechanism in addition to an electron-phonon mechanism, significant anharmonicity, or unusual screening (which changes μ^* and thus the isotope effect.[17)])

Rigid muffin tin calculations gave very low Tc's for both La_2CuO_4[18)] and YBa2Cu_3O_7.[19)] These calculations make two assumptions, both of which are unlikely to hold in these materials. The first assumption is that the local environment of each atom is isotropic. This is certainly not true in the highly anisotropic copper oxide superconductors. The second assumption is that only local changes in the potential are included; all long-range interactions are presumed to be completely screened. However, we find that the potential changes on atomic sites that are stationary for a given atomic motion, whereas the rigid muffin tin approximation ignores this contribution. Jarlborg has examined this approximation using the LMTO method and concluded that the effective electron-phonon matrix elements are at least twice as large as those obtained using the rigid muffin tin approximation.[20)]

Fig. 4 shows Tc versus λ for different values of the mean frequency $<\omega>_{F/\omega}=<\omega^{-1}>^{-1}$, and Fig. 5 shows what combination of λ and frequency moment is required to obtain a Tc of 94 K within Eliashberg theory. We find, as did Jarlborg, that the higher frequency modes have large Madelung contributions to the change in potential, and we find that the very low frequency, anharmonic modes have relatively small couplings. Thus we expect the effective frequency moment to be larger than that one would obtain simply from neutron scattering data. For the electron-phonon interaction to give T_c's of above 90K, enhancements of η on the order of 10 times the rigid muffin tin values are

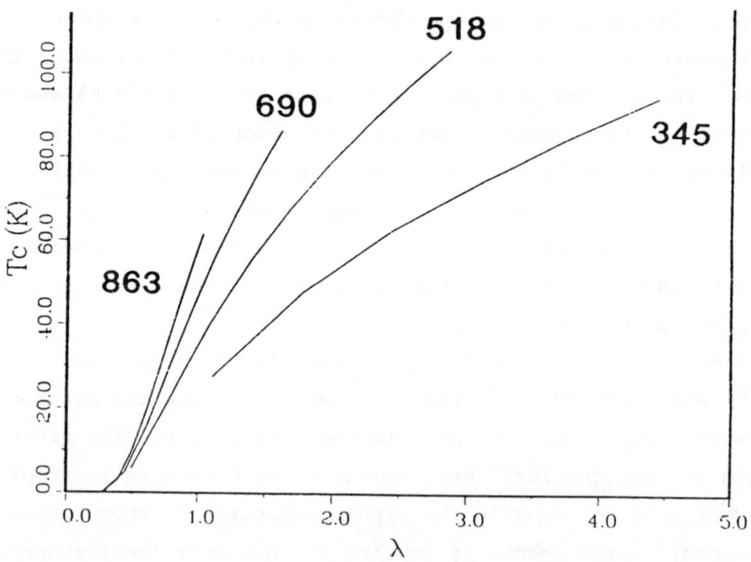

Figure 4. Tc versus λ for different mean phonon frequencies (in K) for $\mu^{*}=0.1$ using Eliashberg theory.

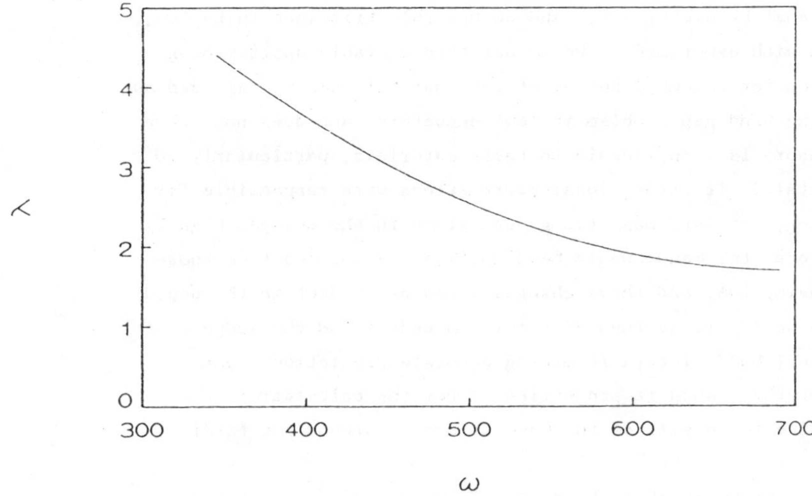

Figure 5. Combinations of frequency (in K) and λ that give Tc of 94K.

required.

Such large enhancements of the electron-phonon matrix elements over the rigid muffin tin values do not seem impossible, considering the ionic nature of these materials that leads to Madelung contributions to electron-phonon interactions that are not present in "normal" metals. The fact that these materials are ionic metals may be the most important aspect that leads to high Tc, rather than the presence of other sorts of excitations. We are now performing self-consistent calculations of electron-phonon matrix elements using the LAPW method, without making the rigid muffin tin approximation, and will find out whether phonons can indeed be responsible for high T_c.

CONCLUSIONS

We find that state-of-the-art electronic structure calculations within a conventional band framework do quite well in calculating phonon frequencies and several aspects of transport properties in the metallic regime. This suggests that the local density approximation (LDA) is an excellent starting approximation for describing the electronic structure of high Tc superconductors. We find a high frequency planar breathing mode, and an unstable tilt mode in La_2CuO_4, in agreement with experiment. We do not find a stable antiferromagnetic solution for La_2CuO_4, but point out that this can be regarded as similar to the band gap problem in semiconductors, and does not prove that band theory is inapplicable to these materials, particularly to the doped metals. If strong local correlations were responsible for the insulating, antiferromagnetic ground state in the undoped high T_c superconductors, the bands would be significantly changed from those calculated using LDA, and these changes would be evident in the doped materials as well. It is then difficult to understand the success of the traditional band picture in making accurate predictions, particularly for the transport properties, since the calculations are quite sensitive to details of the band structure around the Fermi level.

We would like to thank L. L. Boyer, J. W. Serene, J. L. Feldman, W. H. Weber, J. Hardy, and B. M. Klein for helpful discussions. Computations were carried out on the IBM 3090 at the Cornell National Supercomputer Facility, on a Multiflow Trace 7 at NRL, and on Cray X-MPs at the Pittsburgh Supercomputing Center and at NRL. This work was supported in part by the Office of Naval Research, and H.K. was supported by NSF grant DMR-87-19535.

References:

1. T. C. Leung, X. W. Wang, and B. N. Harmon, Phys. Rev. B 37, 384 (1988).

2. W. M. Temmerman, Z. Szotek, and G. Y. Guo, J. Phys. C 21, L867 (1988).

3. W. E. Pickett and C. S. Wang, Int. J. Quant. Chem.: Quant. Chem. Symp. 20, 299 (1986).

4. P. B. Allen, W. E. Pickett, and H. Krakauer, Phys Rev. B 37, 7482 (1988).

5. M. Suzuki and T. Murakami, Jpn. J. Appl. Phys. 26, L524 (1987).

6. S. W. Tozer, A. W. Kleinsasser, T. Penney, D. Kaiser, and F. Holtzberg, Phys. Rev. Lett. 59, 1236 (1987).

7. W. Weber, Phys. Rev. Lett. 58, 1371 (1987).

8. R.E. Cohen, W. E. Pickett, L. L. Boyer, and H. Krakauer, Phys. Rev. Lett. 60, 817 (1988).

9. R. E. Cohen, W. E. Pickett, H. Krakauer, and L. L. Boyer, Physica B 150, 61 (1988).

10. S.H. Wei and H. Krakauer, Phys. Rev. Lett. 55, 1200 (1985).

11. R. E. Cohen, W. E. Pickett and H. Krakauer, unpublished.

12. P.W. Anderson, in Novel Superconductivity, ed. S.A. Wolf and V.Z. Kresin (Plenum, New York, 1987), p. 295.

13. I. Ohana, M. S. Dresselhaus, Y. C. Liu, P. J. Picone, D. R. Gabbe, H. P. Jenssen, and G. Dresselhaus, unpublished.

14. W. H. Weber, C. R. Peters, B. M. Wanklyn, Changkang Chen, and B. E. Watts, Phys. Rev. B 38, 917 (1988).

15. W. H. Weber, C. R. Peters, B. M. Wanklyn, Changkang Chen, and B. E. Watts, Sol. State Commun. (to be published).

16. B. Batlogg, R. J. Cava, L. W. Rupp, Jr., A. M. Mujsce, J. J. Krajewski, J. P. Remeika, W. F. Peck, Jr., A. S. Cooper, and G. P. Espinosa, Phys. Rev. Lett. 61, 1670 (1988).

17. L. C. Bourne, M. F. Crommie, A. Zettl, H. zur Loye, S. W. Keller, K. L. Leary, A. M. Stacy, K. J. Chang, M. L. Cohen, and D. E. Morris, Phys. Rev. Lett. 58, 2337 (1987).

18. W. E. Pickett, H. Krakauer, D. A. Papaconstantopoulos, and L. L. Boyer, Phys. Rev. B 35, 7252 (1987).

19. H. Krakauer, W. E. Pickett, and R. E. Cohen, J. Supercon. 1, 111 (1988).

20. T. Jarlborg, Sol. State Comm. 67, 297 (1988).

CRYSTAL-FIELD-MODEL INTERPRETATION OF MAGNETIC RESONANCE DATA IN HIGH-T_c SUPERCONDUCTORS: LOCATION OF CHARGE-CARRYING HOLES

Frank J. Adrian
Applied Physics Laboratory, The Johns Hopkins University
Laurel, Maryland 20832

ABSTRACT

A crystal field model is used to interpret nuclear magnetic resonance (NMR) and nuclear electric quadrupole (NEQ) resonance experiments on high-T_c superconductors, with emphasis on ascertaining the nature of the Fermi-level states. Previously reported results of calculations of the electric field gradients and NEQ coupling constants for La in La_2CuO_4 and Cu and Fe (substituted for Cu) in $YBa_2Cu_3O_7$ are reviewed briefly. Analysis of the Knight shift (K_Y) and relaxation data ($1/T_1$) for ^{89}Y in $YBa_2Cu_3O_7$ is reviewed and extended to consider the implications of very recent reports that K_Y is negative. It is shown that consideration of K_Y and the related relaxation can provide considerable information about the the Fermi-level states in this case provided that it can be definitely established what part of the observed ^{89}Y NMR frequency shift is a Knight shift.

I. The Crystal Field Model

Useful insights into the structure of the high-T_c superconductors have been obtained using a model which regards these materials as largely ionic crystals in which the charge-carrying holes move under the influence of a potential determined primarily by the Madelung potential at the various ions and the intra-atomic potentials of these ions.[1] These holes, which are inherently present in $YBa_2Cu_3O_{7-\delta}$ because it is nonstoichiometric for $\delta \leq 0.5$, and are introduced into La_2CuO_4 by alkaline earth doping, are henceforth denoted nonstoichiometry holes to distinguish them from the hole inherently present in the Cu^{2+} 3d shell. This model predicts that pure La_2CuO_4 is an antiferromagnetic insulator because of the large energy required to move an electron from an O^{2-} ion to a Cu^{2+} ion or, equivalently, move the Cu^{2+} hole to an O^{2-}. The nonstoichiometry holes are mobile, however, because little or no energy is required to move them between Cu and O ions. Furthermore, these nonstoichiometry holes are predicted to occupy states formed from Cu 3dπ and O 2pπ orbitals rather than the Cu $3d\sigma_{x^2-y^2}$ and O 2pσ orbitals as is generally believed. (Here, σ denotes orbitals whose maximum overlap is along their bond axis while π denotes orbitals oriented away from this bond axis.) Placing the nonstoichiometry holes in O 2pπ orbitals is favored by crystal-field energy considerations, while the lower crystal-field energy of an additional Cu $3d\sigma_{x^2-y^2}$ hole is overridden by the large intra-atomic energy required to place two holes in the same Cu 3d orbital.[1]

II. Nuclear Electric Quadrupole Coupling Constants

The crystal field model is well suited to calculating the NEQ coupling constants (e^2Qq) from the semi-empirical formula

$$(e^2Qq)=(e^2Qq)_{val}+(e^2Qq)_{ionic}. \tag{1}$$

Here, $(e^2Qq)_{val}$ is the contribution from unfilled valence shells in the subject atom or ion, and $(e^2Qq)_{ionic}$ is the contribution from the surrounding ions. For the spherically symmetric La^{3+} ion in La_2CuO_4 only the ionic term is nonzero, which term is readily calculated by assuming point charges for the surrounding ions.[2] The result is $(e^2Qq)=-85$ MHz, which agrees well with experiment (∓88.4 MHz),[3] and thus supports the crystal field model.

$(e^2Qq)_{ionic}$ calculated for Fe at the two Cu sites in $YBa_2Cu_3O_7$ is too large to explain the observed NEQ splittings,[2] and thus suggests that the iron is not present exclusively as the spherically symmetric Fe^{3+} species but has some Fe^{2+} character (Fe^{4+} is believed less likely because of the very high ionization potential of Fe^{3+}), resulting in a valence contribution to (e^2Qq). An Fe^{2+}/Fe^{3+} ratio of 0.25 to 0.50, depending on the d orbitals occupied in the Fe^{2+} state, yields agreement with experiment assuming that the two predominant NEQ splittings of 13 and 23 MHz observed in Mössbauer experiments are due to Fe substituted at the plane and chain sites of $YBa_2Cu_3O_7$, respectively.[2] It should be noted, however, that these assignments are currently the subject of much controversy.[4]

Two Cu NEQ frequencies of 22.05 and 31.48 MHz are observed in

$YBa_2Cu_3O_7$, and their assignment to the chain and plane sites, respectively, has only very recently been finally resolved.[5] Single crystal studies have further shown that the chain site is extremely axially anisotropic ($\eta \simeq 1$) whereas the plane site is axially symmetric ($\eta \simeq 0$).[6] The calculated $(e^2Qq)_{ionic}$ again does not agree with experiment, which is expected because the Cu should be predominantly Cu^{2+} which has a large $(e^2Qq)_{val}$ due to its $3d\sigma_{x^2-y^2}$ hole. A Cu^{2+} character of 93 and 73% for the chain and plane Cu^{2+} ions, respectively, yields excellent agreement with experiment both as regards the the NEQ frequencies and the asymmetry parameters.[2]

III. ^{89}Y Knight Shift and Relaxation

NMR investigations of ^{89}Y in $YBa_2Cu_3O_7$ are of great potential importance because they show, uncomplicated by quadrupole interactions which complicate the Cu NMR, two effects due to the in-plane conduction electrons: a small Knight shift (K_Y) and relaxation rate satisfying a Korringa relation, i.e., $T_1T=3200$ s K.[7] It has been shown that the positive K_Y, reported by the first studies,[8,9] requires that the Cu-O_2-plane conduction band be formed from the in-plane, π-bonding Cu $3d\pi_{xy}$, $O^{(x)}$ $2p\pi_y$ and $O^{(y)}$ $2p\pi_x$ orbitals depicted in Fig. 1b, rather than the σ-bonding Cu $3d\sigma_{x^2-y^2}$, $O^{(x)}$ $2p\sigma_x$ and $O^{(y)}$ $2p\sigma_y$ orbitals shown in Fig. 1a, because, as also shown in Fig. 1, only the π orbitals can mix with the Y^{3+} s atomic orbitals (AO's) to give the postive conduction-electron spin density at the Y nucleus required by a positive K_Y.[10] It is even possible in this case to specify that the Fermi-level states are antibonding states formed largely from the O $2p\pi$ orbitals because any other assignment makes K_Y much too large.[10]

Very recent reports have indicated K_Y is negative,[7,11] however, which case is more ambiguous. $K_Y<0$ results not from direct interaction of the conduction electrons with the nucleus but rather from the spin polarization they induce in closed-shell orbitals that have a nonzero density at the nucleus. If we exclude contributions from admixture of conduction electron density into the Y^{3+} 4p AO's, which calculation shows is small and furthermore leads to orientation-

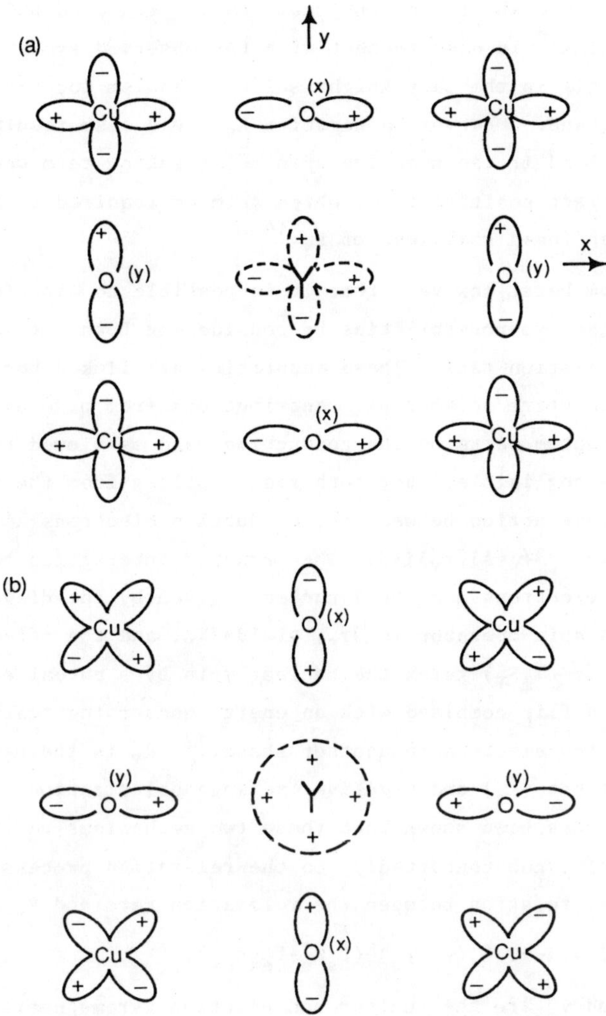

Fig. 1. Illustration of different types of conduction band orbitals for the CuO_2 plane in $YBa_2Cu_3O_7$ and their overlaps with the Y^{3+} AO's. (a) σ-bonding Cu $3d_{x^2-y^2}$ and $O^{(x)}$ $2p_x$ and $O^{(y)}$ $2p_y$ AO's. (b) π-bonding Cu $3d_{xy}$ and $O^{(x)}$ $2p_y$ and $O^{(y)}$ $2p_x$ AO's.

dependent Knight shifts and unsymmetrically broadened NMR lines in powder samples[10] in disagreement with the observed symmetrical lines.[7,8] this is the only Knight shift mechanism for the σ conduction band. However a negative K_Y could also result from the π conduction band if the negative spin polarization term were larger than the direct positive term, which term is required to be small by the aforementioned smallness of K_Y.[10]

We show here, however, that it is possible to discriminate between these two possibilities by considering both the Knight shift and the relaxation rate. These quantities are linked because, in the present case where anisotropic contributions from p, d and higher angular momentum parts of the conduction band as viewed from the Y nucleus are negligible, they both result solely from the isotropic hyperfine interaction between the conduction electrons and the Y nucleus, i.e., $\mathcal{H}_F = A\delta(\mathbf{r_Y})\mathbf{I}\cdot\mathbf{S}$. The magnetic interaction between the conduction electrons and the Y nucleus given by the diagonal (I_zS_z) part of the spin operator in $\mathcal{H}_F$, yields K_Y, and the off-diagonal spin terms ($\frac{1}{2} I_\pm S_\mp$) relax the nuclear spin by a mutual electron-nuclear spin flip combined with an energy-conserving scattering of the conduction electron to another state.[12] K_Y is the net sum of the positive direct (K_s) and negative exchange polarization (K_{ex}) terms, however, it has been shown that these two mechanisms contribute independently, not concertedly, to the relaxation process,[13] so that the Korringa relation between the relaxation rate and K_Y is:

$$1/T_1=(4\pi kT/\hbar)(\gamma_n/\gamma_e)^2(f_sK_s^2+f_{ex}K_{ex}^2), \qquad (2)$$

where γ_n and γ_e are the nuclear and electron gyromagnetic ratios and usually $f_s=1$ and $f_{ex}<1$.[13] An additional complication here is that the Y nucleus interacts with conduction bands in two neighboring CuO_2 planes, and although these two bands contribute equally to K_Y, they may act either independently or concertedly in producing the relaxation. An important point, which we next prove is that their action is concerted for $(1/T_1)_s$ and independent for $(1/T_1)_{ex}$.

It is convenient to write the isotropic hyperfine Hamiltonian in the many electron form

$$\mathcal{H}_F = A \sum_\kappa \delta(\mathbf{r}_{Y\kappa})\, \mathbf{S}_\kappa \cdot I \qquad (3)$$

where $\mathbf{S}_\kappa$ and $\mathbf{r}_{Y\kappa}$ are, respectively, the spin and distance from the Y nucleus of the κ'th electron. The small admixture of Y^{3+} s AO's into the conduction orbitals can be estimated by orthogonalizing the latter to the former, or equivalently, and more useful here, including all pertinent orbitals in an antisymmetrized product.[10] For a single pair of Y^{3+} s AO's (the important inclusion of all the s AO's in actual calculations is straightforward[10]) and a single conduction orbital, this product is: $\mathcal{A} s\bar{s}\mathbf{k}_j$, where $\mathcal{A}$ is the antisymmetrization and renormalization operator and the unbarred and barred orbitals have "up" and "down" spins, respectively. The resulting expectation value of $\mathcal{H}_F$ is $\langle \mathcal{A} s\bar{s}\mathbf{k}_j | \mathcal{H}_F | \mathcal{A} s\bar{s}\mathbf{k}_j \rangle = -A \langle s|\mathbf{k}_j\rangle\langle\mathbf{k}_j|s\rangle\langle\bar{s}|\delta(\mathbf{r}_Y)S_z|\bar{s}\rangle$, which yields a positive conduction-electron spin density at Y because $S_z\bar{s} = -\frac{1}{2}\bar{s}$. The matrix element for relaxation via electron-nuclear spin flip scattering between the conduction band states $\mathbf{k}_j$ and $\bar{\mathbf{k}}_\ell$ is: $\langle \mathcal{A} s\bar{s}\bar{\mathbf{k}}_\ell | \mathcal{H}_F | \mathcal{A} s\bar{s}\mathbf{k}_i \rangle = A\langle\bar{s}|\bar{\mathbf{k}}_\ell\rangle\langle\mathbf{k}_j|s\rangle\langle\bar{s}|\delta(\mathbf{r}_Y)S_-|s\rangle$. Since the same conduction band orbital in either of the two CuO_2 planes next to the Y^{3+} ion has the same overlaps with the Y^{3+} s AO's, this matrix element will be the same no matter whether $\mathbf{k}_\ell$ and $\mathbf{k}_j$ are in the same or different planes. Thus, the direct term can produce relaxation with scattering of the conduction electron either within or between the two planes, so the contributions of the two conduction bands add constructively and $f_s=1$ in Eq. (2).

It has been shown that the exchange polarization mechanism leading to metal cation hyperfine interactions with an electron spin predominantly on an anion such as O^{2-} does not operate by direct polarization of the cation orbitals, because these orbitals are very strongly bound with correspondingly high excitation energies, but rather involves polarization of a filled pair of anion orbitals which overlap with these cation orbitals.[14,15] For the present case, the negative electron spin density at the Y nucleus due to the exchange polarization mechanism operating on the state: $\mathcal{A} s\bar{s}o\bar{o}\mathbf{k}_j$, where o denotes an O 2p orbital which has nonzero overlaps with the Y^{3+} s

AO's, can be shown to be:[14]

$$-\langle s|\delta(\mathbf{r}_Y)|s\rangle\langle s|o\rangle^2\langle \mathbf{k}_j o|e^2/r_{12}|o\mathbf{k}_j\rangle/\varepsilon(o\rightarrow o^*) \qquad (4)$$

where $\langle \mathbf{k}_j o|e^2/r_{12}|o\mathbf{k}_j\rangle$ is the exchange integral between $\mathbf{k}_j$ and o and $\varepsilon(o\rightarrow o^*)$ is the average excitation energy of the O 2p orbital, which has been found to be approximately 20 eV in other metal oxide systems.[15] A calculation similar to that leading to Eq. (4) can be carried out for the matrix element of the hyperfine-induced transition, with electron-nuclear spin flip, between the conduction band states $\mathcal{A}s\bar{s}o\bar{o}\mathbf{k}_j$ and $\mathcal{A}s\bar{s}o\bar{o}\bar{\mathbf{k}}_\ell$, in which case, the excited states admixed with $\mathbf{k}_j$ and $\bar{\mathbf{k}}_\ell$ by the exchange interaction are $\mathcal{A}s\bar{s}o^*o\bar{\mathbf{k}}_\ell$ and $\mathcal{A}s\bar{s}\bar{o}\bar{o}^*\mathbf{k}_j$, respectively. The result is similar to Eq. (4) except that the exhange integral now has the form $\langle \mathbf{k}_j o|e^2/r_{12}|o\mathbf{k}_\ell\rangle$, where it is to be noted that in deriving both this matrix element and Eq. (4) terms involving pairs of oxygen orbitals on different oxygens are neglected because they result in small overlap terms and/or large separations of the interacting charge distributions in the exchange integral. Thus. $\mathbf{k}_j$ and $\mathbf{k}_\ell$ must be in the same CuO_2 plane because otherwise the exchange integral will be very small. Consequently, the conduction bands in the two planes contribute independently to exchange-polarization-induced relaxation and $k_{ex}=1/2$ in Eq. (2).

Based on analysis of the NMR frequency shift as a function of oxygen content ($0\leqslant x\leqslant 1$ in $YBa_2Cu_3O_{6+x}$), and the relation between this shift and the macroscopic magnetic susceptibility, Alloul et al. have concluded that the constant shift observed for $x\leqslant 0.4$ is the zero reference for K_Y in $YBa_2Cu_3O_7$, which gives $K_Y=-0.98\times10^{-4}$.[11] Balakrishnan et al. obtain $K_Y=-1.45\times10^{-4}$ from a similar analysis.[7] For the in-plane σ conduction band, K_Y and the relaxation are due solely to the exchange polarization process, and Eq. (2) with $k_{ex}=1/2$ predicts the Korringa relation: T_1T=22800 and 10400 s K for K_Y= -0.98×10^{-4} and -1.45×10^{-4}, respectively, both of which are much larger than the experimental value of T_1T=3200 s K.[7] Thus, the π conduction band is indicated in this case because, with a direct and exchange polarization making positive and negative contributions to K_Y, respectively, a theoretical relaxation rate consistent with the observed Korringa relation can readily be obtained.

On the other hand nonconducting yttrium oxides in which Y^{3+} has eight nearest-neighbor O^{2} typically have chemical shifts of 1.5×10^{-4} to 2.5×10^{-4} with respect to the aqueous YCl_3 reference used in the experiments.[7] If such oxides are taken as the zero reference then $K_Y=-2.5\times10^{-4}$ to -3.5×10^{-4}. In this case the observed Korringa relation is consistent with either a σ or a π conduction band although the σ conduction band is somewhat more likely because anything more than a very small direct contribution to K_Y ($K_s\leqslant0.1\times10^{-4}$) will yield faster relaxation than is observed.

Thus, the question of whether the exchange polarization mechanism operating in a σ conduction band can account for the observed K_Y is of special interest. This can be estimated using the aforementioned calculation of transferred isotropic hyperfine interactions resulting from exchange polarization[14,15] combined with the method used previously to calculate K_Y in the case of a direct Y hyperfine interaction in a π conduction band.[10] The result is:

$$K_{ex}=-0.01\ [\langle o_{2p\pi}\ o_{2p\sigma}|e^2/r_{12}|o_{2p\sigma}\ o_{2p\pi}\rangle/\varepsilon(o\rightarrow o^*)] \times \langle (c^{(x)}_{0,\mathbf{k}})^2+(c^{(y)}_{0,\mathbf{k}})^2\rangle\ \rho(E_F) \quad (5)$$

Here, $\langle o_{2p\pi}\ o_{2p\sigma}|e^2/r_{12}|o_{2p\sigma}\ o_{2p\pi}\rangle=0.91$ eV (Ref. 15) is the exchange integral between an O 2pσ orbital, which is a component of the conduction band wave function, and an O 2pπ orbital on the same atom and pointing at the Y atom. Also, $\langle (c^{(x)}_{0,\mathbf{k}})^2+(c^{(y)}_{0,\mathbf{k}})^2\rangle$ is the average oxygen character of the conduction band at the Fermi surface where the two wave function coefficients correspond to the two oxygens of the CuO_2 plane, $\varepsilon(o\rightarrow o^*)\simeq20$ eV as noted previously, and $\rho(E_F)$ is the density of orbital states per eV per Cu atom at the Fermi surface. Thus, even if $\rho(E_F)$ is small, e.g. $\simeq0.5$ to 1 eV^{-1}, as calculations and experiments indicate,[16 17] this mechanism can account for the observed $K_Y=-10^{-4}$ to -3×10^{-4} if the conduction band has a substantial oxygen character as recent experiments indicate.[18]

In summary, the ^{89}Y Knight shift and relaxation rate in $YBa_2Cu_3O_7$ can provide very important imformation about the nature of the in-plane conduction band states. If K_Y is indeed the frequency shift observed on going from the nonconducting $YBa_2Cu_3O_{6.4}$ to $YBa_2Cu_3O_7$ then a π conduction band formed from the orbitals shown in Fig. 1b is indicated. On the other hand, it is important to resolve the question of why the chemical shift of the nonconducting $YBa_2Cu_3O_{6.4}$ is substantially different from that of other yttrium oxides before a final conclusion is reached.

Work supported by the Department of the Navy, Space, and Naval Warfare Systems Command under Contract No. N00039-88-C-5301.

References

1. Adrian, F. J., Phys. Rev. B37, 2326 (1988).
2. Adrian, F. J. Phys. Rev. B38, 2426 (1988).
3. Lutgemeier, H. and Pieper, M. W., Solid State Commun. 64, 267 (1987).
4. Sedykh, V., Nasu, S. and Fujita, F. E., Solid State Commun. 67, 1063 (1988), and references contained therein.
5. Walstedt, R. E., et al, Phys. Rev. B38, 9303 (1988), and references contained therein.
6. Pennington, C. H., Durand, D. J., Zax, D. B., Slichter, C. P., Rice, J. P. and Ginsberg, D. M., Phys. Rev. B37, 7944 (1988).
7. Balakrishnan, G., Dupree, R., Farnan, I., Paul, D. McK., and Smith, M. E., J. Phys. C: Solid State Phys. 21, L847 (1988).
8. Markert, J. T., Noh, T. W., Russek, S. E., and Cotts, R. M., Solid State Commun. 63, 847 (1988).
9. Kramer, G. J., Brom, H. B., Van den Berg, J., Kes, P. H., and Ydo, D. J. W., Solid State Commun. 64, 705 (1987).
10. Adrian, F. J., Phys. Rev. Lett. 61, 2148 (1988).
11. Alloul, H., Mendels, P., Collin, G. and Monod, P., Phys. Rev. Lett. 61, 746 (1988).
12. Slichter, C. P., Principles of Magnetic Resonance (Springer-Verlag, New York, 1978), 2nd ed., pp. 106-121 and 144-150.
13. Yafet, Y. and Jaccarino, V., Phys. Rev. 133, A1630 (1964).
14. Adrian, F. J. and Jette, A. N., J. Chem. Phys. 81, 2415 (1984).
15. Adrian, F. J., Jette, A. N. and Spaeth, J. M., Phys. Rev. B31,

3923 (1985).

16. Mattheiss, L. F. and Hamann, D. R., Solid State Commun. 63, 395 (1987). Yu, J., Massidda, W., Freeman, A. J. and Koeling, D. D., Phys. Lett. A122, 203 (1987).

17. Mariot, J.-M., Barnole, V., Hague, C. F., Geiser, V. and Guntherodt, H.-J., Solid State Commun. 64, 1203 (1987), and references contained therein.

18. Takahashi, T., et al., Nature 334, 691 (1988).

Role of Rare Earth Substituted for Y in High-T_c $Y_1 Ba_2 Cu_3 O_{7-\delta}$ Superconductor: Polarizing of Fluctuations from Antiferromagnetism; Breakdown of Cooper Pairing at Close Proximity; Hall Effect.

Gary C. Vezzoli*

US Army Laboratory Command
Materials Technology Laboratory
Materials Science Branch
Watertown, Massachusetts 02171

and

B.M. Moon, B. Lalevic, and A. Safari
Department of Electrical Engineering
and Department of Ceramics
Rutgers University
Piscataway, New Jersey

Abstract

We have shown herein that in the new high-T_c superconducting oxides of the form $(R.E.)_1^{3+}Ba_2^{2+}Cu_3O_{7-\delta}$ the paramagnetic central ion has the dual and counter effects of breaking down Cooper pairs in the vicinity (1.0A) of the ion and secondly undergoing indirect exchange causing weak polarization correlation with the Cu^{2+} fluctuations from antiferromagnetism (AFM) that exist at some chain sites normally occupied by Cu^{3+} at O_7 stoichiometry. We believe that these fluctuations are a mixed system having antiferromagnetic and ferromagnetic correlation (and spin density waves) as well as noncorrelated diamagnetic ions. We have also shown that for single crystalline films the Cooper pair concentration is of the order 10^{19} pairs/cm^3 whereas for polycrystalline bulk material this concentration is about 10^{15} pairs/cm^3. Under idealized space filling condition the maximum concentration of Cooper pairs is of the order 10^{20} pairs/cm^3, as compared to the experimental total electron concentration of 10^{21} holes/cm^3. Furthermore, we indicate that quantized flux trapping gives rise to vortice lines of about 50 Å diameter. Hall effect studies have shown an increase in $+R_H$ with decreasing temperature near T_c in the range where resistance vs temperature deviates from linearity.

*Also Visiting Scientist, Department of Material Science The Massachusetts Institute of Technology, Cambridge, Mass 02139 and Adjunct Professor, Department of Electrical Engineering, Rutgers University, Piscataway, N.J.

1. Introduction

The superconductor $Y_1Ba_2Cu_3O_{7-\delta}$ was synthesized by Chu et al (1) whose work was strongly influenced by the work of Bednorz and Mueller (2) on the high-T_c material $La_{2-x}Sr_xCuO_{4-y}$ and also by the earlier work of Raveau et al (3). Because, traditionally, paramagnetic ions were known to break up Cooper pairs, it was at first speculated that the Y ion was magnetically isolated so as not to interfere with the observed superconductivity in each crystalline direction of the unit cell. Similarly it was reported that unlike copper, the yttrium was not indispensable and could be substituted for by many trivalent rare earth ions without causing major changes, and was hence not of first-order significance. Closer scrutiny of these two assertions reveals that they are not totally valid.

In Fig. 1 the work of Chu et al (4) is plotted as a function of the rare earth that is substituted for Y^{3+}. These data show enhancement of the zero resistance temperature and even moreso an enhancement of the equally important temperature which identifies deviation from linear R vs T behavior (with peak enhancement associated with the rare earth ions of maximum spin). It is also shown in this figure that in the deviation from linear R vs T behavior there is asymmetry and anomaly in the Ho^{3+}, Er^{3+} and Lu^{3+} data as compared to the La^{3+} and Nd^{3+} data points. In Fig. 2 the spin and effective magnetic moment are plotted versus the rare earth series and correlated with the decrease in T_c caused when these rare earth elements with 4f electrons are substituted into lanthanum (5). In Fig. 2 we clearly observe that the effective magnetic moment is nonsymmetric for Ho^{3+} and Er^{3+} as compared to Nd^{3+} and Pm^{3+}. In Fig. 3 we additionally plot the paramagnetic Curie Point and the domain transition for Gd^{3+}, Ho^{3+}, and Er^{3+} and note peaked behavior at Ho^{3+} which correlates with peaked μ_e (also plotted but as open circles) (6). In the Laver phases AB_2 where B=Ge,Ru,Os,Ir, or Pt, and A is either a rare earth with 4f electrons (symbolized A') or Y,Sc,Lu or La (which do not contain 4f electrons(symbolized A")), it is known that $A''B_2$ are always superconducting but $A'B_2$ are ferromagnetic. The Curie Temperatures of the $A'B_2$ compounds show a correlation with the number of valence electrons similar to the criterion for the appearance of superconductivity. Comparison of Figs 1 and 2 suggest that the asymmetry and anomaly in the closed circle data of Fig. 1 (deviation from linear R vs T) may be related to the asymmetry of effective magnetic moment and Curie Point and domain transition in Figs 2 & 3. These correlations suggest that the Y or the rare earth substitute may contribute <u>some</u> net factors in high-T_c $Y_1Ba_2Cu_3O_{7-}$ that favor superconductivity, contrasting their <u>destructive</u> influence on the superconductivity of <u>low</u>-T_c materials. The reason why in <u>conventional</u> (low-T_c) superconductors the magnetic impurities, cause very <u>large</u> reduction in T_c, in comparison to the effect of non-magnetic impurities, is related to the rotational freedom of the spin direction of each impurity atom or ion (7). In high-T_c oxide superconductors the effect of magnetic impurities is small and of the same order as the effect of non-magnetic impurities. This seems to imply that the impurity spin in the new high-T_c superconducting oxides is <u>not</u> free to rotate and thus fixed. Hence the further implication is that this fixing is an indication of magnetic ordering which is intrinsic to the host material (such as antiferromagnetism of the $R.E.^{3+}$ sites) and localized behavior rather than impurity - impurity interactions (7) and rather than very long range correlations. One purpose of the present paper is to explore this hypothesis qualitatively and quantitatively.

One manner by which the paramagnetic central ion ($R.E.^{3+}$) can enhance T_c (and thus superconductivity) in a material in which the mechanism for transition to the superconducting state is for the most part non-phonon is via relationship to the spin-fluctuation phenomenon (8a). This form of Boson mediator will be discussed in the next section but is based on a fluctuation from the planar Cu^{2+} idealized antiferromagnetism caused by a stoichiometry which creates uncompensated Cu^{2+} states at chain sites which are ordinarily in Cu^{3+} valence. One contribution, then, of the paramagnetic central ion might be speculated to involve the polarization (8b) of the unpaired Cu^{2+} spins which constitute the deviation or fluctuation from idealized antiferromagnetism. (The antiferromagnetic spin fluctuation mechanism can indeed supplement other mechanisms such as phonon coupling (8c) and bound excitons or internal charge transfer excitations (8d). This supplementing can be additive (8e)).

The interaction which can create such a spin polarization correlation is very complicated and has been addressed in isolated and local terms in a series of well known previous papers (8b). One conclusion from these studies is that in superconductor alloys of $Mo_{0.8}Re_{0.2}$ the addition of Fe causes an abrupt depression of T_c, however when Fe is added to Ti,V,Nb,Zr the net result is a raising or only a slight lowering of T_c (8b). The authors suggest that localized spins exist on Fe when it is dissolved in group (column) VI elements, but not when dissolved in groups IV and V elements.

The modern basis for studying magnetic polarization interactions in solids is through the Rudderman Kittel interaction which will be described herein. However in a complex structure having complicated multi valence non-stoichiometric chemistry these polarization interactions are of extreme complexity, and have not yet been solved in closed form.

Even in materials which are ordinarily regarded as simple paramagnetic substances such as rare earths, there exist some interionic interactions albeit weak interactions, which are also associated with a temperature below which ferromagnetic or antiferromagnetic behavior prevails. In substances which have a Curie or Néel temperature near or above room temperature these interactions are quite large (such as Cu^{2+} in $Y_1Ba_2Cu_3O_{7-\delta}$). It is probable that in many such cases these dipole-dipole interactions are not simple or direct, but are coupled through the electrons of intervening atoms especially in oxides, and halides (9). The ferro or antiferromagnetic interactions are normally decreased when the magnetic species are physically separated from each other. In the case of $RE_1Ba_2Cu_3O_{7-\delta}$ the effect of the planar Cu^{2+} ions as well as the planar and inverted pyramidal apical oxygen ions will influence the dipole-dipole interaction between RE^{3+} and the Cu^{2+} unpaired spin at the formerly Cu^{3+} site. The effect of this superexchange will be assessed herein. The possibility also exists for ferromagnetic correlations.

We also wish to address in this work the capacity of the central paramagnetic ion to break down a Cooper pair (in near proximity) by causing an unparing of the two magnetic spins. Similarly, we wish to examine the counter effect. of the polarization of the copper spins (related to the fluctuation or deviation from idealized antiferromagnetism) due to the influence of the magnetic moment of the $R.E.^{3+}$ central ion on the spin uncompensated Cu^{2+} ions at chain sites normally occupied by Cu^{3+} ions.

Concommitant to this analysis we shall also calculate the concentration of Cooper pairs relative to the total conduction carrier concentration, as well as the maximum concentration of Cooper pairs within a coherence-length topology, and the magnetic-flux-containing topology of the system.

Finally we shall describe recent Hall Effect experiments on $Y_1Ba_2Cu_3O_{7-\delta}$.

2.A. Antiferromagnetic (AFM) Spin Correlations and Spin-Fluctuations in $Y_1Ba_2Cu_3O_{7-\delta}$

Early in the research progress on the new high-T_c oxides it was shown by neutron diffraction that $La_{2-X}Sr_XCuO_{4-y}$ was an antiferromagnetic material (10). Because the even higher-T_c material $Y_1Ba_2Cu_3O_{7-\delta}$ is more closely related to a defect substitutive derivative (11) of the antiferromagnetic K_2MnF_4 structure (12) (see figs 4 and 5) than it is to a true Perovskite, we suggested that it too was ideally antiferromagnetic in its Cu^{2+} correlations. The Cu^{2+} ions are normally confined to the planes in $Y_1Ba_2Cu_3O_7$. We believe this compound to be structurally/chemically described as $(Y_1{}^{3+}Cu_1{}^{3+})_1(Ba_2{}^{2+}Cu_2{}^{2+})_1O_{8-w}$ with $1.1 < w < 1.5$ for superconductivity. This then is an $A_4B_2X_{8-w}$ structure with Cu^{3+} at the chain sites (or an $A_2BX_{4-w'}$ structure). Our preliminary study using the SQUID magnetometer (13) at low magnetic field (<100G) suggested a Néel temperatures of

390K, and preliminary electron capture study indicated at room temperature behavior intermediate between ferromagnetism and paramagnetism (14). Recent neutron diffraction work has shown that for $Y_1Ba_2Cu_3O_{6+x}$ (for X=0 and X=0.15) the structure is indeed 3D antiferromagnetic with respect to Cu spins (10). However for these two oxygen stoichiometries the material is non-superconducting. For the O_6 stoichimetry the most probable chemical structure is $(Y_1^{3+}Cu_1^{1+})(Ba_2^{2+}Cu_2^{2+})_1 O_6^{2-}$. In the superconducting state antiferromagnetism is <u>not</u> observed using neutron diffraction probably because at such stoichiometry the fluctuations from AFM are so severe that a Néel condition cannot be observed. The AFM tendency which we observe using the SQUID at very low B is probably observable because of very low field < H_{c1} and low input energy.

Exactly why this O_6 structure is not superconducting is not clear from a charge excitation or virtual exciton perspective because of the two valence states of copper involved, however, perhaps the charge transfer excitations $Cu^{1+} \rightarrow Cu^{2+}$ and $Cu^{2+} \rightarrow Cu^{3+}$ in the O_6 stoichiometry are <u>not</u> equivalent to $Cu^{2+} \rightarrow Cu^{3+}$ and (possibly $Cu^{3+} \rightarrow Cu^{4+}$) in the $O_{6.5}$ to $O_{6.9}$ stoichiometry from an exciton mediator standpoint. The most probable chemical structure for the $O_{6.15}$ stoichiometry is $Y_1^{3+}Cu^{3+}_{0.15}Cu^{1+}_{0.85}Ba_2^{2+}Cu_2^{2+}O_{6.15}$. On the other hand, the superconducting stoichiometry of $O_{6.9}$ should show the chemical structure $Y_1^{3+}(Cu_{0.8}^{3+}Cu_{0.2}^{2+})_1(Ba_2^{2+}Cu_2^{2+})_1O_{6.9}$. This means that two out of every ten Cu^{3+} chain sites are occupied by Cu^{2+} ions. This deviation from the Cu^{2+} being exclusively at the plane sites (base of the inverted pyramidal polyhedral substructural coordinated species of the unit cell) can constitute a deviation from otherwise medium-range antiferromagnetic order of the planar Cu^{2+} ions. This deviation in the d^9 ions, due to a finite number of $d^8 \rightarrow d^9$ transitions to maintain charge balance when the stoichiometry changes from, for example, $O_{7.0}$ to $O_{6.9}$, can give rise to a contributing Boson to the formation of Cooper pairs. This mediator arises from the attraction that two conducting electrons have to the unpaired spin (dangling bond and unoccupied hole) deviation from antiferromagnetism. The net result of this attraction is a Cooper pairing of neighboring electrons (with opposite $\vec{m}_s$ an $\vec{p}$ (as a lower-energy reaction than the pairing with the localized d^9 state). This then is Boson mediation via a fluctuation from antiferromagnetism or a "spin-fluctuation" in its most rudimentary form. The most modern formalism to describe this regime is given in ref 8f using supersymmetrical gauge theory of strongly correlated electronic system based on quantization of the Hubbard model and treating topological statistical magnetic excitations.

The above spin fluctuations in the absence of a polarizing effect will not be ordered in the material and in this sense are best described by a fractal mathematics. In an order-regime it seems that the spin fluctuation would be a far more effective mechanism to create Cooper pairing (and hence elevate T_c) than in a disordered-spin regime. We thus must explore sources of internal ordering of these deviant spins and associated holes as well as the possibility of positional processing-induced ordering. The most appealing mechanism for intrinsic ordering is polarization by the $R.E.^{3+}$ ion which is also believed to be antiferromagnetic but with a much lower Néel temperature. In general, this type of polarization effect (15-18) can be described by Rudderman-Kittel (19) theory. The most promising candidate for processing-induced ordering is a low-temperature technique such as by sol-gel, or employment of long anneal times and very slow cooling.

It is generally believed that the kinetic superexchange mechanism is the dominant mechanism for antiferromagnetic correlation ordering (15). A correlation super-exchange mechanism has been invoked to describe high-T_c in La_2CuO_{4-y} and involves virtual charge fluctuation processes (7). If the mechanism is indeed a significant and valid contributor, then antiferromagnetism, superexchange, and high-T_c are all linked together.

B. AFM in $Gd_1Ba_2Cu_3O_z$

The independence of superconductive and magnetic behavior in this material suggests strong anisotropy. Magnetization measurements below T_c at intermediate values of field indicated antiferromagnetism for Gd^{3+} (an f configuration S-state ion) (15). Because Gd^{3+} has zero orbital angular momentum the Gd ion does not interact to first order with the crystal field and retains degeneracy of its half-integral spin

$S=7/2$. Investigations of $Ho_1Ba_2Cu_3O_Z$ indicated that high-T_c superconductivity coexists largely independently with strong Ho^{3+} paramagnetism, suggesting that superconductivity is excluded from layers containing Ho ions and is thus of lower dimensionality. This viewpoint is consistent with calculations of electronic band structure and with the known adverse effect of paramagnetic impurities on low T_c superconductors treated in the previous section. These observations suggest examination of the spatial domain of influence of the spin and effective magnetic moment of the rare earth ion as related to its effect on formed Cooper pairs in its vicinity as well as its effect on ordering the spin fluctuations.

The plan of this calculation then is to : (1) evaluate the Cooper pairing energy or gap (this can be accomplished from scanning tunneling microscopy and from the $J_c(T)$ analysis), and evaluate the spacial dependence of the field of the paramagnetic ion (and its associated energy and pair-destroying capability); and (2) evaluate the level of interaction between AFM-coupled rare earth central ions such as $Gd^{3+}(S_{7/2})$ with the d^9 spin fluctuation ion in the Cu-O chain region.

3. Calculation of Cooper-Pair Decoupling Capability of Paramagnetic Central Ion

We embark upon this calculation under the simplifying condition of zero external magnetic field ($B_{ext}=0$). When Cooper-pair conducting electrons, because of the varying <u>phase</u> of their wave function, suddenly find themselves in a spatial position where they are exposed to a field ($\vec{B}_p$) due to the paramagnetic moment of the central ion (Gd^{3+}), then they must experience a <u>time-changing</u> local magnetic field B_p and therefore submit to Maxwell's third Equation on Faraday's Law:

$$\vec{B}_p = -\nabla \times \vec{E}_p \qquad 1)$$

Thus a local field is superimposed upon the Cooper pair. Even though a superconducting material will <u>not</u> microscopically support an electric field, the microscopic supercurrent (at $T>0K$) interprets this internal field ($-\vec{E}_p$) in terms of an induced linear momentum $\vec{P}_p$ which supplements the additional total linear momentum ($\vec{P}_s = 2m\vec{V}_s$) of a Cooper pair in the supercurrent condition (the pair refers to an electron with $\vec{P} = \vec{p}+m\vec{V}_s$ and an electron with $\vec{P} = -\vec{p}+m\vec{V}_s$). Thus there exists a microscopic occupation of this momentum state, and this occupation is in the form of pairs.

We write the wave function in the London form:

$$\Psi(r_1,r_2,r_3\ldots r_N) = \exp[i\sum_j x(r_j)]\Psi_o(r_1,r_2,r_3\ldots r_N) \qquad 2)$$

where $\chi(r_j)$ = phase of the condensate wave-function as a function of position, and $\Psi_o(r_1,r_2,r_3\ldots.r_N)$ = ground state wave vector with no charge flow. The total local momentum is then

$$\vec{p}_s + \vec{p}_p = \hbar\bar{\nabla}\chi \qquad 3)$$

Since $\bar{\nabla}\times\bar{\nabla}$ vector (or curl grad vector) equals zero, we write $\bar{\nabla}\times(\vec{p}_s+\vec{p}_p)=\emptyset$ and thus the existence of potential flow. The force (F) which gives rise to crystal momentum is thus <u>conservative</u> since $\vec{F}\Delta t = m(\vec{v}_s+\vec{v}_p)$ and since $\bar{\nabla}\times(\bar{F}\Delta t) = \Delta t\bar{\nabla}\times\bar{F}=\emptyset$. Thus the supercurrent regime is independent of path provided that the electrons remain paired.

The kinetic momentum (P) is the parameter which governs how high a current density (J) a Cooper pair can sustain before undergoing pair scission. This is because the kinetic momentum governs the kinetic energy of the pair and when this energy exceeds the binding energy of the Cooper pair, then the pair undergoes scission and the current density exceeds the value of critical current (J_c) for that temperature. However for the case in which pair-break-up is due to proximity of the paramagnetic ion and its de-coupling power we must consider the internal field established by the paramagnetic ion.

We write the kinetic momentum P as follows

$$\vec{P}=\vec{p}_s+\vec{p}_p-\frac{2e\vec{A}(r)}{c} \qquad 4)$$

Where $\vec{A}(r)$ = vector magnetic potential defining the internal field due to the paramagnetic ion such that $\vec{\nabla}\times\vec{A}(r) = \vec{B}_p(r)$. The current density (J) in general terms is expressed as (18).

$$\vec{J}(p) = \frac{\vec{F}_s(P)}{\partial p} \qquad 5)$$

Where Fs= Gibbs Free Energy of the state of macroscopic occupation defined by $\vec{P}$. Thus the critical current (J_c) above which superconductivity is destroyed, or locally above which a Cooper pair breaks apart is written $\vec{J}_c(P) = \frac{\partial F_s}{\partial P}(P)$ evaluated at the value of momentum $P=P_c$ corresponding to local Jc.

At this condition $F_s = F_N$ where Fn is the Gibbs Free Energy of the normal state. Thus we have $J_c(P) = \frac{\partial F_N}{\partial P}(P)$ evaluated at P=Pc.

Because this expression is a function of P, then it is implicity also a function of temperature (T). (The higher T will cause greater energy $(1/2)mv^2 = KT$ and will influence the wave vector $\vec{k} = \vec{P}/\hbar$). Also because the kinetic momentum is influenced by the proximity to the paramagnetic ion (since the paramagnetic ion is the source of the highest internal pertubing field), then a component of Jc is also a function of the position coordinate r. Therefore, in the case being discussed the total critical current density arises from a component (J_f) due to the supercurrents accelerating field and a component due to the paramagnetic ion (J_p) or a virtual component

$$J_c = J_f + J_p \qquad 6)$$

Thus since J_c is fixed by F_s and by T, then the closer a Cooper pair becomes to the paramagnetic ion, the lower is the allowable field current such that $J < J_c$. At a minimum distance r_p the Cooper pair must then break down due to proximity to the paramagnetic perturbing source even at Jf=Ø. It is this distance for which we seek a solution. Then for small changes in P caused by closer proximity to the paramagnetic central ion we can write

$$J_c(P) = \Delta F_N / \Delta P$$

$$\Delta P = \Delta F_N / J_c(P) \qquad 7)$$

Hence from our equation for P in terms of A we can conclude.

$$\Delta \frac{F_N}{J_c} = \Delta\left[P_s + P_p - \frac{2e}{c}\frac{mxr}{r^3}\right] \qquad 8)$$

Where Ps is the linear momentum which is not a function of r but a function of J. (M=the magnetic moment of the paramagnetic ion). For an exact solution the above equation must be solved for r.

Thus when a Cooper pair moves from position r_1 to position r_2 we represent the value of Jc at the terminus of this move as

$$Jc=(F_{2N}-F_{1N})/\ [(p_{s2}-p_{s1})+(p_{p2}-p_{p1})-\frac{2e}{c}\left(\frac{mxr_2}{r_2^3}-\frac{mxr_1}{r_1^3}\right)] \qquad (9)$$

evaluating parameters with subscript 1 at r_1 and subscript 2 at r_2.

The above equation is difficult to presently evaluate because of the free energy terms. We can however evaluate the value of r_2 at which the Cooper pairs decouple due to the paramagnetic central ion by equating the magnetic energy (at $J_f=0$) to the gap energy Δ which is itself equal to the binding energy of a Cooper pair. From recent scanning tunneling microscopy work (18) we observe that

$$2\Delta = 3.53k(T_c)$$

$$\text{or } \Delta = 225 \times 10^{-16} \text{ erg}$$

$$\text{But we know } \Delta = -(1/2)\int M\cdot B dV = -(1/2)m_1\cdot B = (1/2)m_1\cdot \mu\nabla\phi)$$

Where m_1 = Magnetic moment of Cooper pair at moment of decoupling (M_1=2e= 3.46BM); B= Magnetic field due to paramagnetic central ion of magnetic moment m_2.

μ = Magnetic permeability; Ø= scalar magnetic potential

$\phi = (m_2\cdot r)/r^3$

Taking the parallel or antiparallel moment condition of θ =0 and differentiating with respect to r we have

Where χ= magnetic susceptibility

$$\Delta = (1/2)\mu_0(2\mu_e)(1+x)m^2/r^3(-2) \qquad (10)$$

We evaluate this for the paramagnetic ion Gadolinium where M_2= 10 Bohr magnetons; $\chi=775000\times10^{-6}$ cgs units and derive

$$r_2 = 0.52\ \text{Å}$$

This is a reasonable magnitude because we know the value must be considerably less than the distance from the central ion to the Cu^{2+} planes which is essentially the Y-O(2) and Y-O(3) distances which are 2.409Å and 2.386Å. Thus, if a Cooper pair approaches Gd^{3+} (radius = 0.938Å) to within about one angstrom, it will undergo scission and de-couple its formerly antiparallel paired spin. For Er number is 0.54 A. Although Y is not expected theorectically to have a magnetic moment, $Y_1Ba_2Cu_3O_{7-\delta}$ shows a moment of 0.5 B.M.-formula unit. If the Y has some covalency, then it may be associated with a small moment and then break up pairs in its immediate environment.

(4) Frequency and time Considerations of Interaction Between Cooper pair and Paramagnetic Ion.

Now we must determine whether the Cooper pair "sees" the magnetic moment of the center ion ($R.E.^{3+}$) as oscillating or as stationary. If the spin of the central ion is oscillating at such a high frequency that during the time of critical proximity of the Cooper pair, the pair experiences an oscillating moment of the paramagnetic ion, then there will be no net force to de-couple the pair.

The maximum velocity of the Cooper pair is about 10^8 cm/sec or 10^{16} Å/sec. This means that the pair is within 1Å of the paramagnetic center ion for about $t \simeq 0.1$fs (10^{-16} sec) The magnetic spin of the center ion oscillates at $\nu = 10^{11}$ Hz in the host material (this frequency may be even four orders of magnitude higher in an isolated condition). Thus within the crystal grain the period of oscillation of the paramagnetic ion is of the order of 10^4 fs, thus appears stationary to the passing Cooper pair which remains in its vicinity for less than 1fs. We can conclude that the calculation of section 3 is valid for he real time regime.

5. Polarization of Cu^{2+} Spin Fluctuations by Paramagnetic Central Ion.

a. General Considerations. Normally the Rudderman-Kittel,(19-22) theory is employed to study the effect of paramagnetic impurities such as Mn. Briefly, an impurity such as Mn retains some of its Hunds Rule Magnetization in the solute state and by this identical mechanism polarizes the spins of the conduction electrons in its vicinity. The resulting spin polarization, however, is curiously not well localized in the vicinity of the impurity but, remarkably, is long-ranged and oscillatory. This means that a second impurity atom at an arbitrary distance from the first admits to ferromagnetic or antiferromagnetic interaction with the first impurity atom depending on whether the second atom is positioned at the trough or the peak (crest) of the polarization wave.

In the case which is addressed herein we must note that the number of paramagnetic central ions is about five times the number of spin fluctuations (There exists one R.E.$^{3+}$ ion per unit cell, and one Cu^{3+} per unit cell; however only one Cu^{3+} site in five undergoes transition from d^8 to d^9 to become a <u>spin-fluctuation</u> from antiferromagnetism due to charge excitation rearrangement (to institute charge balance after stoichiometry change from an O_7 to $O_{6.9}$ chemical formula). Thus if Cu^{2+} is viewed as the impurity, although Copper is diamagnetic in the bulk metal configuration, the Cu^{2+} spin fluctuation ion becomes either ferromagnetic or antiferromagnetic in the context herein when polarized as a fluctuation.

The calculation of the indirect exchange coupling constant according to Rudderman-Kittel theory is extremely complicated and derives from second order perturbation theory. This is written in abbreviated form (19^{-22}) as

$$J(R_{ij}) \text{ IND EX} = +\left[\frac{J^{Hu}}{2N}\right]^2 \left\{ \sum_{\substack{k<k_F \\ |k+q|>k_F}} \frac{2\cos q\cdot Rij}{E(k+q)-E(k)} \right\} \quad 11)$$

These interactions can be represented to link the spins and as such be utilized in the effective Heisenberg Hamiltonian only under very specific conditions. Two of these conditions are: a) the Fermi Sea remains in the ground state and is <u>not</u> substituted for by a long-range order state; and b) the elementary excitations must be free particle-like (thus Coulomb repulsion must <u>not</u> cause incipient long-range ordering of the conduction gas).

If the basic mechanism which causes high-T_c in the (Y or R.E.)$_1$ Ba_2 Cu_3 $O_{7-\delta}$ material is based on bound excitons or internal charge transfer excitations and assisted by spin fluctuations, then although long range ordering of the elementary excitations in the superconductor may not necessarily exist in a single crystal, an excited paired state ($1S^1$-$2S^1$) may actually indeed exist. Hence the applicability of the Rudderman-Kittel theory may not be totally appropriate.

If the correlation energy in the electron gas <u>can</u> be neglected, then the Hartree-Fock ground state is not the conventional Fermi Sea but is apt to be a spiral configuration described as a spin density wave. This possibility would render the Rudderman-Kittel approach less appropriate, and thus applicable only to non-ordered magnetic materials.

It has been shown (23), however, that indirect exchange theory can be applied to metals containing elements of the rare earth series (lanthanides) because the f-shell radii of the rare earth atoms are so small (additional electrons enter inner orbits) that even nearest neighboring atoms do not have significant direct overlap.

One considers the Hamiltonian

$$\mathcal{H}_{\text{IND EX}} = -\sum J_{ij} S_i S_j \qquad 12)$$

and calculates the Eigenstates where the spins S_i are situated on locations on a uniform lattice. The term $q_\emptyset$ is defined as the wave vector for which the interaction constant $J(q_\emptyset)$ attains a maximum. Then for metals containing lanthanides, when $q_\emptyset=0$, all spins are parallel and the ground state is ferromagnetic. However when $q_\emptyset$ is a wave vector on one or more points of symmetry of the Brillouin zone boundary such as $\frac{\pi}{a}(\pm 1, \pm 1, \pm 1)$ in the simple cubic Bravais lattice, then the ground state is an antiferromagnetic configuration. If $q_\emptyset$ is not any of these unique wavevectors, the ground state is a spiral spin variation. Since the location of the Cu^{2+} fluctuations from antiferromagnetism are not likely to be definitively ordered on the lattice chain sites, we should expect some spiral configurations and some antiferromagnetic correlations, with the $q_\emptyset=\emptyset$ ferromagnetic case being most rare. Thus if the ions which host the spin fluctuations are ordered on a sublattice, their antiferromagnetic (or ferri-magnetic) interactions correlation with the Gd^{3+} ion should foster a very high-T_c near room conditions. This is because these ordered mediators are of the order of 20% of the concentration of the Cu^{3+} charge transfer excitations (bound exitons) which should cause an increase of about 7 to 20% of the maximum T_c based on bound exitons (220K), thereby driving T_c to as high as ~270K. This is based on the belief that T_c is proportional to the number of boson mediators, and that the Cu^{3+} ions or sites are more influential as boson-mediators, from both the exiton and spin fluctuation perspectives than are the antiferromagnetically balanced Cu^{2+} sites.

The Rudderman-Kittel interaction must be solved numerically and is cast in terms of k_F which is defined as $2\pi/\lambda_F$ where λ_F = wavelength of conduction electrons when, constrained by the Pauli exclusion principle, they become polarized by the magnetic impurity. The important parameter $n_{c/a}$ (dimensionless) = number of conduction electrons/magnetic atom is associated with k_F. In the cubic lattice case an antiferromagnetic correlation exists for $1/4<n_{c/a}<3/2$ (for bcc and fcc) or $<5/2$ (for simple cubic, sc). In $Gd_1 Ba_2 Cu_3 O_7$ there is one magnetic ion per unit cell (Gd^{3+}) but there are 1.7 conduction electrons per unit cell. Thus the parameter c/a = 1.7, and would indicate antiferromagnetic indirect exchange for a simple cubic unit cell. The value of J(o) is found to be proportional to the Curie Temperature, hence antiferromagnetic indirect exchange correlations may be a maximum for Gd^{3+} as the rare earth constituent relative to Ho^{3+} or Er^{3+}. To determine whether the type of correlation is the stable ground state at low temperature it is necessary to examine the spin wave spectrum $\hbar\omega(r)\, 2S[J(\emptyset)-J(k)]$.

Thus the above analysis suggests (but does not prove) the strong possibility that in $RE^{3+}_1 Ba_2Cu_3O_{6.9}$ the RE paramagnetic ion polarizes some of the Cu^{2+} fluctuations from antiferromagnetism either antiferromagnetically or ferromagnetically while others are correlated by a spin density wave. Crucial to this analysis is that neither the Cu^{2+} plane ions nor the Cu^{3+} chain ions should show characteristics of conventional BCS superconductivity. This is borne out by recent NMR and NQR studies including spin-lattice effects (23b) which also show degradation of the sharp NMR spectrum of the Cu chain ions due to oxygen loss and a strikingly different electron dynamics between the Cu plane and Cu chain ions. Appreciable disorder is reported in this work (23b) in the Cu^{2+} planes which is probably due to puckering and the effects of 5-coordination. The $RE_1 Ba_2 Cu_3 O_{7-\delta}$ unit cell is itself elongated orthorhombic, however for the purpose of the present analysis due to its symmetry we can divide the cell in half and still retain all of the essential chemical physics of the polarization interaction. As such, the cubic approximation is not destructive to the analysis and inferences. The polarization of the spin fluctuation is illustrated in the structural schematic shown in Fig. 4.

b. Calculations of Approximate Energy and Torque for Polarization.

The geometry and structure of the $Y_1Ba_2Cu_3O_{7-\delta}$ cell shows that the distance from Y^{3+} to the Cu2+ at a Cu^{3+} site is 6.45Å with no directly on-diagonal intervening atoms (or ions situated directly between them).

We now calculate the torque T

$\vec{T}=\vec{m}\times\vec{B}$

$=mB\sin\theta$

Taking θ = 90 degrees for a worst case requirement for alignment we have

$$T=mB=-m_{Cu}(2+)\mu\nabla\left\{\frac{\vec{m}_{R.E.}\cdot\hat{r}}{r^3}\right\} \qquad 13)$$

We take $m_{cu}2+\tilde{-}0.5$ B.M. (Ref. 10B)

$m_{Gd}3+\tilde{-}8.0$B.M.

Then we have

T=0.27 MeV

This value has an order of magnitude slightly less than that for spin orbit coupling and slightly more than that for hyperfine electron-nuclear coupling, hence is of the correct order for magnetic spin-spin coupling (22).

6. Ferrimagnetic Correlations

The case must be at least mentioned for correlations between the $R.E.^{3+}$ central ion and the Cu^{2+} antiferromagnetic spin correlations that are neither ferromagnetic nor antiferromagnetic, but ferrimagnetic(24). The interactions between magnetic ions in some materials are negative rather that positive, and hence in such cases antiparallel alignment of neighboring or interacting magnetic moments is favored. Ferrimagnetic compounds are characterized by exchange interactions and two or more interpenetrant sublattices which are spontaneously magnetized in different directions. The two most fully studied crystal types which are known to be ferrimagnetic are classified as spinels and garnets. The spinel structure is an A_2BX_4 compound such that X is a non-magnetic divalent ion such as oxygen, sulfur, or selenium and B is a divalent metal such as Mn, Fe, Ni, Co, Cu, Pb, or Mg (in the case of $RE_1^{3+}Ba_2Cu_3O_7$ the B ions are Cu^{3+} and Y^{3+}); the A ion in a spinel is a trivalent ion of a metal such as Mn, Fe, Co, Al, or Ga (in the case of the superconductor it is Ba^{2+} and Cu^{2+}). If the "B" ions are not all of the same metal (as is the case in the superconductor), the net result can be a mixed material. In the spinel structure it is shown that one-third of the A and B ions are in the spin-up orientation and the other two-thirds are in the spin-down direction.[30]

In the RE^{3+} $Ba_2Cu_3O_7$ structure, interpreted as a $(RE_1^{3+}Cu_1^{3+})_1(Ba_2^{2+}Cu_2^{2+})_1\ O_{8-w}$ derivative of the K_2MnF_4 structure, there are several ways to envision the sublattices. The sublattices can be divided into that which encompasses the CuO_3 ions in the basal planes and that which encompasses the CuO_2 ions in the slightly puckered central planes. In addition the Cu2+ ions which constitute the fluctuations from antiferromagnetism can be considered a quasi sublattice. In the case of spinel the structure can be divided into octants at each of the corners of the cubic cell, and the ferrimagnetism is described by the Néel molecular field model. The elongated orthorhombic cell of $Y_1Ba_2Cu_3O_{7-\delta}$ can easily be envisioned as subdivided into a triple Perovskite structure of three nearly cubic sublattices, and is for this reason often mistakenly referred to as a Perovskite.

Based upon analogy the possibility for ferrimagnetic correlations between the ions (that constitute the spin fluctuation systems) with themselves and with the rare earth central ion is very reasonable. The question then which naturally arises is what is the magnetic state of those antiferromagnetic fluctuations which are <u>not</u> correlated? Here, we are perhaps assisted by noting that in the RE_1^{3+} $Ba_2Cu_3O_7$ superconductor as well as the more recently discovered superconducting oxides containing bismuth and those containing thallium, the copper, bismuth and thallium in atomic form are <u>diamagnetic</u> constituents (with susceptibilities of -5.46, -280.1 and -50.9×10^{-6} cgs units). These elements are excellent candidates for high-T_c non-phono-mechanism superconductivity because of their multivalence and the manner with which they can coordinate oxygen. It is interesting to note that of all the divalent anions which can participate in the A_2BX_4 structure, only oxygen is <u>not</u> diamagnetic, and only oxygen seems at present to be compatible with high-T_c superconductivity. This suggests as mentioned above that a diamagnetic anion may interfere with superexchange and thereby destroy antiferromagnetic indirect exchange correlations. From the above, we conjecture that the <u>non-correlated</u> Cu^{2+} ions at Cu^{3+} sites are diamagnetic. The frequency of oscillation of the non-correlated diamagnetic Cu^{2+} fluctuations from AFM is apt to be much greater than the spin-flip oscillation of the R.E.$^{3+}$ ions and even if the period is of the order of the time of proximity of a Cooper pair, the <u>transient paramagnetism</u> would be too weak to break up a pair. Diamagnetism of isolated ions results from the orbital motion of electrons as a consequence to the force which each electron experiences when moving in a magnetic field. Although diamagnetism is a universal property of tight and complete pairing it is outweighed by paramagnetism when electrons undergo transitions to other unique quantum states of differing $\vec{m}$ due to applied H. Since it would be an unrealistic assumption to conjecture that the Cu^{2+} spin fluctuation from AFM does not undergo transition, then we can suspect that those which are uncorrelated undergo oscillations between diamagnetic and paramagnetic states.

We believe that the orbital contributions to the spin indirect exchange interacitons are small in (Y^{3+} or R.E.$^{3+}$), $Ba_2^{2+}Cu_3O_{7-\delta}$ because the orbital moment is essentially quenched due to the intense, non-uniform, electric field arising from the oxygen ions that surround the paramagnetic rare earth and weakly paramagnetic barium. Again in anology with the spinel archtype of ferrimagnetism, the exchange correlation in an A_2BX_4 spinel is influenced by whether the ferromagnetic ion is distributed over only the VI c.n. sites or both the VI and IV sites. The latter case characterizes inverse spinels in which one-half of the A ions are in tetrahedral interstices and the other half and all of the B ions are in the octahedral interstices. This occurs when the preference toward octahedral coordination is greater for the B ions than for the A ions. In (Y^{3+} or R.E.$^{3+}$)$_1$ $Ba_2Cu_3O_{7-\delta}$ it seems clear that Cu^{2+} can exist at plane sites, and depending on δ, chain sites as well. Furthermore, Cu^{3+} can exist at chain sites and may on occasion substitute for Y^{3+} or R.E.$^{3+}$. Because of these added complexities or deviations the indirect exchange and superexchange correlations are effected and the net result causes the spontaneous moment per molecule unit to be equal to the average moment of the B ion, or in the case of the superconductor the weighted average moment of Y^3 and Cu^{3+}. The deviations probably also insure presence of some non-correlations spins and some intrinsic disorder which must ultimately limit T_c. However, the above analysis does not suggest that this ultimate limitation is at a temperature less than room temperature. Here we must realize that for every Cu^{3+}d8 -> Cu^{2+}d9 (AFM spin fluc) site in $\rho_{6.9}$ stoichiometry, there are eight Y^{3+} or R.E.$^{3+}$ central ion nearest neighbors. Since at T_c the temperature is above the Néel Temperature for the Y or R.E. ions then these ions are not antiferromagnetically related at T_c and we suspect that their average paramagnetism (majority spin) strongly enforces parallel or antiparallel correlation with the spin fluctuation unpaired state. This means that relative to each other the spin fluctuation sublattice will have essentially a similar relationship as the polarizing paramagnetic central ions have to each other (either parallel to the RE spins, or antiparallel to them). The second and further-away nearest neighbor should have only minor effects on this polarization.

7. Quantized Flux Trapped in Topological Region of High-T_c Superconductor When $H>Hc_1$: Vortice lines of quantized flux.

We, in general, believe that in the new high-Tc oxide superconductors correlations are more localized and of a shorter range than in low-T_c conventional superconductors such as Nb. This is certainly suggested from the vast difference in coherence length of the former (10-17A) as contrasted to the latter (1000A). However, because of the new oxide high value of H_{c2} we must ask ourselves over what spatial range is flux quanta trapped when $H_{c1} < H < H_{c2}$ (re-admission of flux beneath the penetration depth from the external field)?

The London (25) value for flux quanta enclosed in a topological region of closed loop cross sectional perimeter dl is $\Phi = A\cdot dl = n(hc/e)$. This relationship is independent of material and is derived from the quantum mechanical requirement or condition for a single valued solution to the wave function in the case when the current density J is Gauge-invariant. The value for hc/e is $4X10^{-7}$ Gauss cm^2. Experimental evidence (26) shows quantization in the form $n(\frac{hc}{2e})$ in which the factor of 1/2 arises from pairing.

From our own study (23) of resistance versus B at$T=T_c=81.7K$ we find that $Hc_2 > 15T = 150000$ gauss, however Hc_1 is much less than 1 kilogauss. The quantization evidence (experimentally proven for tin (26) and lead (27) cylinders in three octhogonal Helmholz coils) is interpreted by Onsager (28) and by Byers and Yang (29) as direct proof of Cooper pairing.

We then must ask ourselves when H exceeds Hc_1 and flux[30] is being readmitted into the superconductor (because the high-Tc materials are Type II superconductors with a low value of Hc_2) what topological surface will trap it? This surface will admit to circulation currents and will be described by fractal mathematics. From Stokes Theorem $\oint \vec{A}\cdot d\vec{l} = (\vec{\nabla}x\vec{A})\cdot\vec{n}ds$. We can write $\nabla x\vec{A}=\vec{B}$

We take $Hc_1 = 150$ gauss, and if we can approximate A as independent of ℓ, and B as independent S, then

$$\int_S dS= n\ (2X10^{-7}\ \text{gauss}\ cm^2)/12T$$

$r = 25$ Å

for n=1.

8. Calculation of Cooper Pair Concentration

a. Calculation of Cooper Pair Concentration for Ideal Topological Space Filling.

The coherence length in a high-T_c material is believed to be about 15Å, contrasting as previously mentioned the 1000Å coherence length of conventional superconductors such as Nb.

The volume of a sphere of radius 15Å can be envisioned as the domain of freedom of a single Cooper pair

$$V=\frac{4}{3}\pi r^3=4(15)^3 A^3=13.5\times10^{-21} cm^3$$

Based on space filling this indicates a carrier concentration of superconducting electrons (c.p.) of 7.4×10^{19} carriers/cm^3 or 3.7×10^{19} Cooper pairs/cm^3. Based on a closest packing of c.p. spheres, $n=2.6\times10^{19}$. Measured (23) total carrier concentration (from Hall studies) in (~90K) $Y_1Ba_2Cu_3O_{7-\delta}$ is about 1×10^{20} - 1×10^{21} /cm^3. Thus there is approximately one Cooper-paired electron per 3.8 conducting electrons. Using "pancake" geometry (r=17.5Å; h=3.5Å) yields $n_{c.p.}\approx 1\times10^{20}$ pairs cm^{-3}.

b. Calculation of Real Cooper Pair Concentration Based on J_c for Poly- and Single Crystal Regimes

We should be able to re-calculate the Cooper pair concentration from a binding energy and J_c analysis, and approach the idealized condition by employing the value of J_c for the best crystalline films of $Y_1Ba_2Cu_3O_{7-\delta}$ ($J_c\approx10^5$ A/cm^2).

We can write the critical current in terms of the critical momenta (p_c) of the Cooper pair, the values of p_c being associated with an energy that exceeds the binding energy (or gap, Δ) of the Cooper pair:

$$J_c=\frac{(\rho_{c.p.})p_c}{m}$$

Where m=effective mass of Cooper pair.

The gap energy can then be described as

$$\Delta=\frac{p_c^2}{2m}=\frac{(mJ_c/\rho_{c.p.})^2}{2m}=\frac{m}{2}\left\{\frac{J_c}{\rho_{c.p.}}\right\}^2$$

From scanning tunneling microscope studies (18) we have $2\Delta=3.53kT_c$ or $=22.5\times10^{-15}$ erg=14meV.

For excellent single crystalline films we take $J_c=10^5$ A/cm^2; the effective mass of the electron at this energy is $m_e=265\times10^{-28}$ gm which we multiply by two in order to consider the pair. Using the conversion factor of about 0.62×10^{19} electrons/coulomb we calculate

$$\rho_{c.p.} = 6.5 \times 10^{17} \text{ Cooper pairs/cm}^3$$

$$\approx 10^{18} \text{ c.p./cm}^3$$

This value is almost within an order of magnitude of the idealized concentration given. Previous work[23] indicates $J_c = 10^2 A/cm^2$ which yields $\rho_{c.p.} \simeq 6.5 \times 10^{14}$ c.p./cm^3. In the former case (single crystal) the results yield 1 Cooper pair per about 10^2 electrons, and the latter case (polycrystalline), because of the effect of grain boundaries, yields 1 Cooper pair per 10^6 electrons. The reasons why the non-superconducting electrons exceed the superconducting electrons by a factor of approximately 10^2 is due to point defects including magnetic impurities, thermal scattering (because T_c is greater than absolute zero), exclusion of Cooper-pairing near the paramagnetic RE ion (and in the <u>very</u> near vicinity of the Ba ion), and scattering with electrons undergoing charge transfer excitations. Thus space filling does not occur, and Cooper pairs are excluded from a non-negligable volume of the unit-cell.

9. Hall Effect Studies

Hall Effect experiments were performed at values of magnetic field to 15T. Data taken at 6.8 and 5.0T are given in Fig. 6 and 7. These data show a rising positive Hall voltage with decreasing temperature, this effect beginning at temperatures somewhat above T_c (at approximately the temperature corresponding to the deviation fron linear behavior in R vs T). the results bear <u>strong</u> similarity with the peaked signal reported by Hundley et al[32]. We performed the experiment at 6.8T in a point by point manner reversing both the current and the magnetic field at each equilibrium temperature. However, so as to preclude missing a maximum or minimum R_H turning-point-temperature we also monitored V_H continuously as a function of temperature for a single polarity of current and magnetic field. This continuous data showed essentially the same trend as the point by point method, proving that stray voltages due to any very-slight mismatch of Hall contacts or thermal emf's were small. Each of our independent measurements (at several temperatures) of V_H at B=0 were less than 1%

of our measured Hall signal with B>5T. In the lower portion of Fig. 6 the resistance data on both the undoped and the tin-added samples are given. An anomaly is shown in the tin-added sample at about 1Ø3K where a pre-onset effect was observed. EDAX-SEM experiments identified the presence of tin within the grain boundaries. At concentrations greater than about 3 weight % tin there is a clear indication of a second phase. In general, the tin-added material is more machineable than the undoped samples. In Fig. 7a we show four-terminal R vs T data indicating onset in the Rutgers sample at 82.ØK and near-zero resistance at 8Ø.7K, using slow cooling rate < Ø.3ØK/min. In this figure, we also give the recovery of resistance corresponding to the intermediate region between the super-conducting and normal states at 81K as a function of increasing B to 15T. These data indicate that recovery to the normal state has not occurred even at fields as high as 15T at temperatures near T_c, and a very strong B dependence of resistance up to 5T.

In Fig. 8a we give our Hall data taken at 15T which suggest a positive anomaly spike just above T_c, followed by a lowering of $+V_H$ and including some oscillatory or irregular behavior. In Fig. 8b we reproduce the data of Ref. 33 which shows similar behavior for $Ba_{1.2}Y_{Ø.9}V_{Ø.9}Cu_3O_{7-}$ but also includes a region of changes of sign to $R_H<Ø$. The data in Ref. 33 are explained by the authors as being due to differences between the interior of grains and the grain boundary itself. (The latter can still support a Hall voltage even though the grain "islands" are superconducting). An additional explanation is needed.[34)]

If an exciton-mediated mechanism is indeed present as an initiator of high-T_c superconductivity, one would expect an <u>increase</u> in exciton concentration as the temperature were decreased to the T_c region. This would be expected because of the <u>increase</u> in exciton lifetime as heat is withdrawn from the sample. The core of a bound exciton is a positive hole which is either bound on an oxygen, or is due to a charge excitation or fluctuation from one valence state of copper to a second valence state. Thus if exciton concentration increases, there should be a <u>decrease</u> in positive local bound carriers which would be reflected by an <u>increase</u> in $+R_H$ (as we observe). The decrease in bound holes due to exciton formation could cause a decrease in measured carrier concentration by interfering with the mobility of free holes via a bucket-brigade hopping process[12]. This reasoning, however, might imply a monotonically rising $+V_H$ with decreasing temperature rather than the relatively sharp increase that we observe near T_c in Figs. 6-8. One possible explanation is that Cooper-pairing should cause a decrease in electron-exciton scattering. Such scattering would otherwise be a means to promote recombination of the constituents of the exciton. Further Cooper-pairing attraction exceeds the weak exciton binding energy, then the exciton should ionize. If such a phenomena occurs, it must do so near T_c where Cooper-pairing is rapidly building-up. The result of such an ionization will furnish bound holes and free electrons into the conduction system just prior to reaching the superconducting state. This may account for why there is an observed maximum and a rapid fall-off in the positive Hall signal in the work of Hundley et al[32], and in the Hall study that we performed at 15 & 6.8T, Hundley's[32] work and Ref. 10 show the Hall signal becoming negative before returning to zero which may imply that so many excitons become ionized that electrons overcome holes in the conduction mechanism. In our work the approximate carrier concentration at the inception of the transition to the superconducting state is $p=1x10^{20}$ holes/cm^3.

References

1 C.W. Chu, P.H. Hor, R.L. Meng, L. Gao, Z.J. Huang, and Y.Q. Wang, Phys. Rev. Lett 58 (4), 4Ø5 (1987); M.K. Wu, J.E. Ashburn, C.J. Torng, P.H. Hor, R.L. Meng, L. Gao, Z.J. Huang, Y.Q. Wang, L. Gao, Z.J. Huang, J. Becktold, R. Foster, C.W. Chu, Phys. Rev. Lett 58 (18). 1891 (1987).

2 J.G. Bednorz and R.A. Mueller, Z. Phys. B64, 189 (1986).

3 C. Michel, L Er Rakko, and B. Raveau, Mat. Res. Bull. 2Ø, 667 (1985).

4 C.W. Chu, APS Meeting March 1987, New York City.

5 B.T. Matthias, H. Suhl, and E. Corenzwit, J. Phys. Chem. Solids 13, 156 (1959) and Phys. Rev. Lett 1, 92 (1958).

6 I. Pop Stud. Cercetari Fig. (Rumania) 17 (7), 755 (1965).

7 E.W. Fenton, Solid State Comm. 65 (9) 1Ø13 (1988).

8a P.W. Anderson Science 235, 1196 (1987); G. Baskaran, F. Fow, and P.W. Anderson Solid State Comm. G3, 973 (1987); M. Cyrot Solid State Comm. 63,1Ø15 (1987); H. Fukuyaamd and K. Yosida Japan J. Appel Phys. 26, L371 (1987), M.M. Mohan and N. Kumar, J. Phys. C: Solid State Physics 2Ø, L527 (1987); A.E. Ruckenstein, PJ. Hirshfeld and J. Appel Phys. Rev. B6, 857 (1987).

8b B.T. Matthias, M. Peters, H.J. Williams, A.M. Clogston, E. Corenzwit and R.C. Sherwood, Phys. Rev. Lett 5 (12), 524 (196Ø); T.H. Geballe, B.T. Matthias, A.M. Clogston, H.J. Williams, R.C. Sherwood. J.P. Maita, J. Appel Phys. 37 (3), 1181 (1986); T.J. Geballe, B.T. Matthias, J.P. Remeika, A.M. Clogston, V.P. Compton, J.P. Maita, H.J. Williams, Physics 2 (6) 293 (1966), P.W. Anderson Solid State Physics 14,99 (1963); H. Frohlich, Phys. Rev. 79 (5), 845 (195Ø), Proc. Roy Soc. A215 (1952); J.Bardeen, L.N. Cooper, and

J.R. Schreiffer, Phys. Rev. 1Ø8 (15) 1957; Proc. Roy Soc. A223 (1954).

8c V.Z. Kresin, Phys. Rev. B 35 8716 (1987); C.M. Varna, S. Schmitt-Rink and E. Abrahams Solid State Comm. 62, 681 (1987); J. Ruvalds Phys. Rev. B35, 8869 (1987); R. Jagadish and K.p. Sinka Curr. Sci. 56, 291 (1987); Pramana 28, L317 (1987b); see also A.J. Freeman & J.J. Yu, APS Meeting New Orleans, M7-9, Mar 1988 and private communication.

8d J. Ruvalds, Univ. of Virginia and Harvard private communication (1987). J. Ruvalds, Adv. in Phys. 3Ø (5), 677 (1981).

8e P.B. Wiegman, Phys. Rev. Lett 6Ø (9), 821 (1988).

9 F.A. Cotton and G. Wilkinson, Advanced Inorganic Chemistry, New York, Interscience (1966) pp. 642,643.

1Øa D. Moncton, Science 236, 78Ø (1987); J.R. Thompson, S. Sekula, D. Christen, B. Sales, L. Boatner, Y. Kim, Phys. Rev. B36 (1), 718 (1987); R.T. Collins, Z. Schlesinger, M.W.Shafer and T.R. McGuire, Phys. Rev. B39 (1Ø), 5817 (1988).

1Øb J. Tranquada, D. Coz, W. Kunnman, H. Moulden, G. Sirane, M. Suenaga, P. Zolliker, D. Vaknin, S. Sinka, M. Alvarez, A. Jacobson and D. Johnson Phys. Rev. Lett 6Ø (2), 156 (1988).

11 G.C. Vezzoli, R. Benfer, and W. Spurgein, Proc. of Conf. on Novel Mechanisms of Superconductivity, Ed by S. Wolf and V. Kresin, New York, Plenum 1987 pp. 1Ø22-1Ø32.

12 R.J. Bergeneau, H.J. Guggenheim, G. Sirane. Phys. Rev. B5, 2211 (1979).

13a A. Austin, R. Frankel, G.C. Vezzoli, work conducted in Sep. 1987 at National Magnet Lab, MIT, unpublished;

13b G.C. Vezzoli, B.M. Moon, T. Burke, B. Galevic, A. Safari, A. Sundyar, submitted to Phys. Rev. B.

14 K. Waters, C. Raw, A. Austin, G.C. Vezzoli, work conducted in Oct 1987 at Dept. of Physics, Rice Univ. (unpublished); also Ref. 13b.

15 K.P. Sinha and M. Singh, J. Phys. C.; Solid State Phys. 21, L231 (1988); K.P. Sinha Ind. J. Phys. 35, 434 (1961).

16 J.R. Thompsom, S.T. Sekula, D.K. Christen, B.C. Sales, L.A. Boatner, and Y.C. Kim, Phys. Rev. B36 (1), 718 (1987); J.R. Thompson, D.K. Christen, S.T. Sekula, B.C. Sales, and L.A. Baotner, Phys. Rev. B (Rapid Comm) 36 (1) 836 (1987).

17 J. Bardeen in Superconductivity in Science and Technology Ed by M.H. Cohen, Chicago, Univ. of Chicago Press, 1968, pp. 8-11.

18 R. Meservey, The Massachusetts Institute of Technology, 1987, private communication.

19 M.A. Rudderman and C. Kittel, Phys. Rev. 96,99 (1954); N. Bloembergen and T.J. Rowland, Phys. Rev. 97, 1679 (1955).

20 A.W. Overhauser, Phys. Rev. 128, 1437 (1962).

21 W. Mashall, Rev. Mod. Phys. 36, 399 (1964); G.S. Rushbrooke, J. Math. Phys. 5, 1106 (1964).

22 D.C. Mattis, The Theory of Magnetism, Harper and Row, New York (1965), pp. 193-207.

23a G.C. Vezzoli, T. Burke, B.M. Moon, A. Safari, and B. Lalevic in review for Phys. Rev. B.

23b W.W. Warren, Phys. Rev. B 36 (10), 5727 (1987).

24 D.H. Martin, Magnetism in Solids. The MIT Press, Cambridge, Mass, 1967, pp. 240-270.

25 F. London, Superfluids, New York, John Wiley and Sons,Inc, 1950, p. 158.

26 B.S. Deaver and W.M. Fairbank, Phys. Rev. Lett F (2), 43 (1961).

27 R. Doll and M. Nabauer, Phys. Rev. Lett 7(2), 51 (1961).

28 L. Onsage, Phys. Rev. Lett 7(2), 50 (1961).

29 N. Byers and C.N. Yang, Phys. Rev. Lett 7 (2), 46 (1961).

30 A.A. Abrikosov, Zh. Eksperim i Teor. Fig. 32, 1442 (1957); translation Sov. Phys. JETP 5, 1174 (1957).

31 Structural Ignorance Chemistry by A.F. Wells, Oxford, Claredon Press 1984 pp. 208, 318-319, 593, 594.

32 M.F. Hundley, A. Zettl, A. Stacy, and M.L. Cohen, Phys. Rev. B35 (16), 8800 (1987); For Hall studies see also N. Ong, Z.Z. Wang and J. Clayhold in Novel Superconductivity, ed by S. Wolf & V. Kresin, Plenum, New York, pp. 1061-1066 and A.I. Braginski in same volume, pp. 935-949.

33 Y. Zhao, J. Xia, Zhenhui He, Shitang Sun, Qirui Zhang, Y. Qian, Z. Chen, G. Pan, Chinese Phys. Lett 5 (5), 221 (1988).

34 Zhenhui He, private communication.

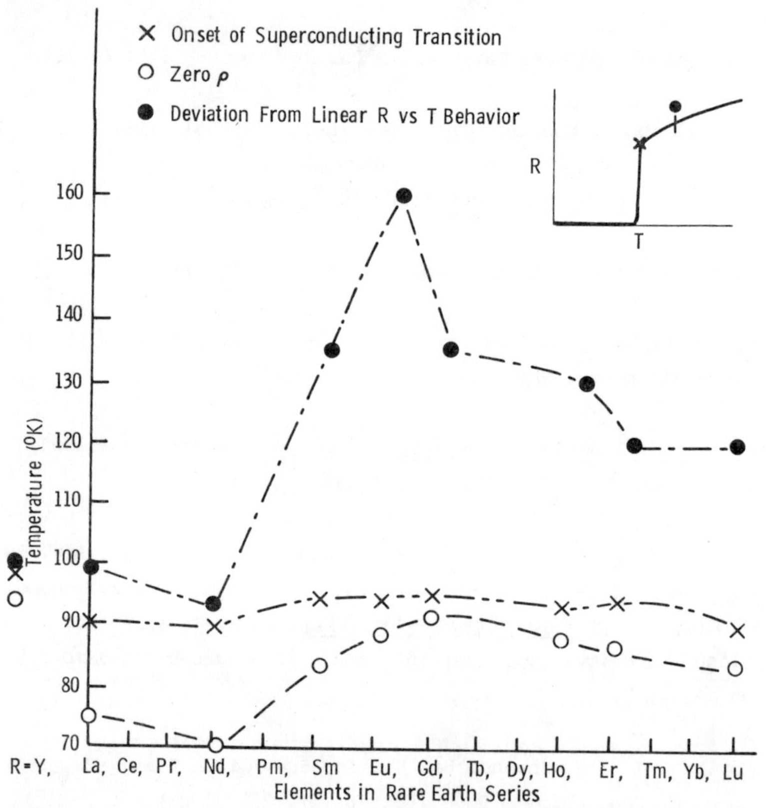

TRANSITION TEMPERATURES VERSUS ELEMENT (R) IN $R_1Ba_2Cu_3O_{7-8}$ (After Chu et al.)

Fig.1

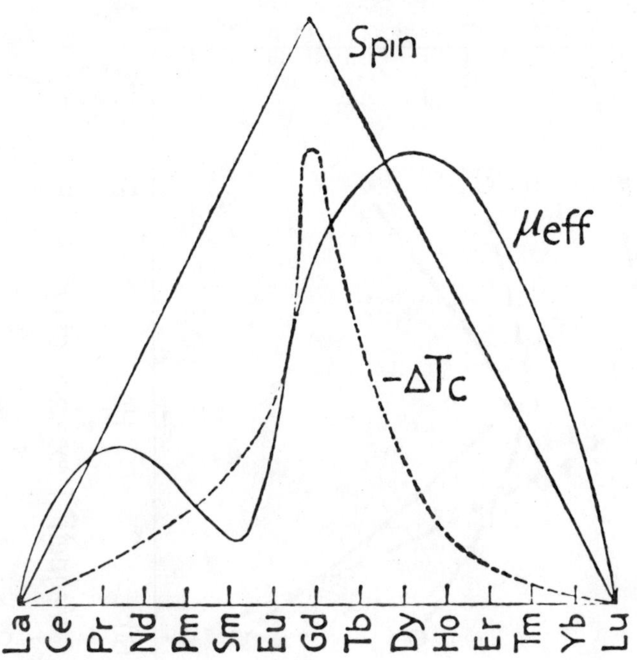

Fig. 2

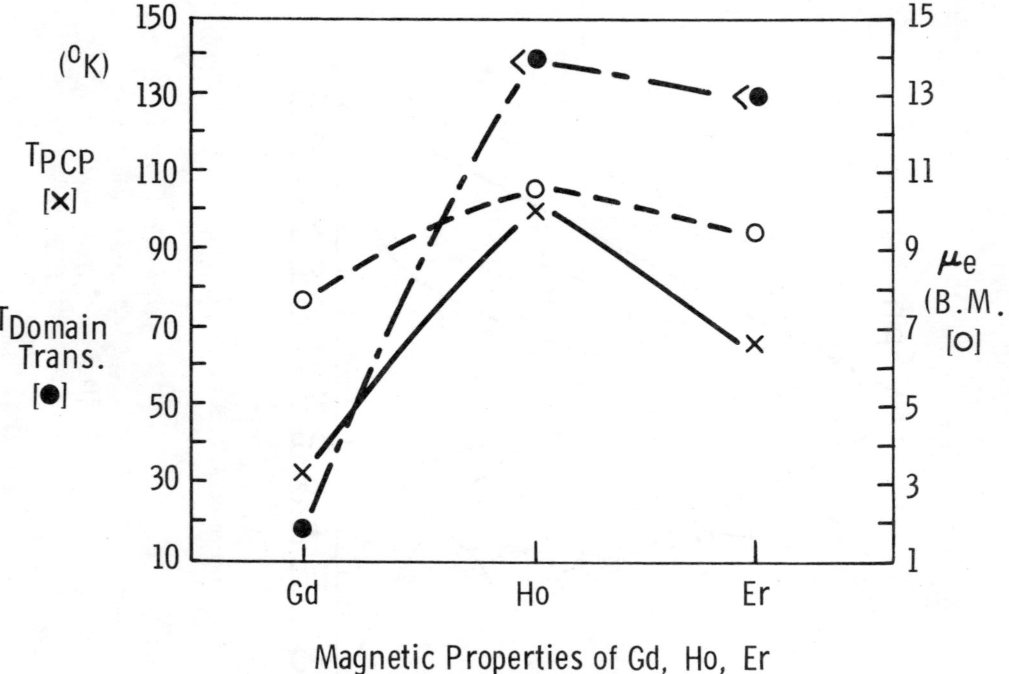

Magnetic Properties of Gd, Ho, Er

Fig.3

SPIN-FLUCTUATION CONTRIBUTING MECHANISM FOR HIGH-T_c SUPERCONDUCTIVITY

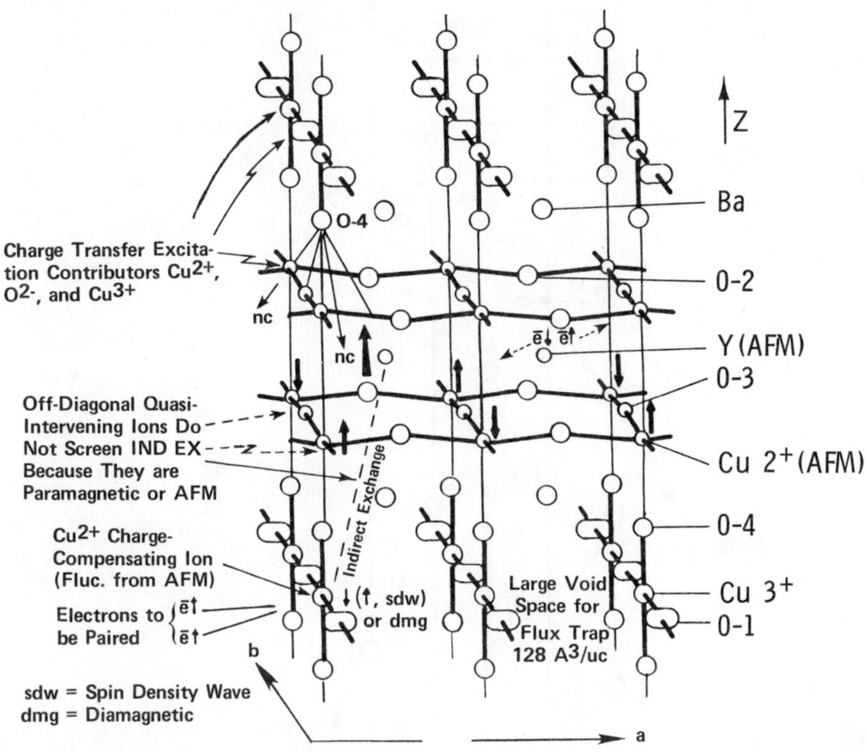

$Y_1Ba_2Cu_3O_{6.9}$
(Indirect exchange can be ferrimagnetic, spin density wave, or ferromagnetic)

O-1 - Oxygen Chain Ion
O-2 - Oxygen Planar a-Axis Ion
O-3 - Oxygen Planar b-Axis Ion
O-4 - Oxygen Apical Ion of 5 Coordinated Cu^{2+} Pyramidal Polyhedral

nc = Next Cell
uc = Unit Cell
⇠ ⇢ Pair Breakup
➡ Refers to Paramagnetic Central Ion

Fig.4

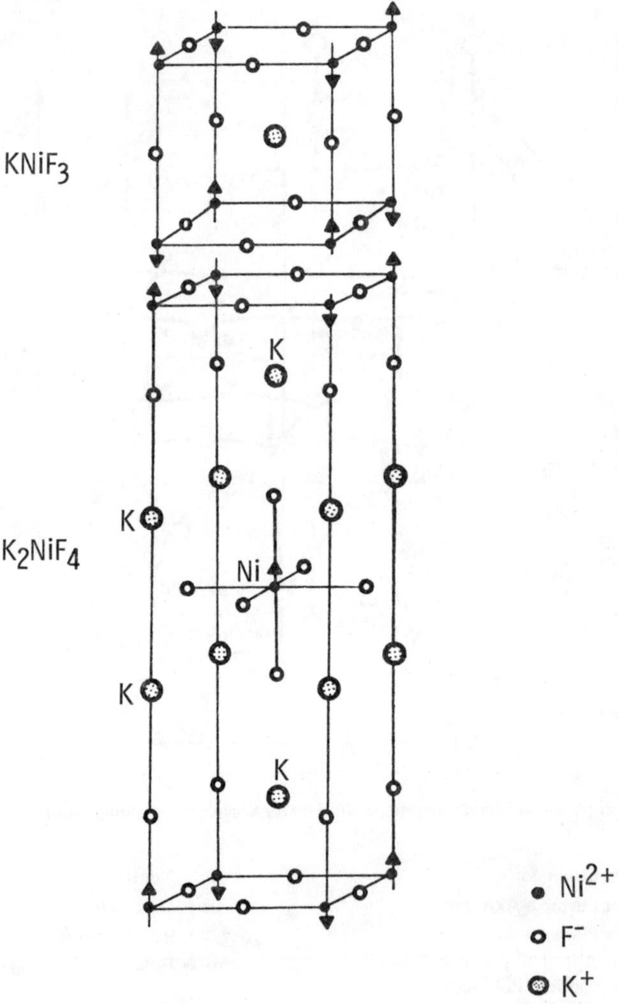

Fig. 5

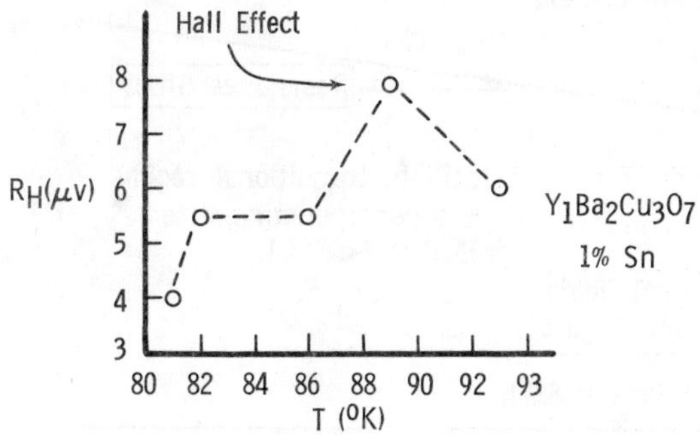
Hall Effect
$R_H(\mu v)$
8
7
6
5
4
3
80 82 84 86 88 90 92 93
T (^{0}K)
$Y_1Ba_2Cu_3O_7$
1% Sn

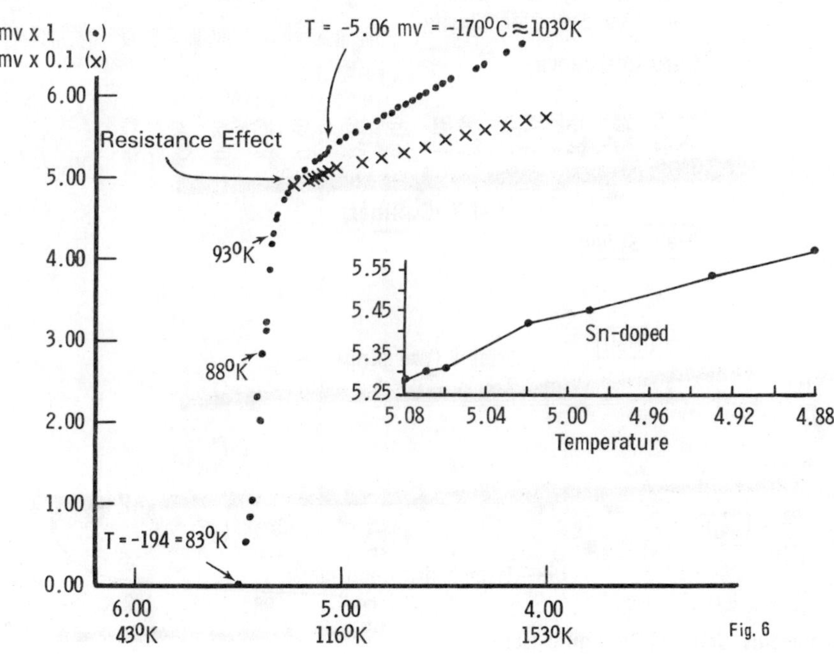
mv x 1 (•)
mv x 0.1 (x)
T = -5.06 mv = -170^{0}C ≈ 103^{0}K
Resistance Effect
93^{0}K
88^{0}K
T = -194 = 83^{0}K
6.00
5.00
4.00
3.00
2.00
1.00
0.00
6.00 43^{0}K
5.00 116^{0}K
4.00 153^{0}K
5.55
5.45
5.35
5.25
Sn-doped
5.08 5.04 5.00 4.96 4.92 4.88
Temperature

Fig. 6

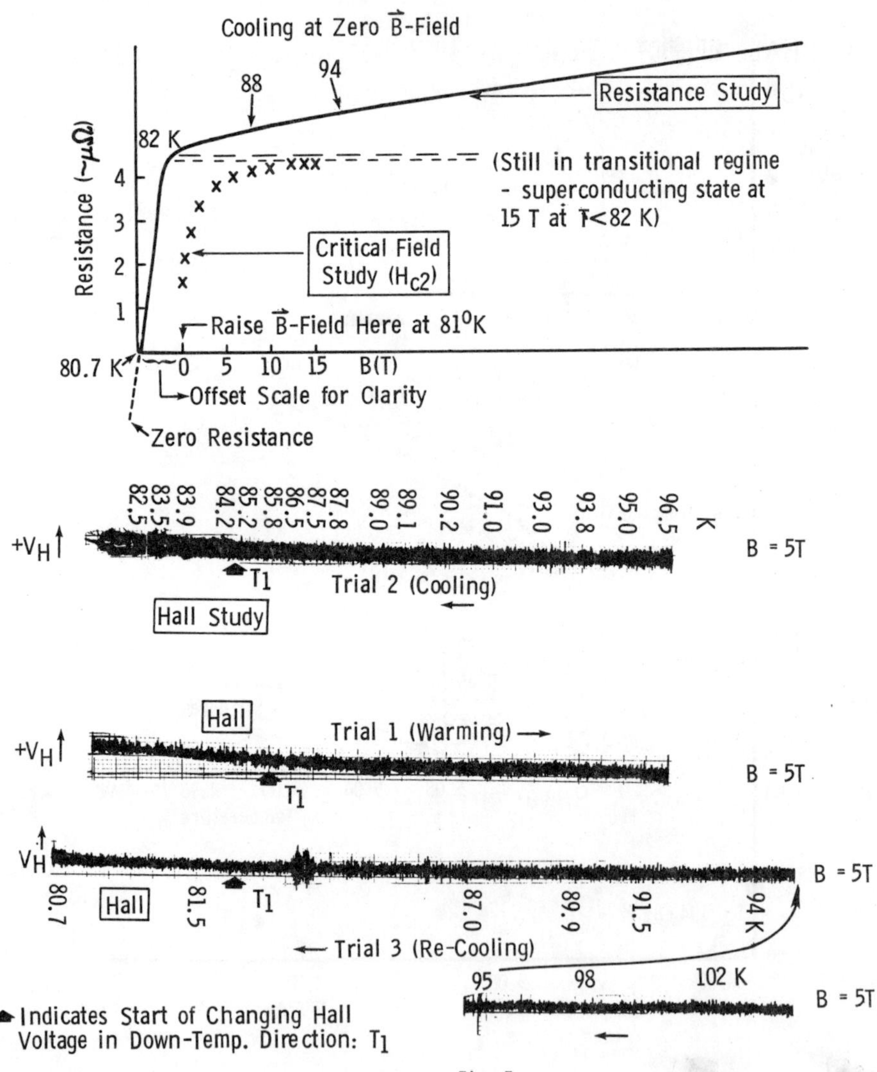

Y1Ba2Cu3O7-δ SUPERCONDUCTOR
(Polycrystalline)
Cooling at Zero B-Field
88
94
Resistance Study
82 K
(Still in transitional regime
- superconducting state at
15 T at T<82 K)
Resistance (~μΩ)
4
3
2
1
Critical Field
Study (Hc2)
Raise B-Field Here at 81°K
80.7 K
0 5 10 15 B(T)
Offset Scale for Clarity
Zero Resistance
82.5 83.5 83.9 84.2 85.2 85.8 86.5 87.5 87.8 89.0 89.1 90.2 91.0 93.0 93.8 95.0 96.5 K
+VH
B = 5T
T1
Trial 2 (Cooling)
Hall Study
Hall
Trial 1 (Warming)
+VH
B = 5T
T1
VH
B = 5T
80.7
Hall
81.5
T1
87.0
89.9
91.5
94 K
Trial 3 (Re-Cooling)
95
98
102 K
B = 5T
Indicates Start of Changing Hall
Voltage in Down-Temp. Direction: T1

Fig. 7a

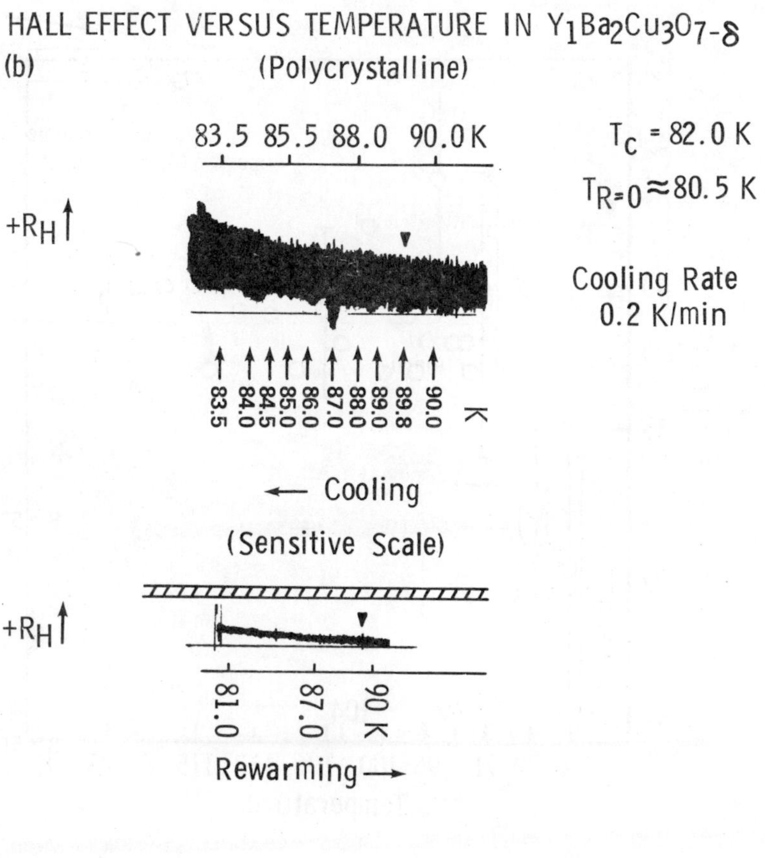
HALL EFFECT VERSUS TEMPERATURE IN $Y_1Ba_2Cu_3O_{7-\delta}$
(b)
(Polycrystalline)
83.5 85.5 88.0 90.0 K
T_c = 82.0 K
$T_{R=0} \approx 80.5$ K
$+R_H$
Cooling Rate
0.2 K/min
83.5
84.0
84.5
85.0
86.0
87.0
88.0
89.0
89.8
90.0
K
← Cooling
(Sensitive Scale)
$+R_H$
81.0
87.0
90 K
Rewarming →
Temperature

Fig. 7b

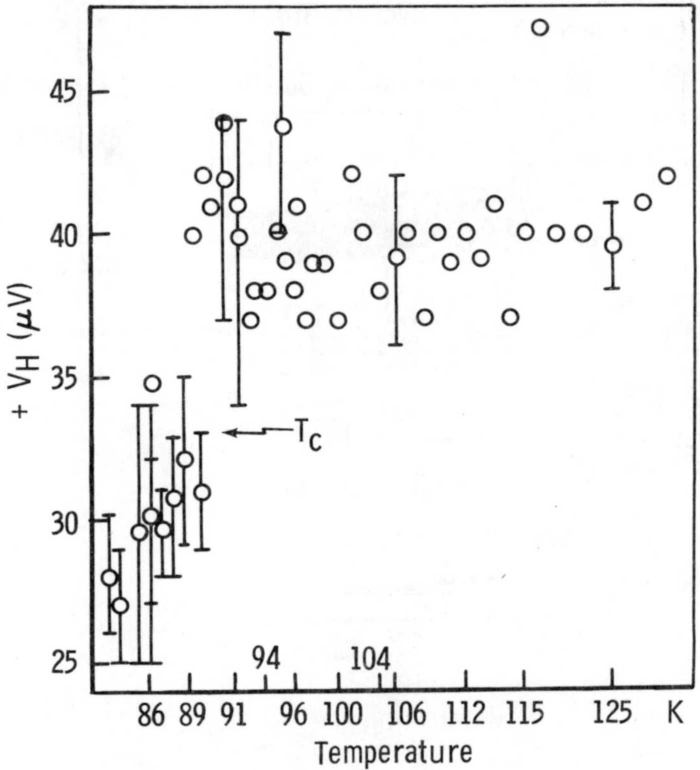
45
40
35
30
25
+ V_H (μV)
T_c
94
104
86 89 91 96 100 106 112 115 125 K
Temperature

Fig. 8a

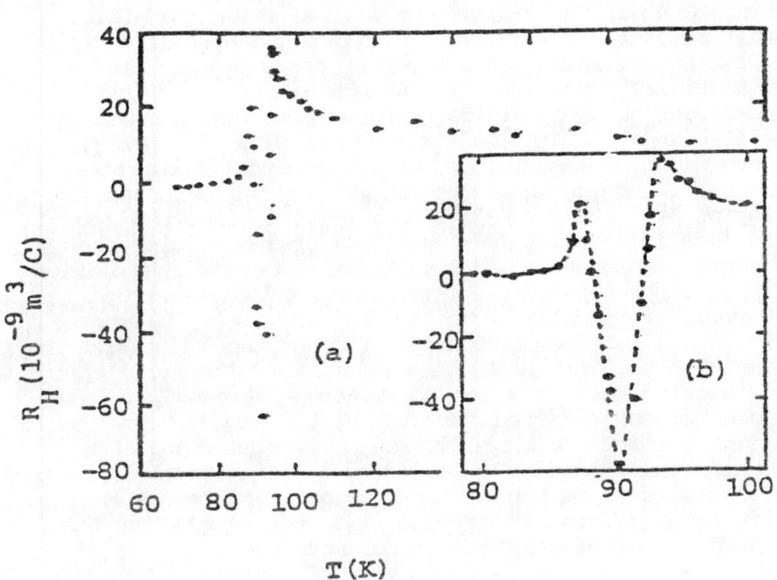

Fig. 8b

ROLE OF CU-O PLANES IN SUPERCONDUCTORS

J. Ruvalds, C.Y. Lin, M. Rilee and A. Virosztek
Physics Department, University of Virginia
Charlottesville, VA 22901

ABSTRACT

The surprising variation of the superconducting transition temperature from 0 to 125 K in numerous cuprates is examined by considering the occupation of two energy orbitals emanating from the d-states of copper and hybridized with the oxygen p-states. The broader energy $d_{x^2-y^2}$ band is substantially occupied over a wide range of oxygen concentrations and yields a two-dimensional plasma oscillation with maximum energy near $\Omega_\ell \sim 1\text{-}2$ eV in superconducting as well as in other metallic cuprates. However, a variation of the oxygen content or selective alloying may shift the Fermi energy to intersect a second more localized energy band and thus create a lower energy acoustic plasmon mode which provides a favorable mediation of the electron pairing interaction. A logical choice for the narrower band would be the d_{z^2} orbital. We present calculations of the plasmon spectrum and the superconducting transition temperature T_C as a function of the Fermi energy and thereby obtain a close correlation between T_C and the oxygen content observed for $YBa_2Cu_3O_X$. Similar physical insight predicts a large change in T_C caused by atomic substitutions, such as Pr replacing Y. Experimental probes of this plasmon mechanism are proposed, and recent Knight shift data is shown to provide evidence for the two-band density of states expected for our model.

I. Experimental Background

A remarkable feature of superconductors which are known to have transition temperatures in the liquid nitrogen range and above, is the evidently ubiquitous presence of planar structures containing copper atoms coupled strongly to nearby oxygen sites. As the list of newly discovered cuprates proliferates, however, the finding of several similar planar structures in metallic, but not superconducting, compounds poses a vital challenge for any mechanism of superconductivity.

The sampling of oxides shown in Figure 1, illustrates the evolution of interest in their properties and vividly points out the theoretical challenge to distinguish high T_C materials from the others.

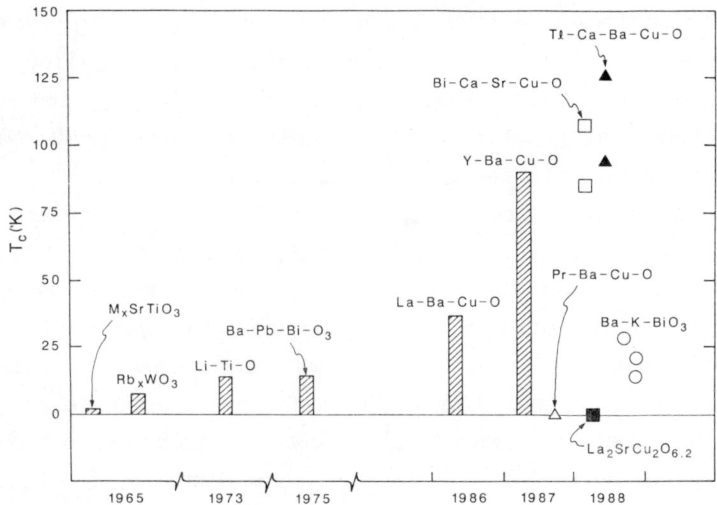

Fig. 1. Historical comparison of superconducting oxides showing a wide range of transition temperatures T_C.

Some interesting common features of these oxides include the following:

1. They tend to have low electron densities of order 10^{21} electrons/cm^3 and therefore are unlikely candidates for electron pairing in the BCS formalism.
2. Many of the oxides exhibit a metal-insulator transition at a given composition.
3. Their transition temperatures are highly sensitive to structure, oxygen content, and alloy variations.

Bednorz and Muller[1] discovered the $La_{2-X}Ba_XCuO_4$ high T_C superconductor. Their findings show that T_C increases suddenly when the Ba content exceeds a threshold value of $X \simeq 0.07$ and then reaches a maximum near 35 K before subsequently dropping rather sharply for $X \geq 0.3$. This mysterious behavior has been replicated in more recent studies which correlate the Ba content with the concentration of hole charge carriers in the alloy. The corresponding changes in the Fermi energy may indicate a preferential population of a localized energy level E_f in the superconducting cases, which could yield the above T_c variation on the basis of an acoustic plasmon mechanism.[2] As E_f is lowered, by adding Sr for example, a threshold occurs when E_f first intersects the narrow band edge and then the acoustic plasmon energy Ω_h grows as the number of hole states increases; the optimum value of T_c is obtained when $\Omega_h \simeq 0.1\ E_f \simeq 2.0$ eV. Decreasing E_f further creates a plasmon whose energy is too high for an effective electron pairing interaction and lowers T_c in agreement with the experimental data on Sr or Ba substitutes in La_2CuO_4. Incidentally, the same physical effect can be achieved by changing the oxygen content in pure La_2CuO_X.

The discovery[3] of the $YBa_2Cu_3O_X$ class of superconductors with $T_c \sim 90$ K for $X \simeq 7.0$, generated widespread interest in compounds with the planar Cu - O structure. However, very similar structures such as $YBa_2Cu_3O_{6.5}$ and $PrBa_2Cu_3O_7$ have $T_c = 0$, with a challenging series of intermediate T_c cases for oxygen content $6.5 < X \leq 7.0$.

Professor W.A. Little has presented in this workshop an insightful and comprehensive review of experimental constraints that oxides pose for theorists. His conclusions reveal that electrons in some cuprates are paired in states having s-wave symmetry, and the well-defined energy gap conforms to a weak coupling version of the BCS theory. Of course the isotope effect data, and the high T_c values relative to such low electron concentrations essentially rule out the phonon-mediated interaction for the pairing. In brief, the key experimental evidence for an s-state follows from measurements of the flux quantization, AC Josephson effect in $YBa_2Cu_3O_7$ - insulator - Pb, Sn junctions, muon spin precession, small isotope effect, insensitivity of T_c to ordinary impurity scattering, specific heat, and magnetic field attenuation as a function of temperature. A weak coupling formulation would allow T_c values near 100 K by the exchange of an electronic excitation with a typical energy in the range of tenths of an eV. Excitons provide a mechanism which fulfill all of the above criterion,[4] as do the acoustic plasmon modes in a two-band model originally proposed by Pines[5] and studied by many groups over the last thirty years.

Our goal is to decipher those clues and theoretical attributes which may distinguish the superconducting cuprates from the others shown in Figure 1. It is particularly interesting to note that well-defined Cu - O planes exist in metals such as $La_2SrCu_2O_{6.2}$ and a temperature variation of their normal state resistivity appears linear and of similar magnitude to $YBa_2Cu_3O_7$, despite the fact that the former cuprates do not exhibit superconductivity.[6] In brief, our expectation is that the non-superconductiviting alloys should not have the partial filling of two energy bands corresponding to orbital states with disparate localization and symmetry. Thus we intend to display those characteristics of the two-band structure which influence T_c and may be verified experimentally.

II. Electronic Structure

We shall consider here the case of $YBa_2Cu_3O_X$, although the general features of the Cu-O orbitals should apply to the other oxides as well. The first requirement is a partially filled band corresponding to the Cu $d_{x^2-y^2}$ orbital. It should be broadened by the substantial overlap of this state with the oxygen p_x, p_y orbitals. Band structure calculations are expected to provide a reasonable estimate of the width of this band which we approximate by a tight-binding model

$$E_\ell = \ell_0 + \ell_1 \ [\cos (a_x k_x) + \cos (a_y k_y)], \tag{1}$$

with $\ell_1 \simeq 1.25$ eV. The position of the Fermi energy E_f relative to this majority band is uncertain, but nevertheless we use the band structure calculations[7-9] as a guide to yield the partial occupation of states shown in Figure 2.

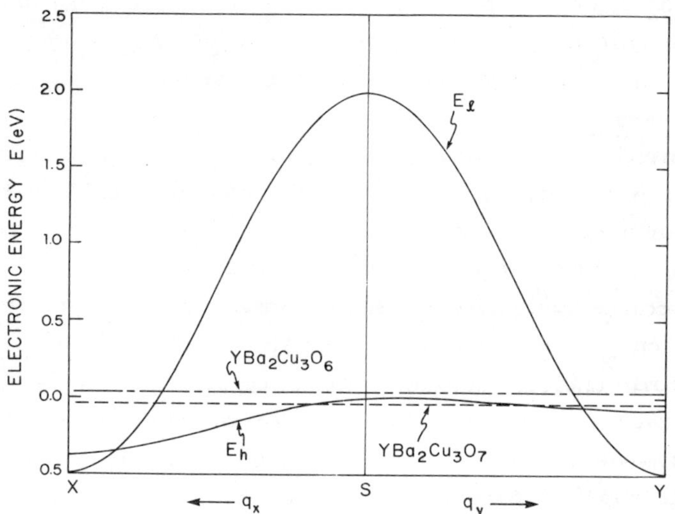

Fig. 2. Electronic structure model for $YBa_2Cu_3O_x$ showing the localized orbital state energy E_h as a function of momentum in comparison to the majority carrier band E_ℓ corresponding to the in-plane $d_{x^2-y^2}$ orbital mixed with the oxygen p_x, p_y states. The Fermi energy E_f is shown by the dashed line in the superconducting phase, and by the dot-dash line in the case of non-superconducting $YBa_2Cu_3O_6$.

Our reliance on the band structures is based on the bandwidths ℓ_1, t_1 and t_2. Although the positions of these bands are surely shifted by correlation effects, the final resolution of their existence near E_f must be provided by experiment, and we show below the Knight shift data which provides support for the model chosen here. For the narrower, "heavy"-mass band we use the energy,

$$E_h = t_1 \cos (a_x k_x) + t_2 \cos (a_y k_y). \quad (2)$$

with $t_1 = 0.2$ eV and $t_2 = 0.04$ eV. The bandwidth anisotropy in the $YBa_2Cu_3O_x$ case is related to the peculiar chain structure of one group of Cu-O atoms. In tetragonal compounds we expect $t_1 = t_2$ with roughly similar magnitudes. The essential point in the acoustic plasmon mechanism is that one band should be fairly broad, like the E_ℓ band shown here, and the other band should be narrower and more sparsely populated by virtue of the slight intersection of the Fermi energy with the E_h band.

A common objection to band structure calculations is their prediction of a metallic state for $YBa_2Cu_3O_6$, in spite of the real insulating antiferromagnetic structure. Although future investigations of the Fermi surface nesting in these two-dimensional systems may elucidate the role of exchange and correlation effects from the band starting point, the insulating phase represents an inadequacy of the band structure in its present form.

The metallic state of $YBa_2Cu_3O_7$ is a far better case from the band structure point of view. Resistivity and Hall effect data on single crystals yields a mean free path extending roughly 70 A within the Cu-O plane,[10] and positron annihilation data presented by S. Berko at this workshop reveal a Fermi surface in good agreement with the Freeman computations.[7]

III. Plasmon Spectrum

Using the tight-binding model described by Equations 1 and 2, we have computed[11] the dielectric function of $YBa_2Cu_3O_x$ as a function of the Fermi energy whose value is determined by the oxygen content x. The zeroes of the real part of the dielectric function yield the dispension of the plasmon mode shown in Figure 3.

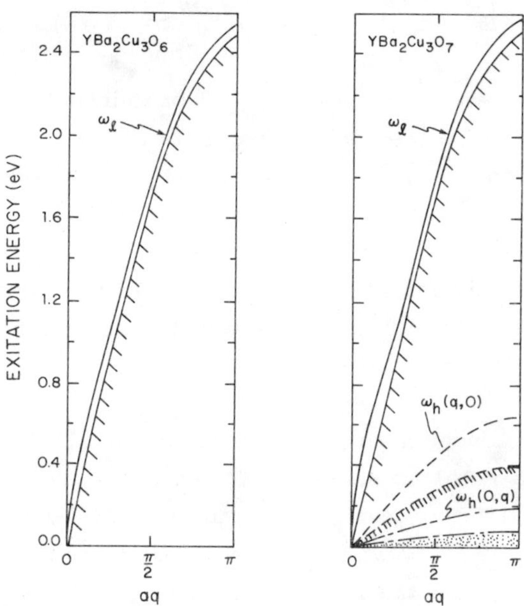

Fig. 3. Calculated plasmon energies ω_ℓ and ω_h as a function of momentum for superconducting $YBa_2Cu_3O_7$ in comparison to the oxygen depleted case with only one plasmon branch ω_ℓ. The corresponding electron-hole continuum regions are shown by the shaded regions.

Several features of these plasmon states are pertinent to the superconducting cuprates. The majority hole plasmon is "acoustic" in character because of the two-dimensional character of the Cu-O planes, but its maximum energy in the tight-binding model $\theta_\ell \simeq 2.5$ eV is much lower (by an order of magnitude) than the results of a free-electron approximation for the holes with $m^* = |m_e|$. Infrared reflectivity data typically shows structure near ~ 2 eV in the cuprates but the analysis of the spectrum in terms of a two-dimensional model is yet to be done. Interlayer interactions further complicate the data analysis because they yield a band of plasmon energies: the plasmons in Figure 3 correspond to oscillations which are out of phase in adjacent Cu-O layers, whereas in-phase motion will raise the shoulder in ω_ℓ at small q to higher values until ultimately a gap at finite $\omega(q=0)$ values opens for perfect phase matching.[12] Our calculations vividly show that a single acoustic plasmon ω_ℓ by itself yields $T_c = 0$, in accord with the known non-superconducting phases $YBa_2Cu_3O_x$ with $x < 6.5$.

Superconducting electron pairing may be generated by the exchange of the lower energy plasmon ω_h which we find to be present in $YBa_2Cu_3O_x$ in the range $6.5 < x < 7.0$. Several features of the localized band plasmons are favorable for this exchange mechanism. First and foremost is the existence of the ω_h branch as a well-defined excitation over a wide momentum region. As seen in Fig. 3 the combined influence of the tight-binding electronic states and the two-dimensional hole dynamics split the ω_h plasmon branch off well above the localized electron-hole continuum, thereby avoiding a source of damping. Also, the decay of the ω_h plasmons into the lighter mass continuum yields only a small broadening (of order 5% of the energy over the entire q-range) so that the structure factor yields sharp peaks for the ω_h plasmons.[11] The latter feature is vital to achieve a strong attractive region in momentum and energy space for the electron (or hole) pairing interaction.

Analysis of the Eliashberg equation supports the viability of achieving high transition temperatures via the exchange of the ω_h acoustic plasmons. In the weak coupling limit we find

$$T_c \simeq \theta_h \exp\left(-\frac{1}{\lambda_{p\ell} - u^*}\right), \tag{3}$$

where $\lambda_{p\ell}$ is the plasmon-induced pairing interaction and μ^* is the properly screened Coulomb repulsion. The main advantage of the plasmon mechanism is clearly the large value of the prefactor $\theta_h \simeq 0.25$ eV $\simeq 3{,}000$ K corresponding to the maximal point in the plasmon dispersion, i.e. $\theta_h \equiv \omega_h$ $(aq = \pi)$. In the case of $YBa_2Cu_3O_x$, the removal of oxygen raises the Fermi energy E_f and therefore lowers the plasmon energy θ_h by depleting the number of holes in the narrow E_h band. Our calculated[11] scaling of $T_c(x) \propto \theta_h(x)$ shows a close correspondence to experimental T_c values over a wide range of x. Also this simple physical picture applies quite well to the data on P_r substitutions for Y in that the T_c values progressively drop as the P_r content increases in $Pr_zY_{1-z}Ba_2Cu_3O_7$. At this workshop, the lectures by J. Rhyne and J. Crow demonstrate that Pr enters as a tetravalent ion in contrast to its Y host which is trivalent. Therefore, the Pr substitution has the effect of raising the Fermi energy, and, in our view, therefore lowering the plasmon energy θ_h and accordingly depressing T_c.

How can a low density of holes yield a large pairing interaction? This question seems to be in contrast to the predictions of BCS theory, and this basic issue is a common feature of all the oxides shown in Figure 1. From the Eliashberg equation we have decoupled the effects of the electronic screening from the plasmon dynamics which enter in the electron-electron (or hole-hole) pairing interaction

$$V_q(\omega) = \frac{2\pi e^2}{q\epsilon(q,\omega)}, \tag{4}$$

where the dielectric function $\epsilon(q,\omega)$ includes the E_ℓ and E_h bands. The static contribution to the screening is dominated by the lighter mass majority holes emanating from E_ℓ, and a calculation of this screening shows vividly an electron-plasmon coupling, proportional to $\lambda_{p\ell}$, which increases substantially as the carrier concentration is lowered. This screening change is qualitatively similar to the three-dimensional analysis of T_C for Cs_xWO_3 on the basis of the acoustic plasmon model.[13] However, the strength of the electron-plasmon interaction is increased in two dimensions by roughly 30% for densities corresponding to the cuprates, in comparison to three dimensions.

The two-dimensional nature of the plasmon dispersion provides another benefit to enhancing $\lambda_{p\ell}$ and T_C. The tight-binding model results of Fig. 3 show the persistence of a flat region of ω_h near $aq \simeq \pi$, which yield a higher density of states than the free electron counterpart and accordingly increases the pairing interaction $\lambda_{p\ell}$ toward unity, while maintining a relatively weaker value of the μ^* Coulomb repulsion.

IV. Experimental Evidence

Despite theoretical enthusiasm for decades, acoustic plasmons generated by the screening of a localized group of electrons by more itinerant carriers have not yet been observed in a metal. The two-dimensional plasmons in superlatticis of GaAs and similar modes in electrons on He surfaces have been observed, however, and the latter modes resemble closely the majority hole plasmon dispersion $\omega_\ell(q)$ discussed above. Hence light scattering studies of the cuprates may be encouraged to look for both types of plasmons in the superconducting phases. Since the lower energy plasmon is sensitive to oxygen content, it may be feasible to study its effect on acoustic attenuation if a single crystal of the low T_C material could be formed.

At present, there is some interesting NMR data[14] for the ttrium nucleus as a function of oxygen content which we show in Fig. 4 by by comparison to the density of states calculated from our model of two energy bands[11].

Recent analysis[15] of the Knight shift data indicates the presence of a Cu $d_{x^2-y^2}$ band near the Fermi energy with a bandwidth comparable to the value used here. It would be interesting to pinpoint the precise origin of both bands near the Fermi energy, even though the plasmon mechanism is not restricted to a given symmetry. A compilation of other relevant experiments, such as electron loss and photoemission spectroscopy is available elsewhere[11].

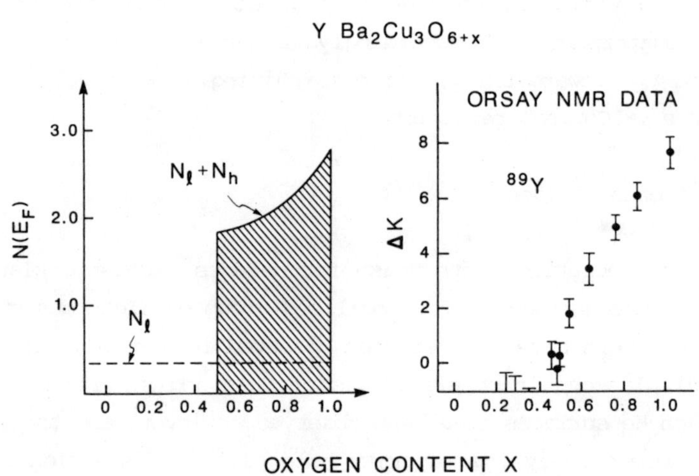

Fig. 4. Computed density of states $N(E_f)$ as a function of oxygen content showing the onset of the narrow band contribution at $x \simeq 0.5$. The Knight Shift data Δ K indicates a similar enhancement in N $(E_f) \propto \Delta$ K in the superconducting regime $x > 0.5$.

In conclusion, we may mention two optimistic prospects on the basis of the plasmon mechanism. The required overlap of two disparate energy bands near the Fermi energy should be present in other planar compounds without copper or oxygen. Thus the potential for finding new superconductors should be widespread. Also, higher transition temperatures could be achieved in principle by relatively minor changes in the plasmon characteristics.

It is a pleasure to acknowledge stimulating suggestions from Y. Ishii and G. Ashe. This research was supported by the U.S. Department of Energy Grant No. DEF605-84-ER45113.

References

1. T.G. Bednorz and K.A. Muller, Z. Phys. B64, 189 (1986).
2. J. Ruvalds, Phys. Rev. B35, 8869 (1987).
3. M.K. Wu, J.R. Ashburn, C.J. Torng, P.H. Hor, R.L. Meng, L. Gao, Z.J. Huang, Y.Q. Wang and C.W. Chu, Phys. Rev. Lett. 58, 908 (1987).
4. W.A. Little, Phys. Rev. 134, A1416 (1964).
5. D. Pines, Can. J. Phys. 34, 1379 (1956).
6. J.B. Torrance, Y. Tokura, A. Nazzal, and S.S.P. Parkin, Phys. Rev. Lett. 60, 542 (1988).
7. J. Yu, S. Massida, A.J. Freeman, and D.D. Koeling, Phys. Lett. 122, 203 (1987).
8. L.F. Mattheiss and D.R. Hamann, Sol. St. Comm. 63, 395 (1987).
9. F. Herman, R.V. Kosawski, and W.Y. Hsu, Phys. Rev. B36, 6904 (1987).
10. N.P. Ong, Z.Z. Wang, S. Hagen, T.W. Jing, J. Clayhold and J. Horvath, Physica C153 (1988).
11. Y. Ishii and J. Ruvalds, Phys. Rev. B. (submitted).
12. A.L. Fetter, Annals of Physics 88, 1 (1974).
13. L.M. Kahn and J. Ruvalds, Phys. Rev. B19, 5652 (1979).
14. H. Alloul, P. Mendels, G. Collin, and P. Monod, Phys. Rev. Lett. 61, 746 (1988).
15. F.J. Adrian, Phys. Rev. Lett. 61, 2148 (1988).

EXPERIMENTAL AND APPLICATIONS

MAGNETIZATION, TRANSPORT AND STRUCTURAL STUDIES OF Tl-Ca-Ba-Cu-O: BULK CERAMICS, SINGLE CRYSTALS, AND POLYCRYSTALLINE THIN FILMS

E. L. Venturini, J. F. Kwak, D. S. Ginley, B. Morosin, and R. J. Baughman, Sandia National Laboratories, Albuquerque, NM 87185

ABSTRACT

A detailed investigation of the Tl-Ca-Ba-Cu-O system of high temperature superconductors is presented including static magnetization, electrical transport, structural and processing data for sintered and melted bulk ceramics, single crystals, and polycrystalline thin films. This system is shown by X-ray diffraction and energy dispersive X-ray analysis to have at least five distinct superconducting phases containing either two or three Cu-O sheets separated by one or two Tl-O planes; $Tl_2Ca_1Ba_2Cu_2O_x$ (Tl-2122) and $Tl_2Ca_2Ba_2Cu_3O_y$ (Tl-2223) are two common phases. Bulk ceramic Tl-2223 has a Meissner effect up to 124 K and zero resistance up to 120 K, but transition temperatures are found to be a strong function of processing even for single crystals. Magnetization versus field data suggest weak flux pinning in all samples above 30 K. Transport properties of the Tl-Ca-Ba-Cu-O system are very different from those of ceramic $YBa_2Cu_3O_{7-\delta}$ (Y-123). Melt processed bulk ceramics of Tl-2223 have transport critical current densities >5000 A/cm^2 at 76 K with modest magnetic field dependence, indicating strong intergranular links which are not present in Y-123. This is also observed for unoriented polycrystalline thin films. Tl-2223 films have critical current densities at 76 K up to 240,000 A/cm^2 (110,000 A/cm^2 for Tl-2122 films) with a modest dependence on in-plane magnetic fields, implying negligible effects of grain boundary Josephson junctions. The strong intergranular links and high critical current densities demonstrate the potential of the Tl-Ca-Ba-Cu-O system for applications.

INTRODUCTION

The enormous experimental and theoretical effort on high-temperature copper-oxide-based superconductors began with the La-Ba-Cu-O system[1)], shifted to the Y-Ba-Cu-O system[2)], and recently has focused on the chemically more complex Bi-Ca-Sr-Cu-O[3)] and Tl-Ca-Ba-Cu-O[4)] systems. Here we concentrate on the Tl-Ca-Ba-Cu-O superconductors with reported zero resistance and Meissner effect up to 125 K.[5,6)] Unlike the Y-Ba-Cu-O system which has one predominant superconducting structural arrangement, $YBa_2Cu_3O_{7-\delta}$ (Y-123), the Tl-Ca-Ba-Cu-O system has a multitude of superconducting phases.[7-9)]

Using single crystals grown from the melt by a self-flux technique (described below), five distinct superconducting phases containing either 2 or 3 Cu-O sheets have been identified from energy dispersive X-ray analysis (EDS) and single crystal X-ray diffraction (XRD).[10)] Table I summarizes the compositions of the five phases together with the c-axis length (these are tetragonal crystal structures with very similar a-axis lengths of 3.85 Å) and the <u>average</u> transition temperature T_c measured on the same as-grown crystals examined by XRD. Also shown are abbreviations for the composition which will be used below for simplicity.

Note that there is only a slight correlation between the number of

Table I. Five Tl-Ca-Ba-Cu-O superconducting single crystal phases.

Composition	Abbreviation	# Cu-O Sheets	c-axis Length	Transition Temperature
$Tl_1Ca_1Ba_2Cu_2O_x$	Tl-1122	2	12.7 Å	107 K
$Tl_2Ca_1Ba_2Cu_2O_x$	Tl-2122	2	29.3 Å	110 K
$Tl_1Ca_2Ba_2Cu_3O_x$	Tl-1223	3	15.9 Å	117 K
$Tl_2Ca_2Ba_2Cu_3O_x$	Tl-2223	3	35.9 Å	114 K
$Tl_1Ca_2Ba_3Cu_4O_x$	Tl-1234	4	19.8 Å	120 K

Cu-O sheets and the corresponding T_c. We have observed significant variations in T_c (not shown in Table I) for crystals with identical structure and nominal composition, particularly following post-growth annealing in vacuum or oxygen. The effects of cation disorder and oxygen vacancies appear to be substantial in this system. Our study of annealing will be discussed in detail elsewhere.[11)]

This paper is divided into three main parts: 1) sintered and melt processed ceramics, 2) single crystals, and 3) unoriented polycrystalline thin films. Based on the magnetic and transport properties, particularly for the thin films, the Tl-Ca-Ba-Cu-O system of superconductors shows great potential for applications at liquid nitrogen temperature.

SINTERED AND MELTED CERAMICS

Ceramic Preparation and Structure

Sintered ceramic pellets were prepared[5)] by grinding dry powders of Tl_2O_3, CaO, BaO, and CuO, sieving (50 μm), regrinding and sieving again, and pressing the resulting powder into pellets at 5 kbar. The optimum resistivity and magnetization results were obtained by air-sintering a pellet for 10 minutes at 850°C and subsequently annealing in flowing oxygen (640 Torr) for 12 hours at 850°C followed by a slow cool. Sintered pellets are porous with densities between 60 and 85% of theoretical density. Although all pellets prepared to date have exhibited minority phases, ceramics with predominantly Tl-2223 or Tl-2122 are routinely prepared by this processing sequence. In general, the Ca, Ba and Cu concentrations are equal to the desired phase, but excess Tl_2O_3 is used since it decomposes during the air sinter to Tl_2O which melts above 600°C. It appears that this Tl decomposition to a liquid phase is integral to grain growth and grain boundary properties, but optimizing the synthesis presents a challenge.

Dense melted ceramic was prepared[12)] by the same double grinding and sieving of the dry oxide powders just described. The resulting powder was loaded into a 2.5×2.0 cm diameter Pt crucible with a tight-fitting Pt lid wired in place. The crucible was rapidly heated (5 minutes) to 950°C in a vertical tube furnace under 1 atmosphere of pure oxygen. The melt was held at 950°C for 1 hour, then cooled slowly to 700°C over 12.5 hours and rapidly to room temperature over 5 hours. This process produces both single crystals and a dense "melt flux" ceramic with over 90% of theoretical density. The melted ceramic typically exhibits numerous shiny black faceted sub-millimeter regions suggestive of very small embedded single crystals. The polycrystalline melt has a plate-like habit similar to the single crystals.

XRD patterns were taken using a Siemens D-500 theta/theta automated powder diffractometer equipped with a monochrometer on the detector set so both Cu $K\alpha_1$ and $K\alpha_2$ radiation were obtained. The region between 4 and 60° 2θ was scanned and patterns compared with standards established for these phases using single crystal data. Though peak positions for a particular phase were very reproducible over a range of starting compositions and synthesis conditions, the intensity values for particular diffraction lines differed. This is believed to be a combination of slight preferred orientation effects together with small differences in the cation/vacancy solid solution ratios involving both the Ca^{2+} and Tl^{3+} ions in their two different structural sites.

Static Magnetization

Thin rectangular plates were cut from both sintered and melted ceramic for static magnetization studies using a commercial SQUID magnetometer (Biomagnetic Technologies, San Diego, CA). Most data was recorded with the cut plate oriented parallel to the applied field to minimize demagnetization corrections. However, limited testing of sintered pellets showed considerable shape effects in 2.5 mT at 5 K, suggesting some strong links between grains. In particular, one plate cut from a predominantly Tl-2223 pellet was 5×5×0.5 mm with a predicted demagnetization enhancement of 7.2 for the field normal to the plate. The measured moment was enhanced by 3.7× for a normal field, roughly

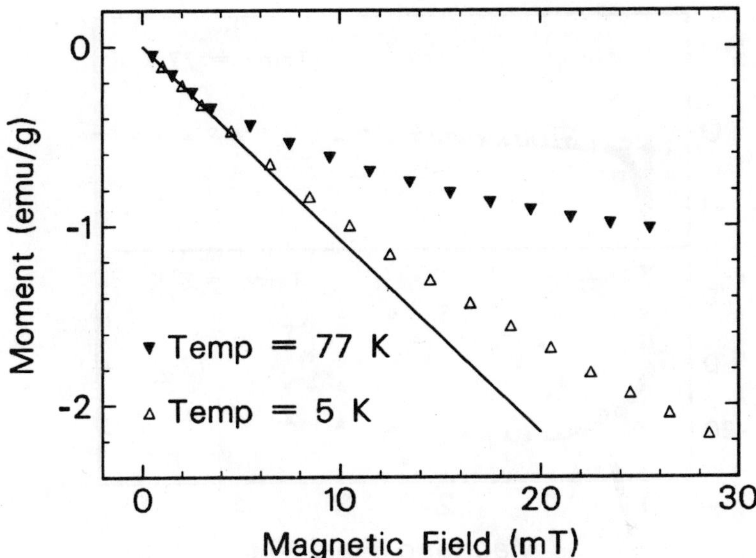

Fig. 1. Low-field diamagnetic shielding response at 5 and 77 K for a thin plate cut from a bulk ceramic containing predominantly Tl-2223.

half the prediction. In contrast, small applied fields (typically 2 mT) substantially decouple the grains[13] in sintered ceramic Y-123, leading to a minimal shape effect. As discussed below for the thin films, these Tl-Ca-Ba-Cu-O superconducting ceramics do not suffer from the severe "weak links" between grains that dominate in randomly oriented Y-123.

Fig. 1 compares the low-field diamagnetic shielding (flux exclusion following cooling in zero field) versus increasing field at 5 (open triangles) and 77 K (closed triangles) for the field parallel to a plate cut from a sintered ceramic pellet containing predominantly Tl-2223 phase (magnetic T_c = 124 K). The solid line represents perfect diamagnetism (volume susceptibility $\chi_v = -1/4\pi$), and the 5 K data fall on the line from 0 to 4 mT while the 77 K data deviate above 2 mT. This is in strong contrast to similar data for ceramic Y-123 where the linear response persists to much higher fields. Traditionally, the

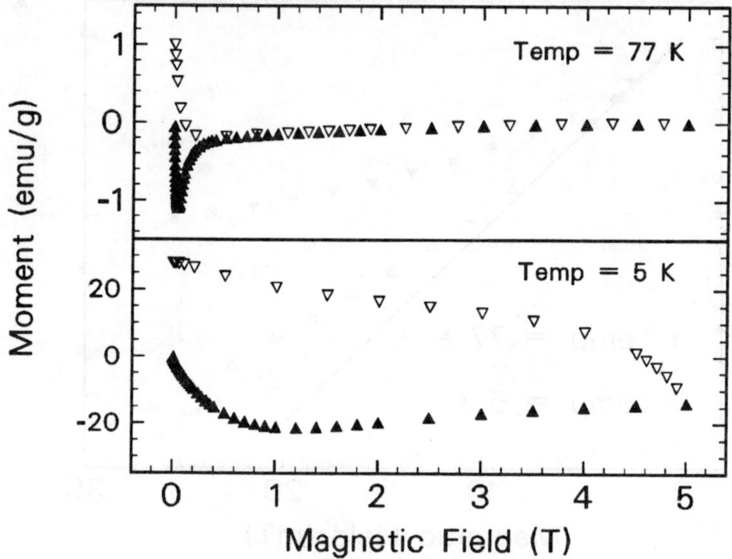

Fig. 2. High-field hysteresis loops at 5 and 77 K for the same cut plate as in Fig. 1. Solid triangles indicate increasing field, open triangles for decreasing field. The negligible hysteresis at 77 K indicates weak flux pinning.

field at which the shielding response starts to deviate from linearity has been defined as the lower critical field, H_{c1}, which has been reported[14] as high as 50 mT at 20 K for Y-123. Hence our data in Fig. 1 suggest an order of magnitude smaller H_{c1} in ceramic Tl-2223 (4 mT at 5 K) along at least some crystallographic directions. This result is in marked disagreement with a recent report[15] of $H_{c1} \approx 90$ mT at 4.2 K for ceramic Tl-2223.

Fig. 2 compares the high-field magnetization to 5 T at 5 and 77 K for the same cut plate of sintered ceramic containing primarily Tl-2223. The solid triangles represent increasing field strength (following zero field cooling) while the open triangles show decreasing fields. There is no measurable hysteresis above 0.5 T at 77 K, while the open loop at 5 K suggests a hard (strongly pinned) type II

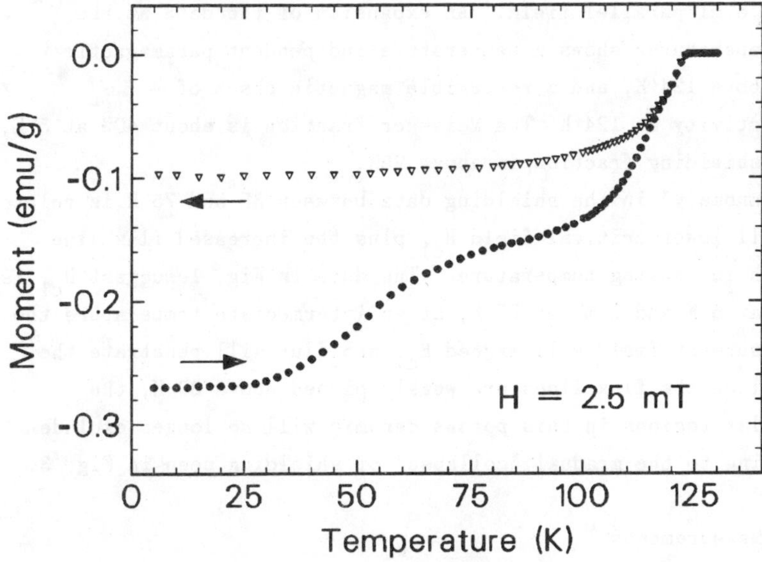

Fig. 3. Shielding (flux exclusion, solid circles) versus increasing temperature and Meissner effect (flux expulsion, open triangles) versus decreasing temperature measured in 2.5 mT for a cut ceramic plate. The "anomaly" above 35 K in the exclusion data is discussed in the text.

superconductor. Note that the magnetization scales differ by 20X. It is clear from this data that the Tl-2223 sintered ceramic has relatively weak flux pinning at liquid nitrogen temperature. We have recently initiated flux creep studies which show that thermally activated flux motion in the Tl-Ca-Ba-Cu-O superconductors will be a problem of the same magnitude as that reported for Y-123[16]. The open magnetization hysteresis loop at 5 K in Fig. 2 closes rapidly with increasing temperature beginning near 30 K. Dense melted ceramic plates exhibit a very similar lack of hysteresis due to weak flux pinning above 30 K.

Fig. 3 shows the shielding signal versus increasing temperature (solid circles, after zero field cooling) and the expulsion or Meissner signal versus decreasing temperature (open triangles, cooling in a field). The data were taken with the same Tl-2223 sintered ceramic cut plate in 2.5 mT parallel field. An expansion of the data at the highest temperatures shows a temperature-independent paramagnetic response above 124 K, and a reversible magnetic onset of superconductivity at 124 K. The Meissner fraction is about 40% at 5 K, while the shielding fraction is above 95%.

The "anomaly" in the shielding data between 35 and 75 K is related to the small lower critical field H_{c1} plus the increased flux line motion with increasing temperature. The data in Fig. 1 suggest H_{c1} is near 4 mT at 5 K and 2 mT at 77 K; at an intermediate temperature the 2.5 mT measurment field will exceed H_{c1} and flux will penetrate the grains. Since the flux lines are weakly pinned above 30 K, the intergranular regions in this porous ceramic will no longer exclude flux, leading to the gradual "collapse" of shielding seen in Fig. 3.

Transport Measurements

Resistivity measurements were made at low frequencies using a linear 4-probe arrangement of silver-paint contacts on roughly rectangular samples. The resistivities versus temperature for representative sintered and melted ceramic samples are illustrated in Fig. 4. The resistivity of the polyphase sintered sample is comparable to that for good quality ceramic samples of Y-123. The resistivity of the polyphase melt material is highly variable, but can be as low as our best Y-123 ceramics. For both sintered and melt-processed material, transitions with zero resistance close to 120 K are observed.

Critical current data were obtained by linear 4-probe pulsed measurements using low-resistance current contacts. Routine checks were made for non-ohmic or high-resistance contacts, non-ideal current paths, and heating effects. The critical current was measured using 600 μs square pulses with < 1% duty cycle in order to minimize heating effects. The critical current was defined by the observation across the inner contacts of a 2 μV signal (the minimum quantifiable signal) on an oscilloscope with differential input. Measurements at 76 K in fields from 0 to 80 mT were made with the samples immersed directly in liquid nitrogen using a copper coil for field generation. The samples

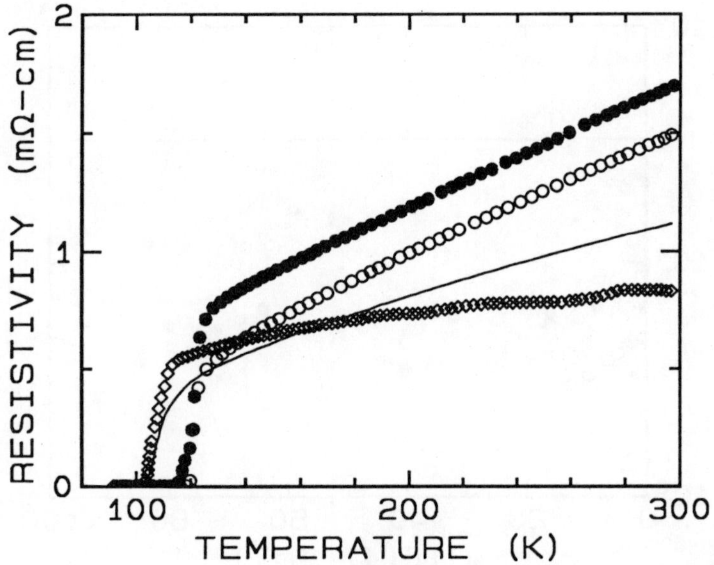

Fig. 4. Resistivity versus temperature for representative samples of sintered ceramic (open circles), melt-processed ceramic (closed circles), single crystal (diamonds), and thin film (solid line) material.

were routinely warmed above T_c after field cycles in order to eliminate trapped flux.

An expected benefit of having T_c at 120 K rather than 90 K is significantly higher critical currents at liquid nitrogen temperature, for both ceramic and crystalline material. Critical current results versus magnetic field below 0.1 T for representative sintered and melt-processed samples at 76 K are shown on a semi-logarithmic scale in Fig. 5. The critical current is significantly higher and less sensitive to field than our best Y-123 material, particularly for melt-processed material. This suggests that intergranular weak links are not dominant, in contrast to Y-123, and hence that the superconducting phase is not as severely disrupted at the grain boundaries. These results are consistent with magnetization results suggesting stronger

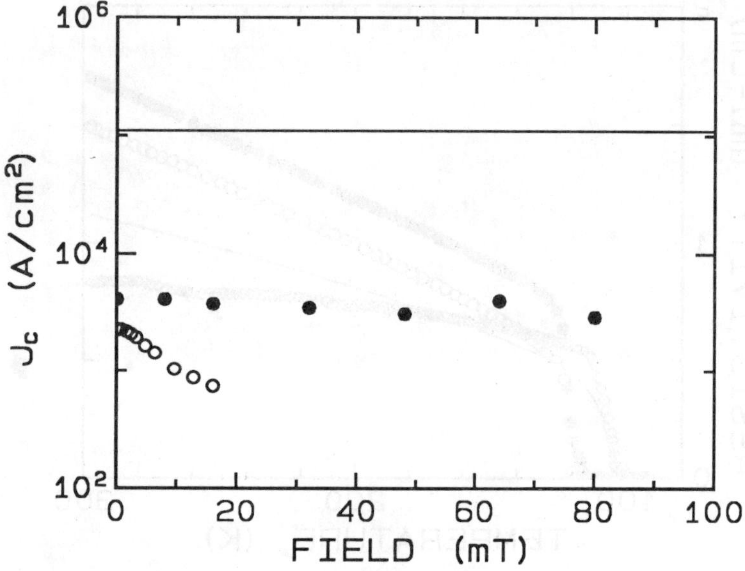

Fig. 5. Semi-logarithmic plot of critical current at 76 K versus magnetic field for representative samples of sintered ceramic (open circles), melt-processed ceramic (closed circles), and thin film (solid line) material. Even for the poly-phase sintered ceramic the fall off with field is significantly less than for the best Y-123 ceramic, suggesting minimal disruption of the superconducting phase at grain boundaries.

grain coupling in the Tl-Ca-Ba-Cu-O bulk material. Considering the fact that these samples represent initial synthesis attempts and are not single-phase, the critical currents measured to date are quite encouraging.

SINGLE CRYSTALS

Crystal Growth

A pseudo flux-type growth technique[12] has been used to produce single crystals of all five Tl-Ca-Ba-Cu-O phases shown in Table I, although the Tl-1234 crystals obtained to date have been very small. The temperature-time cycle is identical to that for preparing "melt flux" ceramic as outlined above. Breaking the shiny black melt exposes melt pockets containing well formed platelets from 1 to 3 mm across and from 0.01 to 0.2 mm thick. A single melt produces both single crystals and syntactic polycrystals (crystals containing intergrowths of more than one phase) of several structural types.[8] These are characterized by a combination of EDS and single crystal XRD techniques.

Plates were investigated by X-ray precession photography to determine the phase(s) present and to establish the perfection of the specimens. This technique allows a determination of the various phases present in a polycrystal provided that the minimum intergrowth blocks are coherent and of the order of 50 - 70 Å and represent at least 5% of the entire platelet. Crystals which are not perfectly coherent with respect to the a axes are easily revealed by split diffraction spots and were discarded for the studies reported here.

Static Magnetization

The normal state magnetic susceptibility was measured at 1 T from 80 to 350 K for a Tl-2122 single crystal approximately $1.2 \times 1.6 \times 0.1$ mm^3 (1.4 mg) as shown in Fig. 6.. The susceptibility with the field applied along the c axis (solid triangles) is over an order of magnitude greater than that with the field in the a-b plane (open triangles). The data have been corrected for the plastic sample holder (leading to the estimated errors shown at 350 K) and for the core diamagnetism (estimated from Pascal's constants as -2.4×10^{-7} cm^3/g). The in-plane susceptibility is very small above 200 K, possibly related to the 2-D short range magnetic order[17] seen in other high-T_c copper

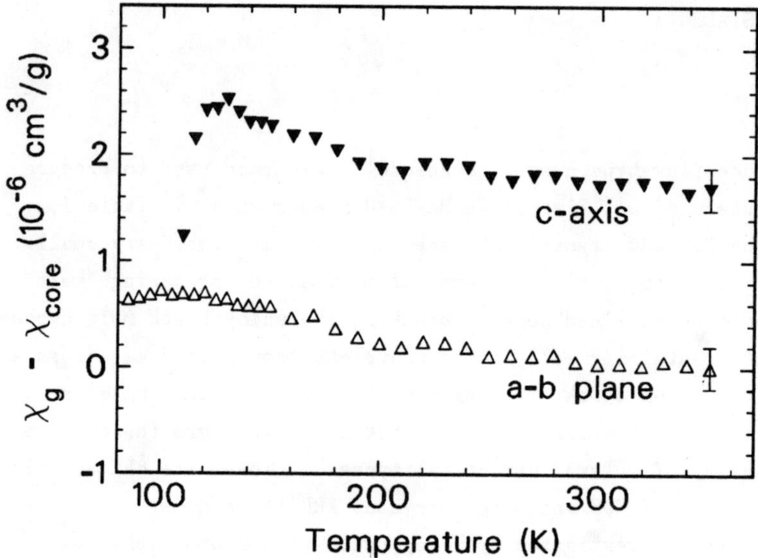

Fig. 6. Normal state magnetic susceptibility measured at 1 Tesla with the field in the a-b plane (open triangles) and along the c axis (solid triangles) for a Tl-2122 single crystal, corrected for the sample holder and the estimated core diamagnetism.

oxides. The drop in c-axis susceptibility below 120 K signifies the onset of superconductivity which is not seen in the a-b plane data, probably due to anisotropy in flux pinning and penetration depth in 1 T at these temperatures.

Fig. 7 compares Meissner effect data (field cooling) in 2.5 mT for three samples: a 0.7 micron thin film on yttria-stabilized zirconia (unoriented polycrystalline film described below), a single crystal, and a sintered ceramic. All three samples are Tl-2223 phase superconductors, but the magnetic onset of superconductivity varies from 103 K in the thin film to 114 K in the single crystal to 124 K in the ceramic. Note that the Meissner signal has been normalized to its value at 5 K. As expected, the transition is sharpest in the single crystal and broadest in the thin film. Since the data were measured

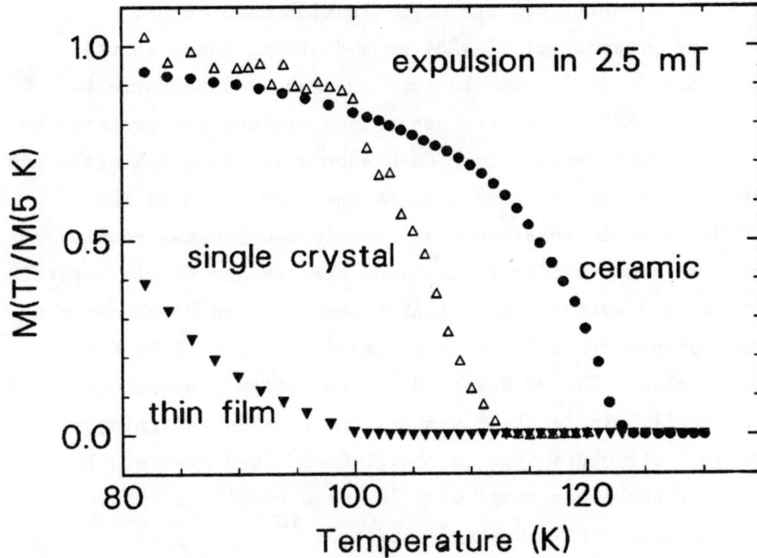

Fig. 7. Flux expulsion in 2.5 mT versus decreasing temperature (normalized to its value at 5 K) near the transition for a thin film (T_c = 103 K), a single crystal (T_c = 114 K), and a sintered ceramic (T_c = 124 K) of Tl-2223 (the ceramic contains other minor phases).

with the field parallel to the film plane, most of the broadening is due to the temperature-dependent penetration depth, although compositional inhomogeneities in the film may increase the transition width.

The substantial difference in transition temperatures between the Tl-2223 single crystal and the ceramic containing predominantly Tl-2223 is surprising. There is no evidence for multiple transitions in the ceramic data in Fig. 7 as might be expected for a mixed phase ceramic. Further, the predominance of Tl-2223 in XRD spectra for the ceramic plus the fact that the ceramic Meissner signal has reached 60% of its low temperature value before the transition in the single crystal begins suggests that T_c for Tl-2223 in ceramic form is considerably higher than that in a relatively large single crystal. This effect is not understood at present, but the magnetization data cannot be attributed to the presence of Tl-1234 as a distinct phase in the ceramic since this phase is not observed in amounts detectable by conventional powder XRD. However, one cannot exclude the presence of intergrowths with 4 or perhaps more Cu-O sheets if the block sizes are sufficiently small to broaden and obscure the corresponding XRD signatures. Hence it is possible that ceramic samples may routinely exhibit higher superconducting transitions than larger single crystals in the Tl-Ca-Ba-Cu-O system. Note that transitions in Y-123 ceramic, thin films and single crystals are reported within 1 K of 92 K for high-quality samples. The 20 K spread in transition temperatures in Fig. 7 for nominally single phase samples complicates assigning a unique T_c to each structure type in the Tl-Ca-Ba-Cu-O system. This is also confirmed by the large range of T_c's for a particular phase measured on well characterized single crystals.[10)]

Transport

The resistivity versus temperature measured in the a-b plane of a representative crystal is shown by the diamonds in Fig. 4. The crystal exhibits a slightly smaller room temperature resistivity than the polycrystalline samples. However, the crystal resistivity has considerably less temperature dependence, suggesting a highly defective structure and perhaps a much smaller intrinsic resistivity than the measured value. Critical current data have not been obtained on crystals due to inadequate electrical contacts.

UNORIENTED POLYCRYSTALLINE THIN FILMS

Thin Film Deposition

Thin films were deposited in a 1.5×10^{-5} mbar oxygen atmosphere by sequential e-beam evaporation of Tl, Ca, Ba, and Cu metals onto substrates of $SrTiO_3$, polycrystalline MgO, sapphire, and yttria-stabilized cubic zirconia (YSZ).[18,19] The best films to date have been prepared on YSZ and $SrTiO_3$. The metals were deposited in a 25-layer sequence of Cu-Ba-Ca-Tl, starting and ending with Cu, to a total thickness of 0.7 μm with each metal layer thickness adjusted to achieve a net stoichiometry of either Tl-2223 or Tl-2122. The substrate temperature was kept below 50°C during deposition.

The key step in thin film processing is proper control of the oxygen and thallium partial pressures during post-deposition sintering and annealing.[19] This control was accomplished by placing the substrate with the deposited metals face down on a sintered Tl-Ca-Ba-Cu-O pellet of the same stoichiometry as the film, air annealing at 850°C for 15 minutes, and then annealing in flowing oxygen at 750°C for 30 minutes. Scanning electron micrographs show a polycrystalline microstructure in all films consisting of plate-like growths up to 20 μm in size. XRD for a number of the 0.7 μm thick films on both YSZ and $SrTiO_3$ shows a random orientation of grains with the intended phase as the predominant one (>80%). EDS indicates that the original metal deposition ratio is maintained in the fully annealed films.

Static Magnetization

Fig. 8 compares the low-field diamagnetic shielding response at 5 K for an unoriented polycrystalline 0.7 μm-thick Tl-2122 film on YSZ for fields parallel and perpendicular to the film. Note that the vertical scale for the perpendicular data (solid triangles) is 200× greater than that for the parallel data (open triangles). This film had a Meissner signal below 103 K. The total film area was 3.0×3.5 mm^2 for a volume of 7×10^{-3} mm^3 or a mass of 40 μg using an estimated 75% of theoretical density based on scanning electron micrographs. The diamagnetic response of the YSZ substrate is negligible at the field strengths in Fig. 8.

The parallel shielding data correspond to a volume susceptibility $\chi_v = -0.87/4\pi$, while the perpendicular data give $\chi_v = -315/4\pi$. This

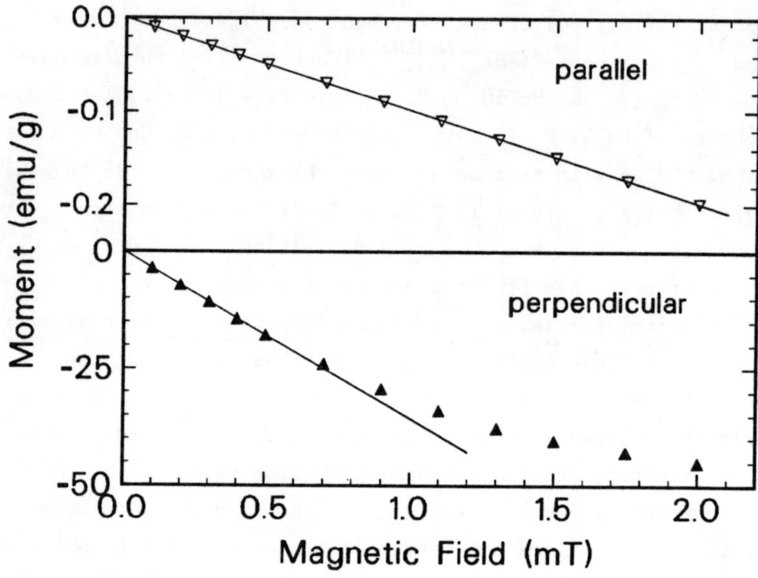

Fig. 8. Low-field diamagnetic shielding response at 5 K with the field applied parallel and perpendicular to a 0.7 μm unoriented polycrystalline Tl-2122 film on a YSZ substrate. The demagnetization enhancement of the perpendicular data demonstrates supercurrent flow over a continuous region > 0.35 mm in diameter.

difference is due to the demagnetization normal to a thin film. The enhancement in the perpendicular direction implies a demagnetization factor D = 0.997, requiring that the shielding currents flow through a region at least 0.35 mm in diameter (500x the film thickness). This demonstrates the ability of the film to transport supercurrent over large continuous areas, confirming the negligible effects of possible weak links between grains in agreement with the transport critical current results discussed below.

Fig. 9 compares high-field magnetization loops to 2 T at 5 and 77 K with the field applied normal to the same 0.7 μm Tl-2122 film on YSZ. The solid triangles represent increasing field strength (following zero field cooling) while the open triangles show decreasing fields. The

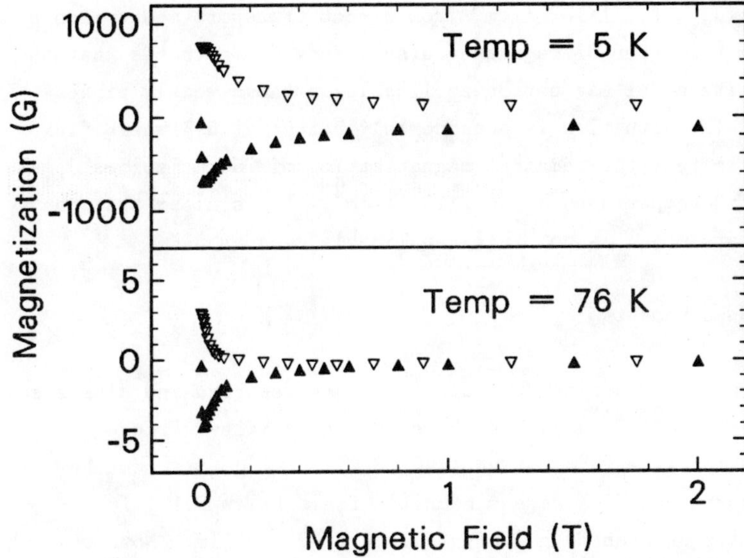

Fig. 9. High-field magnetization loops at 5 and 77 K with the field applied perpendicular to the same Tl-2122 thin film used in Fig. 6. The small hysteresis at 77 K indicates weak flux pinning.

data are corrected for the diamagnetism of the YSZ substrate and sample holder (measured directly). There is no measurable hysteresis above 0.6 T at 77 K, while the hysteresis extends to 5 T at 5 K (not shown). Hence increased flux motion due to thermal activation[16)] at higher temperatures results in weak flux pinning and little hysteresis at liquid nitrogen temperatures.

To demonstrate the problems in applying the Bean critical state model to a superconducting film with weak flux pinning, we can compare the inferred critical current densities J_{cm} from the remanent magnetization at 5 and 77 K following the field loops in Fig. 9 with the direct transport values J_{ct}. The magnetization data were obtained on a Tl-2122 film which was scribed into smaller segments averaging 0.1 cm on a side for critical current studies. Hence at 77 K the remanent magnetization is 3 G, and J_{cm} would be $\approx 30 \times 3/0.05 = 1800$ A/cm^2

compared to J_{ct} of 1.1×10^5 A/cm^2 from direct transport measurements. This nearly two order of magnitude discrepancy demonstrates that the critical state model may not be applicable in these weakly pinned materials. The situation is considerably better at 5 K where flux motion is limited: the remanent magnetization of 750 G implies $J_{cm} \approx 4.5\times10^5$ A/cm^2 compared to $J_{ct} = 2\times10^6$ A/cm^2. It is clear that the critical state model is not fully applicable even at 5 K.

Transport Measurements

The transport results for thin films are reported and discussed in detail elsewhere.[20-21] Fig. 4 shows the resistivity versus temperature for a representative Tl-2223 film. Fig. 5 shows the critical current at 76 K versus magnetic field below 0.1 T for a representative unoriented polycrystalline Tl-2122 film. Note that in zero field, $J_{ct} = 1.1\times10^5$ A/cm^2. Values up to 2.4×10^5 A/cm^2 have been obtained on unoriented polycrystalline Tl-2223 films. The magnetic field is applied in the film plane but normal to the nominal current direction. No drop-off in critical current is apparent at these fields. For fields normal to the film plane, the fall-off is much stronger but still much slower than for bulk material. These results are <u>inconsistent</u> with weak-link dominated critical current. The reason that grain boundaries should be so benign in these films remains a question.

The apparent discrepancy between the magnetization and transport critical currents is understandable in the context of weak flux pinning. The Bean critical state model used to derive critical current from magnetization assumes a uniform current density everywhere in the sample, with pinning forces just balanced so that the flux lines are stationary. In a weakly-pinned system, however, flux will begin to penetrate the sample just above H_{c1} and the screening currents will be confined to the vicinity of the sample surface, with a flux gradient determined by the strength of the pinning. This is the reason that the critical state model is strictly applicable only to highly hysteretic (hard) superconductors.

CONCLUDING REMARKS

The Tl-Ca-Ba-Cu-O system exhibits at least five distinct structural phases with different superconducting transitions above 100 K. Sintered and melted ceramics, single crystals and thin films have been synthesized and characterized by X-ray diffraction, energy dispersive X-ray analysis, static magnetization and electrical transport. Both polycrystalline and single crystal samples exhibit little hysteresis in high-field magnetization loops at liquid nitrogen temperatures, indicating weak flux pinning. This weak pinning prevents use of the critical state model to infer critical current densities from magnetization data. Flux motion decreases at lower temperature, and substantial hysteresis is evident below 30 K. Flux creep measurements are in progress to explore pinning versus field and temperature. Various sample treatments will be explored in an attempt to increase pinning.

Transport critical current densities are large and exhibit only modest magnetic field dependence, particularly in unoriented polycrystalline thin films on a variety of substrates which have J_{ct} over 10^5 A/cm^2 at 76 K and 10^6 A/cm^2 at 4 K. This suggests that transport in the Tl-Ca-Ba-Cu-O system is not dominated by weak links between grains, in contrast to the dramatic effects of weak links in bulk ceramic and polycrystalline thin films of Y-123.

These promising results in the first few months of research on the Tl-Ca-Ba-Cu-O superconducting system clearly demonstrate the potential for applications. Our studies suggest that this material system does not lose oxygen at elevated temperature, greatly simplifying device fabrication. Finally, T_c's above 110 K make operation at liquid nitrogen temperatures very attractive.

ACKNOWLEDGEMENT

This work at Sandia National Laboratories was supported, in part, by the United States Department of Energy, Office of Basic Energy Sciences, under Contract No. DE-AC04-76DP00789. The technical assistance of T. Castillo, R. Hellmer, M. Mitchell, S. Newton, and G. Pannell, Jr. is appreciated.

REFERENCES

1. Bednorz, J. G., and Muller, K. A., Z. Phys. B 64, 189 (1986).
2. Wu, M. K., Ashburn, J. R., Torng, C. J., Hor, P. H., Meng, R. L., Gao, L., Huang, Z. J., Wang, Y. Q., and Chu, C. W., Phys. Rev. Lett. 58, 908 (1987)
3. Maeda, H., Tanaka, Y., Fukutomi, M., and Asano, T., Japan. J. Appl. Phys. 27, L209 (1988).
4. Sheng, Z. Z., Hermann, A. M., El Ali, A., Almasan, C., Estrada, J., Datta, T., and Matson, R. J., Phys. Rev. Lett. 60, 937 (1988); Sheng, Z. Z., and Hermann, A. M., Nature 332, 138 (1988).
5. Ginley, D. S., Venturini, E. L., Kwak, J. F., Baughman, R. J., Carr, M. J., Hlava, P. F., Schirber, J. E., and Morosin, B., Physica C 152, 217 (1988). Further work indicates that the superconducting phase identified as $TlCaBaCu_2O_x$ in this paper is actually $TlCa_2Ba_3Cu_4O_y$ as listed in Table I in the text.
6. Parkin, S. S. P., Lee, V. Y., Engler, E. M., Nazzal, A. I., Huang, T. C., Gorman, G., Savoy, R., and Beyers, R., Phys. Rev. Lett. 60, 2539 (1988).
7. Hazen, R. M., Finger, L. W., Angel, R. J., Prewitt, C. T., Ross, N. L., Hadidiacos, C. G., Heaney, P. J., Veblen, D. R., Sheng, Z. Z., El Ali, A., and Hermann, A. M., Phys. Rev. Lett. 60, 1657 (1988).
8. Morosin, B., Ginley, D. S., Hlava, P. F., Carr, M. J., Baughman, R. J., Schirber, J. E., Venturini, E. L., and Kwak, J. F., Physica C 152, 413 (1988).

9. Parkin, S. S. P., Lee, V. Y., Nazzal, A. I., Savoy, R., Beyers, R., and LaPlaca, S. J., Phys. Rev. Lett. 61, 750 (1988).
10. Morosin, B., Ginley, D. S., Schirber, J. E., and Venturini, E. L., Physica C (accepted).
11. Venturini, E. L., Ginley, D. S., Morosin, B., Baughman, R. J., and Kwak, J. F., manuscript in preparation.
12. Ginley, D. S., Morosin, B., Baughman, R. J., Venturini, E. L., Schirber, J. E., and Kwak, J. F., J. Crystal Growth 91, 456 (1988).
13. Kwak, J. F., Venturini, E. L., Ginley, D. S., and Fu, W., in Novel Superconductivity, edited by S. A. Wolf and V. Z. Kresin (Plenum, New York, 1987), p. 983.
14. Cava, R. J., Batlogg, B., Van Dover, R. B., Murphy, D. W., Sunshine, S., Siegrist, T., Remeika, J. P., Rietman, E. A., Zahurak, S., and Espinosa, G. P., Phys. Rev. Lett. 58, 1676 (1988).
15. Reith, W., Muller, P., Allgeier, C., Hoben, R., Heise, J., Schilling, J. S., and Andres, K., Physica C 156, 319 (1988).
16. Yeshurun, Y., and Malozemoff, A. P., Phys. Rev. Lett. 60, 1676 (1988).
17. Shirane, G., Endoh, Y., Birgeneau, R. J., Kastner, M. A., Hidaka, Y., Oda, M., Suzuki, M., Murakami, T., Phys. Rev. Lett. 59, 1613 (1987).
18. Ginley, D. S., Kwak, J. F., Hellmer, R. P., Baughman, R. J., Venturini, E. L., and Morosin, B., Appl. Phys. Lett. 53, 406 (1988).
19. Ginley, D. S., Kwak, J. F., Hellmer, R. P., Baughman, R. J., Venturini, E. L., Mitchell, M. A., and Morosin, B., Physica C (accepted).
20. Kwak, J. F., Venturini, E. L., Baughman, R. J., Morosin, B., and Ginley, D. S., Physica C 156, 103 (1988).
21. Kwak, J. F., Venturini, E. L., Baughman, R. J., Morosin, B., and Ginley, D. S., Cryogenics (accepted).

MODELING OF CRITICAL CURRENTS IN GRANULAR HIGH-T_c SUPERCONDUCTORS *

R. L. Peterson and J. W. Ekin

Electromagnetic Technology Division
National Institute of Standards and Technology †
Boulder, CO 80303

ABSTRACT The transport critical current density of several samples of bulk sintered high-T_c superconductors was measured at very low magnetic fields and fitted to a model which assumes that the impediments to current at such fields are Josephson weak links. A sample of particular interest was $Y_1Ba_2Cu_3O_x$ made from hydroxycarbonate precursors; the final bulk sintered sample was very fine-grained, having an average grain size of about 1.8 μm as determined by a linear intercept analysis. The fit to the model is excellent if the average linear dimension of the weak links is chosen to be 2.0 μm. We conclude that this sample as well the others has Josephson weak links at its grain boundaries, and that any intragrain defects which may be responsible for flux pinning are not the primary weak links limiting the transport J_c of bulk samples at very low magnetic fields.

Bulk sintered samples of $Y_1Ba_2Cu_3O_x$ show a decrease in the transport critical current density, J_c, of about two orders of magnitude in applied magnetic fields as low as 10^{-2} T.[1-5] This has led to the suggestion that Josephson weak links exist somewhere within the bulk material. The location of these links has been uncertain.

We have previously[6] fitted transport critical current data from a $Y_1Ba_2Cu_3O_x$ sample at very low fields with a statistical model based on Josephson weak links (that is, weak links for which the Josephson $J_c \sin\phi$ equation holds). The fitting parameter is the average effective area of the weak link. Assuming that the total

---

* Contribution of the U. S. Government, not subject to copyright

† Formerly, National Bureau of Standards

junction thickness is twice the London penetration depth, we concluded that the average junction length is comparable to the average grain size, and inconsistent with the dominant junctions being at intragranular twinning boundaries. Other analyses also suggest this.[7]

In this note, we look specifically at a sample of $Y_1Ba_2Cu_3O_x$ prepared by J. Ritter and C. Osterag of the NIST laboratories in Gaithersburg. We use the same averaging procedures as before[6] to determine average linear dimensions of the junctions, although now we use the Airy rather than the Fraunhofer J_c *vs.* B pattern as more appropriate for a typical single link. A detailed explanation of our modeling technique, as well as results on many other samples, will be published elsewhere. For this sample as well as the others, the fit to the theoretical curve is quite excellent, and the average linear dimension of the weak links as deduced from the fit is comparable to the grain size.

The preparation of this sample was different from the more common solid state reaction method in that the starting powder was chemically prepared from hydroxycarbonate percursors. After calcining and annealing, the final bulk sintered sample showed a very fine-grained structure. A linear intercept technique was used to estimate grain sizes from scanning electron micrographs. The average intercept length was corrected for porosity, yielding a final average linear intercept of about 1.8 μm with an uncertainty of about $\pm 20\%$.

The critical current measurements were carried out on specimens cut into bars about 12 mm long, with a cross sectional area of about 1 to 2 mm^2. This sample and all others were tested using a standard four-terminal technique, with liquid nitrogen as a refrigerant. A cryostat was constructed with a mu-metal shield to exclude the earth's magnetic field so that measurements could be made accurately down to magnetic fields of 10^{-5} T.[6] The applied magnetic field was supplied by a copper solenoid and the system was calibrated with a Hall probe. Considerable care was taken to ensure that the results were not affected by heating at the current contacts. An electric field criterion of 1 μV/mm was used to determine J_c. The sample reported here had a zero-field J_c of 420 A/cm^2.

We assume that the principal barrier to current flow in these granular superconductors, at fields below about 10 mT, is a collection of Josephson weak links of random orientations and sizes. We assume that the current through these weak links can be described by the Josephson $J_c \sin\phi$ relation. The contact area between the grains in this sample is quite irregular. Given the random orientation of the grains relative to a magnetic field, the Airy diffraction pattern, describing approximately how J_c of a single junction varies with magnetic field B, is more appropriate than the Fraunhofer pattern.[9] Thus we use

$$\frac{J_c(B)}{J_c(0)} = \left| \frac{J_1(\pi B/B_o)}{\pi B/2B_o} \right|, \tag{1}$$

where J_1 is the first-order Bessel function, and the characteristic field B_o is

$$B_o = \frac{\Phi_o}{dL}. \tag{2}$$

Here B is the component of the field in the plane of the junction, Φ_o is the flux quantum, L is the average linear dimension of a given junction, and $d = 2\lambda + t$ is the effective junction thickness, where $\lambda(T)$ is the London penetration depth and t is the barrier thickness.

At large B, the Airy pattern varies as $B^{-3/2}$, apart from the oscillations, in agreement with the data. Each junction will pass current according to its own pattern, and interference among these patterns will smear out the structure. The results presented here have been obtained by averaging Eq. (1) over junction lengths with a skewed triangular probability distribution. The deduced values of junction area are rather insensitive to the specific type of length distribution. For all the samples, significant skewness is usually required for the best fit. We also average over the angle between the direction of the field and the normal to the junction, using a cutoff angle as an additional parameter. The data show that only those angles near 90° should be considered, consistent with the concept of a weakest link in each percolative path.

Figure 1 shows the measured J_c data together with a theoretical curve for the sample under consideration. For the triangular probability distribution over lengths, we used a minimum length of zero and a maximum length of 10 times the length at the peak. The average of this distribution is 3.7 times the length at the peak. To fit the data as shown, we chose B_o to have the value 1.85 mT, where B_o is defined in Eq. (2) relative to the average length L_{av}. The product dL_{av} thus has the value 1.1 μm^2. Using $d = 2\lambda = 0.56$ μm, we obtain $L_{av} = 2.0$ μm, which is comparable to the average linear intercept value of 1.8 μ m. We estimate an overall uncertainty of about ±10% in the value of B_o. Thus the average linear dimension of the junctions is comparable to the average grain size as estimated by linear intercept analysis.

Two potential problems need to be mentioned. The measuring current induces a magnetic self field whose magnitude may be about the size of B_o. However, the self field decreases with applied field, and detailed analysis shows that its affect on our estimate of B_o is less than the other uncertainties involved in the curve fitting. Second, the percolative network of superconducting paths expected in this material has many closed loops of current which individually give an oscillatory behavior of J_c as a function of B. These loops can be thought of as multi-junction SQUIDs, whose behaviors have been thoroughly analyzed.[8] The loops have areas orders of magnitude larger than the junction areas dL, and the periods of the oscillations are correspondingly very much smaller than B_o. Further, the J_c amplitudes, although irregular, do not show an overall decrease with B. This effect can also be safely ignored.

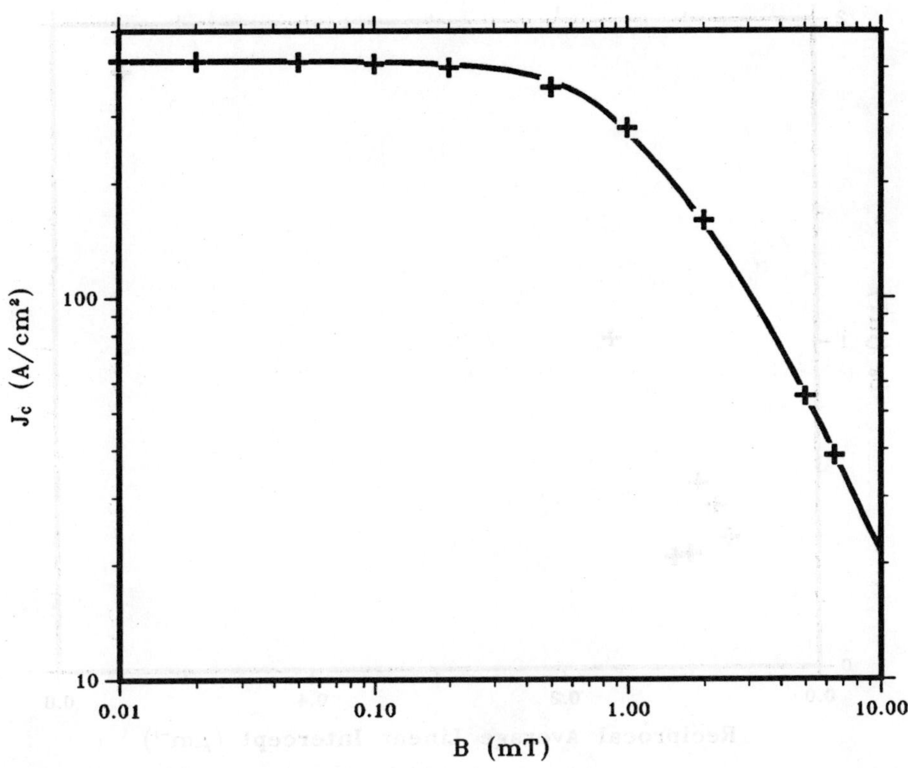

Fig. 1. Theoretical curves and data for the sample discussed in the text. Measured J_c is plotted *vs.* B on a log-log scale. The curve is calculated with a skewed triangular probability distribution of grain sizes, whose length parameters are L_{min} $= 0$ and $L_{max} = 10.0L_{peak}$. To fit the data to the curve, we give $B_o = \Phi_o/dL_{av}$ the value 1.85 mT.

Comparison of theoretical curves with the experimental transport critical current *vs.* magnetic field data on many different samples of bulk sintered high-T_c superconductors gives convincing evidence that the impediments to transport current at very low fields are indeed Josephson weak links. The analysis yields the average weakest-link cross sectional areas. If we accept 280 nm as a reasonable value for the London penetration depth at 77 K, we infer average junction lengths comparable to the average grain size in all cases. The deduced junction lengths scale roughly with the grain sizes. We illustrate this in Fig. 2, where we plot values of B_o obtained from fitting data for many different samples, against the reciprocal of the average linear intercept.

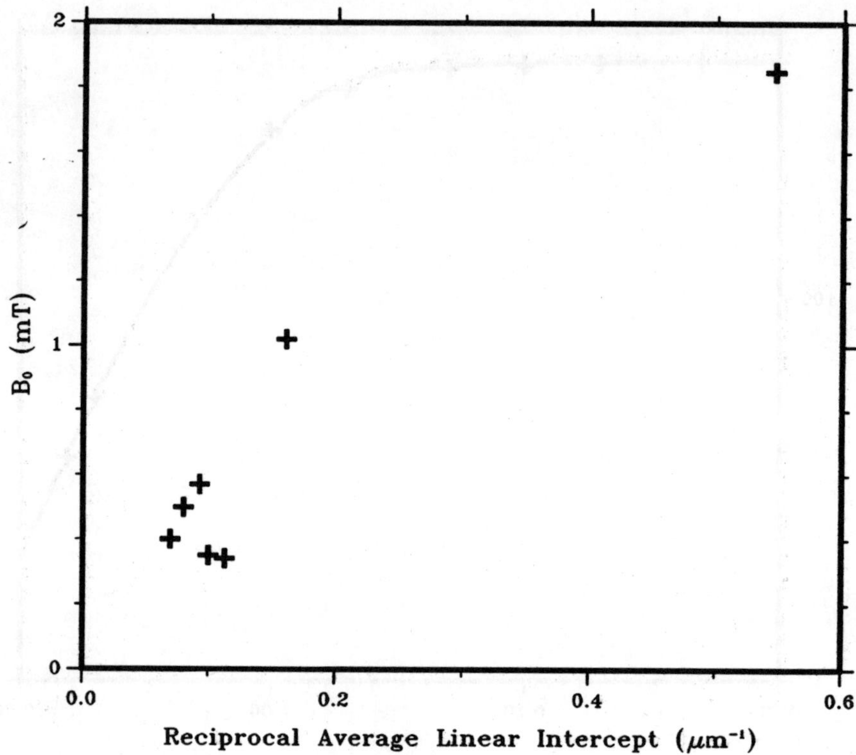

Fig. 2. Plot of B_o *vs.* the reciprocal of the average linear intercept for several samples of bulk sintered superconductors.

We also conclude that any intragrain defects which may exist are not the primary weak links limiting the transport J_c of bulk samples at very low magnetic fields.

Acknowledgments

We thank C. P. Ostertag for sending us the sample used in this analysis, briefing us on the method of preparation, and sending us micrographs.

References

1. Ekin, J. W., Panson, A. J., Braginski, A. I., Janocko, M. A., Hong, M., Kwo, J., Liou, H., Capone II, D. W., and Flandermeyer, B., presented at the Materials Research Society meeting, April 23–24, 1987, Anaheim, CA; *Proceedings of the Symposium on High Temperature Superconductors*, edited by D. U. Gubser and M. Schluter (Materials Research Society, Pittsburgh, 1987), Vol. EA-11, p. 223.
2. Capone, D. W. II, and Flandermeyer, B., in Ref. 1, p. 181.
3. Ekin, J. W., Advanced Ceramic Materials **2**, 586 (1987).
4. Ekin, J. W., Braginski, A. I., Panson, A. J., Janocko, M. A., Capone, D. W. II, Zaluzec, N., Flandermeyer, B., de Lima, O. F., Hong, M., Kwo, J., and Liou, S. H., J. Appl. Phys. **62**, 4821 (1987).
5. Kwak, J. F., Venturini, E. L., Ginley, D. S., and Fu, W., in *Novel Superconductivity*, edited by S. A. Wolf and V. Z. Kresin, (Plenum Press, New York, 1987), p. 983.
6. Peterson, R. L. and Ekin, J. W., Phys. Rev. **B37**, 9848 (1988).
7. Kwak, J. F., Venturini, E. L., Nigrey, P. J., and Ginley, D. S., Phys. Rev. **B37**, 9749 (1988).
8. Peterson, R. L. and Hamilton, C. A., J. Appl. Phys. **50**, 8135 (1979).
9. Peterson, R. L. and Ekin, J. W., Physica C, to be published.

IS THERE A FERMI SURFACE IN HIGH T_c SUPERCONDUCTORS ? A BRIEF REVIEW OF POSITRON ANNIHILATION RESULTS

Stephan Berko
Physics Department, Brandeis University
Waltham, Massachusetts 02254

ABSTRACT

After a brief introduction to the physics of positron annihilation spectroscopy (PAS), the present status of the two-dimensional angular correlation of annihilation radiation experiments (2D ACAR) in high T_c superconductors will be described. The 2D-ACAR technique is being used to study the electronic structure and, in particular, the existence, size and shape of the Fermi surface of the new superconducting materials.

During the last two years a great number of experimental techniques have been used to elucidate the properties of the new high temperature superconductors,[1] and to distinquish experimentally between the many theoretical models proposed to explain their behavior.[1] It is safe to say that, although much information has been obtained, we are far from a clear consensus regarding the nature of superconductivity in these new materials. One of the most critical questions regarding the electronic structure of the high T_c superconductors is: "Do high T_c superconductors have a Fermi surface and is Luttinger's theorem fulfilled?" In other words, is the electronic structure of these materials "Fermi-liquid-like," or are the correlations so strong as to require a new type of many-body description, such as that of Anderson's Resonating Valence Bond (RVB) model?[2] To underline the importance of these questions, we quote from a review of various high T_c theories by C. M. Varma[3]: "A clear experimental test of RVB would consist in mapping the Fermi surface. If the oxide superconductors in the metallic phase have a conventional Fermi surface, which obeys Luttinger's theorem and looks even vaguely as that given by bandstructure calculations, RVB must be abandoned, at least, in the present context."

The classic techniques used to study Fermi surfaces, such as the de Haas-van Alphen effect and magneto-resistance, cannot be applied to the high T_c superconductors, since they require long electronic mean-free-paths. Positron annihilation spectroscopy (PAS) has been shown in the past to be free of such restrictions, and to be able to produce valuable information regarding the electronic structure of various solids.[4] Before discussing the results of the several PAS experiments performed to date on high T_c materials, we briefly outline the fundamental techniques used in PAS. The behavior of positrons in condensed matter has been the subject of intense experimental and theoretical investigation during the last decades, and the use of positrons as a probe in solids has been well documented and reviewed.[4,5] By 1988, the use of positron annihilation spectroscopy (PAS) in solids could be divided into two main categories: a) bulk studies, using mainly fast positrons from radioactive β^+ sources, and b) surface and near-surface studies, using the more recently developed variable energy slow positron beams.[6,7] The information about solids is gained by studying the various properties of the annihilation quanta or, as sometimes is the case with slow positron beams, by detecting directly the scattered positrons.

In bulk measurements, the injected fast positrons lose their energy by ionizing collisions and phonon scattering, and reach thermal equilibrium with the solid prior to annihilation, penetrating the sample well into the bulk to mean depths of 10 - 100μm. In single crystals the thermalized positron takes the form of a delocalized Bloch wave, with maxima at interstitial regions due to core repulsion, while, if defects are present, the positron can localize around lower density defects such as vacancies, dislocations, and microvoids. The annihilation process, well understood theoretically, results in two quanta (2γ) from spin-singlet overlaps, or in three quanta (3γ) from spin triplet overlaps. In metals 2γ decays predominate, since the 3γ decay is $\sim 10^3$ times less probable.

Four typical experiments can be performed with the annihilation γ-s: 1)

2γ angular correlation experiments (ACAR), 2) the measurement of the energy spectrum of the γ-s in 2γ decay (Doppler profiles), 3) the direct measurement of the positron lifetime, and 4) $3\gamma/2\gamma$ yield experiments. The ACAR and the Doppler profile data reflect the momentum distribution $\rho^{2\gamma}(\underline{p})$ of the annihilating $e^+ - e^-$ pair. The positron lifetime depends on the electronic density at the site of the positrons. The measurement of the $3\gamma/2\gamma$ yield is important when the e^+ and e^- spins are relevant, as in studies of magnetically ordered samples, using spin-polarized positrons[4,5] (available in β^+ decay due to parity nonconservation). One can thus obtain detailed information about the electrons annihilating with the positron. The many-body $e^+ - e^-$ correlations importantly influence the local density (thus the lifetimes), but have smaller effect on the momentum density – at least in those metals where the conduction electrons form a "Fermi liquid".

The PAS technique used mostly to study the Fermi surface of metals is the ACAR technique in its high precision two dimensional (2D-ACAR) form.[8] The 2γ angular correlation of annihilation radiation (2D-ACAR) depends, by momentum conservation, on the momentum distribution $\rho^{2\gamma}(\underline{p})$ of the $e^+ - e^-$ pair. Using position sensitive γ detectors, one measures the deviation from (anti)-collinearity of the annihilation quanta; the small angles involved are directly translated to momenta, by $p = \theta mc$ (one milliradian corresponds to a momentum component of $p = mc \times 10^{-3}$). One obtains the two-dimensional projections of $\rho^{2\gamma}(\underline{p})$; 2D ACAR spectra yield $N(p_y, p_z) = \int \rho^{2\gamma}(\underline{p})\, dp_x$ (convoluted by the resolution of the apparatus). The integration direction p_x depends on the orientation of the crystal studied with respect to the axis of the apparatus. In the independent particle model (IPM),

$$\rho^{2\gamma}(\underline{p}) = \sum_{\underline{k},j} n_j(\underline{k}) \mid \int \psi_{\underline{k},j}(\underline{r})\psi_+(\underline{r}) exp(-i\underline{p}.\underline{r}) d\underline{r} \mid^2 , \tag{1}$$

where the $\psi_{\underline{k},j}(\underline{r})$ is the $\underline{k}$ electron in band j, $\psi_+(\underline{r})$ is the $\underline{k}^+ = 0$ Bloch wave positron and $n_j(\underline{k})$ restricts the summation to the occupied electron states. From

Eq. 1 it is clear that the Fermi surface appears as a discontinuity in the momentum density via $n_j(\underline{k})$. One can either search for these discontinuities in the 2D projections $N(p_y, p_z)$ or, by obtaining several projections, one can reconstruct the full 3D $\rho^{2\gamma}(\underline{p})$ prior to Fermi surface studies. Instead of analyzing the full momentum density, one can first form the "folded momentum density" $\tilde{\rho}^{2\gamma}(\underline{p}) = \sum \rho^{2\gamma}(\underline{p} - \underline{G})$, where the $\underline{G}$'s are the reciprocal lattice vectors. This procedure[4,8] (the "LCW folding"[9]) leads to a Fermi surface dominated distribution in the periodic zone, since the folded densities of filled bands produce (in the IPM) a nearly constant contribution. The applicability of Eq. 1 in the analysis of ACAR data has been tested in many well known metals,[8] and 2D ACAR studies have been performed on many systems. For example, the Fermi surface of the "pre-high-T_c" A15 superconductors such as V_3Si[10,11] and of the tungsten bronze[12] Na_xWO_3 (with an incomplete perovskite structure) has been mapped out by the 2D ACAR technique.

The problems facing the experimentalists in performing 2D ACAR experiments in high T_c materials are, unfortunately, much more cumbersome than was the case with the A15 superconductors, for example. First, the availability of "good" single crystals for positron experiments is clearly a problem. Most 2D ACAR experiments would require samples of a much larger size than are available to date. Defects, which are also a problem in other systems, but can be annealed out, play of course an important role in high T_c materials. The question of e^+ trapping at oxygen vacancies is more subtle than in metal vacancies, because of charge effects. The $YBa_2Cu_3O_{7-x}$ single crystals available are too thin (<0.1mm) to stop all positrons; in addition, the larger the sample diameter (2-3mm), the more uncertain its bulk oxygen concentration, because of the well known oxygenation problem. In addition, the number of filled bands is much higher than in other, simpler systems, leading to a large "background" in 2D ACAR Fermi surface studies, and the large real space cell leads to correspondingly small Brillouin zones, increasing the relative resolution of the apparatus.

Two groups have published to date 2D ACAR results in $YBa_2Cu_3O_{7-x}$ single crystal samples. The University of Geneva results[13] are based on a crystal orientation where the integration direction, p_x in Eq. 1, is in the a-b plane. They use a sample with $x = 0.30 \pm 0.05$. One notes that in all single crystal measurements to date the crystals are twinned, thus resulting in an effective summing between the a and b direction, i.e. yielding a "pseudo single crystal with a tetragonal structure."[13] The other group, at the Argonne National Laboratory, performed their experiment[14] on a sample with $x < 0.05$, with the integration direction along the c axis. This geometry required the suspension of the thin crystal with its normal along the axis of the apparatus; a magnetic field is used to spiral the positrons into the sample. Since the shape of the Fermi surface is expected to be nearly cylindrical along the c axis, reflecting the two-dimensionality of the Cu-0 planes and the one-dimensionality of the chains in $YBa_2Cu_3O_{7-x}$, the 2D ACAR spectra $N(p_y, p_z)$ with p_x along the c direction are predicted to show most of the relevant shapes, as computed by band theories using the local density approximation (LDA).[15] In both the Geneva and Argonne experiments, the 2D ACAR spectra show small, but distinct anisotropies ($\sim 0.7\%$ and $\sim 3\%$), presumably due to positron annihilation with oriented conduction electrons. In order to search for Fermi surface breaks, both groups perform momentum density reduction into the first zone (i.e. LCW reduction).[8,9] The Geneva group reconstructs $\tilde{\rho}^{2\gamma}(\underline{p})$ from three $N(p_y, p_z)$ spectra; their results indicate a rough agreement with the main topological features of the Fermi surface predicted by Freeman et al.[15] The analysis by Smedskjaer et al[14] of the Argonne results yields "four nearly cylindrical (Fermi surface) sheets, in substantial accord with band theory predictions". Preliminary temperature dependent measurements by the Geneva group indicate no essential change of the Fermi surface topology with temperature, albeit within rather large error bars. The Argonne data were obtained only at room temperature. The data is not sufficiently precise to set a limit on the "sharpness" of the Fermi surfaces observed. The total counts accumulated by the Argonne group is about a factor of ten less than the usual high quality 2D ACAR data obtained in other materials. To compensate for this, Smedskjaer et al use the symmetry of the crystal to "symmetrize" their data. This procedure has its dangers because of a "kaleidoscope effect" due to noise propagation, leading to Fermi surface-like shapes; some details of the real Fermi surface topology might also be distorted by such effects. The reconstruction of the highly complex 3D reduced momentum density from only three projections, as performed by the Geneva group, also has its questionable merits. These drawbacks could account for the lack of detailed agreement between the two sets of data. We conclude therefore that more, higher precision work has to be performed before the 2D ACAR data can test in detail the validity of the theoretical predictions based on local density band computations. Additional 2D ACAR results from other experimental groups are expected soon.[16] Only with much higher precision data can one hope to test in detail the agreement with Luttinger's theorem regarding the volume enclosed by the observed Fermi surfaces.

In order to analyze the details of the full momentum density anisotropies (rather than only the folded densities), one needs the theoretical computation of the positron wavefunction. Such computations have been performed recently[17], indicating that in $YBa_2Cu_3O_7$, the positrons form "chains" along the ordered row of vacancies, chains that couple strongly through the (010) face of the cell. Surprisingly, the positron is almost completely absent in the Y plane, yielding a large anisotropic positron "effective mass". A partial theoretical $e^+ - e^-$ momentum density in $YBa_2Cu_3O_7$ has been computed recently by Bansil et al[18] based on local-density band theory. They compare their result (along the zone edge) with the data by Smedskjaer et al[14], and find reasonable agreement with experiment, thus concluding that "conventional LDA band-theory framework provides a reasonable description of the ground state and the Fermi surface of $YBa_2Cu_3O_7$". Since their comparison is based on the differentiation of the (symmetrized) data, higher statistics data would be most desirable for a more detailed and conclusive check of the theoretical predictions.

2D ACAR experiments are also in progress to study the momentum density as a function of oxygen concentration in $YBa_2Cu_3O_{7-x}$. Should oxygen be removed randomly from the chains, one expects a smearing of the Fermi surface due to disorder, thus making the identification more difficult. Localization of the positrons due to chain disorder could also become a problem.[17]

Experiments on the semiconducting La_2CuO_4 material have been performed by a Livermore-AT&T Bell Laboratories collaboration.[19,20] Data along several crystallographic orientations show clear anisotropies in the 2D ACAR spectra, due to valence electrons. These spectra are reasonably well reproduced by a theory using a localized ion scheme and a LCAO-MO approach, as developed by Chiba[21,12] to explain earlier PAS data on metal oxides. A similar scheme was also used by the Livermore group to obtain the e^+ charge density[22] in $YBa_2Cu_3O_7$. A surprising result was reported by Tanigawa et al[23] on the folded momentum density $\tilde{\rho}^{2\gamma}(\underline{p})$ from La_2CuO_4. In spite of the known semiconductor-like temperature dependence of the resistivity, they obtain a set of Fermi surface sheets from their analysis. This potentially interesting problem clearly needs independent confirmation from other experiments on different samples.

The question of the temperature dependence of PAS data has been the subject of many studies, mainly performed on sintered powder samples – for a partial list of references, see the literature cited in Ref. 17. These studies use the low resolution, but high efficiency Doppler profile technique for momentum density measurements, and lifetime experiments for the total real space positron-electron charge density measurements. Since the first experiments,[24] all results indicate that the positron lifetime, which in most samples is multivalued, indicating some e^+ trapping at defect sites, follows an "anomalous" behavior below T_c. This change below T_c is also apparent in the overall shape of the momentum density. The many experiments on the temperature dependence show a great variation from sample to sample, indicating the sensitivity of the positrons to sample preparation. The only lifetime results in single crystal samples also show a "temperature effect" below T_c, but in opposite directions from the sintered powder results.[25]

In view of these large variations, it is premature to assign a unique physical interpretation to this potentially important effect. In particular, one should study in detail the temperature dependence of the 2D ACAR spectra in crystalline samples. Several possible causes producing the temperature effect below T_c were reviewed in Ref. 17; they include possible changes in e^+ thermalization, structural changes below T_c, the effect of large anharmonic motions of the O(4) oxygens, as well as possible changes in electronic density near the Cu-O chains. The importance of possible e^+ trapping at twin boundaries has also been mentioned[26] - such trapping could affect even the single crystal results.

In conclusion, the present 2D ACAR results indicate that $YBa_2Cu_3O_{7-x}$ superconductors do exhibit Fermi surfaces, albeit the exact shape of the various Fermi surfaces cannot be determined with the present accuracy of the experiments. The extent to which Luttinger's theorem is obeyed cannot be ascertained from the data. Much higher precision PAS data is therefore required before one could definitely conclude that LDA based band theories can properly account for the ground state of the high T_c superconductors.

I am grateful to many colleagues in the positron field for important discussions regarding their data and to the superconductivity experts who have helped me with the present status of high T_c physics. I thank E. C. von Stetten for a critical reading of this manuscript. The present work was partially supported by NSF Grant DMR-8814185.

REFERENCES

1. See for example the Proceedings of the "Interlaken Conference ," edited by J. Müller and J. L. Olsen, Physica C, 153-155, 1988; see also papers in this volume.
2. P. W. Anderson, Science 235, 1196 (1987); see also R. B. Laughlin, Science 242, 525 (1988).
3. C. M. Varma, "International Meeting on High Temperature Superconductors," Schloss Mantendorf, Austria, February 1988; to be published by Plenum Publ. Corp., NY (1988), edited by H. W. Weber.
4. See the reviews in "Positron Solid-State Physics - Proc. Int. School E. Fermi," edited by W. Brandt and A. Dupasquier (North-Holland, Amsterdam, 1983).
5. See also the reviews in the Proceedings of the International Conferences on Positron Annihilation (ICPA) published every three years - "Positron Annihilation - ICPA-7," edited by P. C. Jain, R. M. Singru and K. P. Gopinathan (World Scientific, Singapore, 1985); "Positron Annihilation - ICPA-8," edited by M. Dorikens, L. Dorikens - van Praet and D. Segers (World Scientific, Singapore) to be published 1989.
6. Review by A. P. Mills, Jr., in Ref. 4, page 432.
7. Review by P. J. Schultz and K. G. Lynn, Rev. Mod. Phys. 60, 701 (1988).
8. See review by S. Berko in Ref. 4, page 64.
9. D. G. Lock, V. H. C. Crisp, and R. N. West, J. Phys. F3, 561 (1973).
10. W. S. Farmer, F. Sinclair, S. Berko, and G. M. Beardsley, Sol. St. Comm. 31 481 (1979); S. Berko, W. S. Farmer, and F. Sinclair, in "Superconductivity in d- and f-band Metals", edited by H. Auhl and M. B. Maple (Academic Press, NY, 1980).
11. T. Jarlborg, A. A. Manuel, and M. Peter, Phys. Rev. B27, 4210 (1983); see also the review by A. A. Manuel, Helvetica Phys. Acta, 61, 451 (1988).
12. T. Akahane, K. R. Hoffmann, T. Chiba, and S. Berko, Sol. St. Comm. 54, 823 (1985).
13. L. Hoffmann, A. A. Manuel, M. Peter, E. Walker, and M. A. Damento, Europhys. Lett. 6, 61 (1988); Physica C153-155, 129 (1988); M. Peter, L. Hoffmann, A. A. Manuel, Physica C153-155, 1724 (1988); see also the general review by M. Peter, in "Positron Annihilation - ICPA-8," edited by M. Dorikens, L. Dorikens - van Praet and D. Segers (World Scientific, Singapore) to be published 1989.
14. L. C. Smedskjaer, J. Z. Liu, R. Benedek, D. G. Legnini, D. J. Lam, M. D.

Stahulak, H. Claus, and A. Bansil, Physica C156, 269 (1988).
15. J. Yu, S. Massidda, A. J. Freeman, and D. D. Koeling, Phys. Lett. A122, 203 (1987); H. Krakauer, W. E. Pickett, and R. E. Cohen, J. of Supercon. 1, 111 (1988).
16. Work is in progress on $YBa_2Cu_3O_{7-x}$ by a collaboration between the University of Texas Arlington and Livermore National Laboratories, as well as a collaboration between AT&T Bell Laboratories, Brandeis University and Brookhaven National Laboratory.
17. See for example the positron charge density as shown in E. C. von Stetten, S. Berko, X. S. Li, R. R. Lee, J. Brynstad, D. Singh, H. Krakauer, W. E. Pickett, and R. E. Cohen, Phys. Rev. Lett. 60, 2198 (1988).
18. A. Bansil, R. Pankaluoto, R. S. Rao, P. E. Mijnarends, W. Dlugosz, R. Prasad and L. C. Smedskjaer, Phys. Rev. Lett. 61, 2480 (1988).
19. A. L. Wachs, P. E. A. Turchi, Y. C. Jean, K. H. Wetzler, R. H. Howell, M. J. Fluss, D. R. Harshman, J. P. Remeika, A. S. Cooper, and R. M. Fleming, Phys. Rev. B38, 913 (1988).
20. P. E. A. Turchi, A. L. Wachs, Y. C. Jean, R. H. Howell, K. H. Wetzler, and M. J. Fluss, Physica C153-155, 157 (1988).
21. T. Chiba and N. Tsuda, Appl. Phys. 5, 37 (1974); T. Chiba, J. Chem. Phys. 64, 1182 (1976).
22. Y. C. Jean, J. Kyle, H. Nakinishi, P. E. A. Turchi, R. H. Howell, A. L. Wachs, M. J. Fluss, R. L. Meng, H. P. Hor, J. Z. Huang, and C. W. Chu, Phys. Rev. Lett. 60, 1069 (1988).
23. S. Tanigawa, Y. Mizuhara, Y. Hidaka, M. Oda, M. Suzuki, and T. Murakami, MRS fall meeting Boston 11/30-12/4 (1987).
24. Y. C. Jean, S. J. Wang, H. Nakanishi, W. N. Hardy, M. E. Hayden, R. F. Kiefl, R. L. Meng, H. P. Hor, J. Z. Huang, and C. W. Chu, Phys. Rev. B36, 3994 (1987).
25. D. R. Harshman, L. F. Schneemeyer, J. V. Waszczak, Y. C. Jean, M. J. Fluss, R. H. Howell, and A. L. Wachs, Phys. Rev. B38, 848 (1988).
26. S. G. Usmar, K. G. Lynn, A. R. Moodenbaugh, M. Suenaga, R. L. Sabatini, Phys. Rev. B38, 5126 (1988).

MUON SPIN ROTATION STUDIES OF MAGNETIC ORDER AND STRONG MAGNETIC CORRELATIONS IN MAGNETIC AND SUPERCONDUCTING SYSTEMS BASED ON THE HIGH T_c COPPER OXIDE STRUCTURES

J.I. Budnick and B. Chamberland
Physics Department, University of Connecticut, Storrs, CT 06268, USA

A. Weidinger, Ch. Niedermayer, A. Golnik, R. Simon and E. Recknagel
Fakultät für Physik, Universität Konstanz, D-7750 Konstanz, FRG

C. Baines
Paul-Scherrer-Institut, CH-5234 Villigen, Switzerland

We have carried out a series of muon spin rotation (μSR) studies down to millikelvin temperatures in order to explore the existence of magnetic correlations below T_C in the $La_{2-x}Sr_xCuO_4$ system. Evidence is presented for the presence of local magnetic fields believed to originate from Cu electronic moments in both superconducting $La_{2-x}Sr_xCuO_4$ and in superconducting oxygen deficient $YBa_2Cu_3O_{6.6}$. μSR results will also be presented for oxygen deficient and superconducting $GdBa_2Cu_3O_{7-x}$ samples. Some discussion of the relevance of these results to recent proposals for pairing mechanisms will be presented.

INTRODUCTION

Soon after the discoveries of High-Temperature-Superconductivity [1] it was recognized initially in $La_{2-x}Sr_xCuO_4$ and subsequently in the $YBa_2Cu_3O_X$ system that the parent compounds La_2CuO_4 and $YBa_2Cu_3O_6$ were strongly antiferromagnetic [2-9]. It is now well established that the antiferromagnetism is due to the coupling of the Cu moments in the CuO_2 planes of these structures. Theoretically the close relationship between antiferromagnetism and superconductivity was explored in the early stages of High-T_c superconductor research [10-14]. Recently magnetic fluctuations were introduced explicitly as the cause for the pairing of charge carriers in the superconducting state [11]. Aharony et al introduced in their picture for spin pairing a local ferromagnetic coupling between copper spins which is introduced by the hole doping produced by Sr addition to La_2CuO_4 [11]. Probing this suggestion of an inhomogeneous magnetic structure provides a clear challenge for local probes such as μSR and NMR.

Gutsmiedl et al [15] concluded from their low temperature specific heat studies the existence of a magnetic contribution to the heat capacity at the copper site which was the first indication of magnetic correlations in superconducting $La_{1.85}Sr_{0.15}CuO_4$.

To answer the still open question of what the relationship is between antiferromagnetism and superconductivity in these systems it is important to study the magnetic properties of $La_{2-x}Sr_xCuO_4$ as a function of Sr concentration and similarly the properties of $YBa_2Cu_3O_x$ as a function of the oxygen content x.

In this paper we present an overview of some recent muon spin rotation studies for a variety of High T_c Oxide related systems. A more complete discussion of some of the research results presented here is to be found in the references contained in this paper. We have also attempted to indicate briefly some points of contact between μSR results on magnetic correlations and results obtained by other methods.

METHOD

μSR provides an ideal probe to study the evolution of weak magnetism in complex systems and to explore whether magnetic correlations exist well below the superconducting transition temperature.

Our μSR experiments were performed at the Paul-Scherrer-Institut (PSI) in Switzerland at the low momentum beam. In these experiments positive muons are implanted into the sample with an energy of approximately 4 MeV. The stopped muons decay by emitting a positron and two neutrinos. The emission probability of the positrons is anisotropic with respect to the muon spin direction and therefore allows the detection of the spin precession. The positron count rate in a fixed detector as a function of time after the arrival of the muon in the sample is given by

$$N(t)=N_0 \exp(-t/\tau_\mu)\{1 + A\, P(t) \cos(\omega t+\Phi)\} + B$$

where N_0 is a normalization constant, A an effective anisotropy (A = 0.2-0.3), Φ a phase reflecting the direction of the initial polarization with respect to the positron emission direction and B a time-independent background. The precession frequency $\nu = \omega/2\pi$ is related to the internal field B_μ by

$$\nu(\text{MHz}) = 135.5\ B_\mu(\text{T})$$

Usually for the time-independent polarization P(t), a gaussian or exponential function is assumed

$$P(t) = \exp\ (-\sigma^2 t^2/2)$$
$$P(t) = \exp\ (-\lambda t)$$

In a typical μSR spectrum only the anisotropic part

$$R(t) = A\ P(t)\ \cos\ (\omega t + \Phi)$$

is shown.

LA_2CuO_4-BASED SYSTEMS

Early μSR experiments [5,6]confirmed the neutron results which established the local antiferromagnetism present in undoped La_2CuO_4.

In Fig. 1 a muon spin rotation spectrum and its Fourier transform at 4.2 K is shown for La_2CuO_4. The observation of a unique μSR frequency unambiguously demonstrates magnetic ordering in this sample. The measured frequency at 4.2 K is 5.6 MHz corresponding to an internal field of 41 mT. A rigorous interpretation of the magnitude of the internal field is difficult since the site of the muon in the crystal lattice is not known. A discussion of possible muon sites is given in reference 5. A fairly open site is formed by the distorted tetrahedral arrangement of two La and two O atoms. Assuming the antiferromagnetic structure as reported in reference 4 reasonable agreement of the calculated and observed dipolar fields is obtained for a copper moment of 0.5 μ_B. This is consistent with the neutron data.

The temperature-dependence of the μSR frequency is shown in Fig. 2 together with the results obtained in Sr doped samples [16]. It can be seen that the Sr doping changes the Neel-temperature dramatically from about 250 K for the undoped sample to less than 6 K for x = 0.05. This destruction of the antiferromagnetic phase by a few percent of Sr-doping is rather striking. On the other hand the measured frequencies are not very different at low temperatures for the different samples indicating that the magnetic moments of the ions are not altered appreciably [16,17]. Thus the low Neel-temperatures are caused by a weakening of the effective interaction strength rather than by a diminishing of the magnetic moments themselves. In this article we focus on the low temperature results in $La_{2-x}Sr_xCuO_4$ with Sr concentrations at which the samples are superconducting.

Our samples were prepared in the usual way by sintering the appropriate oxide constituents. The final pellets of 12 mm diameter and 2 mm thickness were single phase and had superconducting transition temperatures (midpoint) at 14 K for x = 0.07, at 26 K for x = 0.10 and 32 K for x = 0.15. The Meissner fractions of the three samples determined in an external field of 2 mT were 0.08, 0.22, 0.23, respectively. These are reasonable values for this type of material [18]. The Meissner fractions represent conservative lower limits of the superconducting volume fractions as trapped flux will reduce the actual signals [19].

Representative data for $La_{1.93}Sr_{0.07}CuO_4$ for three different temperatures are displayed in Fig. 3. At the left side the time spectra and on the right side the corresponding Fourier transforms are presented. The spectrum at 4.2 K represents the situation without an internal field from electronic moments. In this case the asymmetry R(t) is almost constant in the time range shown in this figure. The slight decay of R(t) at 4.2 K is due to nuclear moments. The spectra at 35 mK and 2.2 K clearly show a fast decay of the asymmetry corresponding to internal fields much higher than those arising from nuclear moments.

Spectra for different Sr-concentrations measured at 35 mK are shown in Fig. 4. A clear relaxation, although much slower than for the $x = 0.07$, is seen also for $x = 0.10$ and $x = 0.15$. Thus in these samples clearly magnetic fields due to electronic moments are present. It should be emphasized that the observed effect is not due to magnetic impurities or to minority phases since essentially all muons (at least more than 70%) which stop in the sample participate in this process. This can be deduced from the observed asymmetry at $t = 0$ which reaches the expected value. Thus the observed phenomenon is indeed an intrinsic property of the sample.

The main feature of the spectra in Fig. 3 and 4 is clearly the large variation of the decay times of the asymmetry function with Sr-concentration and temperature and correspondingly the variation of the magnitude of the μSR frequencies (see Fourier transforms). As a measure of the characteristic decay time of the anisotropy in Fig. 3 and 4 we take the value T_e at which the anisotropy has fallen to 1/e of its value at $t = 0$ and relate it to the average magnitude of the μSR frequency $<|B_\mu|>$ by

$$1/T_e = <|\omega|> = \gamma_\mu <|B_\mu|>$$

where $\gamma_\mu = 851.4$ MHz/T is the gyromagnetic ratio of the muon. In this way all data can be represented on a common graph irrespective of the detailed shape of the spectra. The results are shown in Fig. 5.

Two features can be seen: i) The values of $1/T_e$ (and correspondingly the magnitudes of the internal fields) depend strongly on the Sr-concentrations. At 35 mK, an order of magnitude difference is observed between $1/T_e$ for $x = 0.07$ and $1/T_e$ for $x = 0.15$. ii): The $1/T_e$ -values are fairly constant below 1 K but fall off steeply at higher temperatures reaching unmeasurable small values between 2 and 3.5 K. The temperature at which the ordering disappears is weakly dependent on the Sr-concentration.

Fig. 6 shows a phase diagram deduced from the data discussed in this chapter. In the upper part of Fig. 6 the temperatures T_N and T_C for the transition to magnetic ordering and superconductivity are displayed as a function of the Sr concentration. For x > 0.07 and temperatures below 2-3 K a region with coexisting magnetic ordering and superconductivity is found. In the lower part of the figure we have plotted the magnitude of the observed magnetic fields at 35 mK. It can be seen that the fields are strongly reduced in the superconducting samples compared to undoped La_2CuO_4. Similar phase diagrams have been reported on the basis of NMR studies [20,21]. Recent neutron experiments provide evidence for strong magnetic correlations which persist in the superconducting regime [22].

In summary the present μSR results (see also reference 23) give clear evidence for the existence of local internal magnetic fields in superconducting La_2CuO_4. Our data are consistent with fluctuating magnetic moments (probably residual moments of Cu) which slow down as one lowers the temperature and correlation times in the order of 10^{-8} s or longer are reached below 3 K.

1-2-3 MATERIALS

$GdBa_2Cu_3O_{6+x}$

μSR experiments [24] were carried out on two superconducting samples of $GdBa_2Cu_3O_{6+x}$. In one sample a sharp transition at 60 K was observed with no evidence for any diamagnetic susceptibility in the region of 90 K. Our second sample showed a transition temperature of approximately 90 K as studied by magnetic susceptibility measurements.

The low temperature muon spin rotation results as shown in Fig. 7 reflect the onset of magnetic ordering in the Gd sublattice and indicate the presence of two distinct μSR frequencies which suggest the existence of two muon stopping sites. The lower frequency (4.6 MHz at T=0 K) is dominant at all temperatures below 2.3 K and was seen in all samples. The intensity of the higher frequency (7 MHz at T=0 K) is sample dependent. In the 60 K sample the 4.6 MHz line is about 2.5 times stronger than the 7 MHz line, in the 90 K sample at least 5 times stronger. Our initial suggestions for the location of the muon site were based on dipolar calculations. We now believe that recent calculations of Dawson et al. [25] provide a much more physical interpretion of our data. They suggest that the lower frequency signal should be associated with a muon bound to an oxygen chain site. If we remove some oxygen from the chain sites, as is believed to be the case for the 60 K sample, we expect an increased occupation of a second muon site with a somewhat higher frequency bound to an apical oxygen site. Our experimental results are therefore consistent with the proposal of Dawson et al.

$YBa_2Cu_3O_{6+x}$

Antiferromagnetism in strongly oxygen deficient $YBa_2Cu_3O_7$ has been first reported by muon spin rotation studies by Nishida et al [7] and subsequently an extensive study by Brewer et al [8] measured the dependence of the onset of magnetic ordering on the oxygen content. Our studies of oxygen deficient $YBa_2Cu_3O_{6+x}$ indicate the existence of 2 muon spin precession frequencies (Fig. 8). A 4 MHz signal is observed which is due to muons stopping near the apical oxygen site, since the chain oxygens are almost completely removed. In addition a second frequency at 18 MHz is observed and is probably due to muons stopping close to the CuO_2 planes, where the dipolar fields should be larger. We would like to mention that at higher temperatures greater than 200 K we find the presence of a small amplitude 2 MHz signal. This 2 MHz site is believed to arise from muons bound to an oxygen in a chain site and is currently being studied by us.

Thorough neutron studies by the Grenoble group [26] and the Brookhaven group [9] clearly proved the existence of long range antiferromagnetic order in samples with an oxygen content in the range from 6 to 6.5. From the neutron results it was determined that there is a strong reduction in the ordering temperature and a transition to a superconducting state at around 6.5 along with a transition from a tetragonal to an orthorhombic structure near this composition.

As in La_2CuO_4 we have searched for magnetic correlations in $YBa_2Cu_3O_{6-x}$ well below the superconducting transition temperature. The measurements shown here were performed on a sample with an oxygen content of x = 0.6. This sample showed a broad resistive transition at 44 K. In Fig. 9 a μSR spectrum obtained at T = 0.9 K in zero external field is shown. It can be seen that the muon decay asymmetry relaxes rapidly at this temperature. This fast decay definitely requires magnetic fields of electronic origin. The temperature dependence of the depolarization rate is displayed in Fig. 10 and shows a steep falloff at temperatures above 2 K. Also shown in Fig. 10 is a similar behavior for a semiconducting sample with x = 0.5. These preliminary results [31] are very similar to those found in $La_{2-x}Sr_xCuO_4$ [23,27] and suggest the presence of strong spin correlations persisting into the superconducting state.

Studies of Cu nuclear spin lattice relaxation [28-30, 32] below T_c suggest the presence of a strong spin fluctuation contribution to the temperature dependent rate which is observed. On the other hand NQR studies of ^{17}O relaxation rates [28-30] indicate a BCS like peak in the relaxation rate. A comprehensive understanding of these results is at present not established.

Our data do not in fact prove that magnetic correlations are responsible for the pairing mechanism but do provide evidence for the persistence of strong local magnetic correlations in these two classes of superconducting High-T_c copper oxides. Further studies to positively establish or to rule out direct coexistence remain as a challenge.

ACKNOWLEDGEMENT

The work of the Konstanz group has been supported by the Bundesminister für Forschung und Technologie. We would like to thank the staff of PSI for their strong technical support. The assistance of M. Rauer is greatefully acknowledged. The work at the University of Connecticut has been sponsored in part by the Department of Higher Education.

REFERENCES

[1] J. G. Bednorz and K. A. Müller, Z. Phys. B 64 (1986) 189; M. K. Wu, J. R. Ashburn, C. J. Torng, P. H. Hor, R. L. Meng, L. Gao, Z. J. Huang, Y. Q. Wang and C. W. Chu, Phys. Rev. Lett. 58 (1987) 908.

[2] R. Saez Puche, M. Norton and W. S. Glansinger, Mater. Res. Bull. 17 (1982) 1429, 1523.

[3] Y. Yamaguchi, A. Yamaguchi, M. Ohashi, H. Yamamoto, N. Shimoda, M. Kikuchi and Y. Syono, Japan. J. Appl. Phys. 26 (1987) L445.

[4] D. Vaknin, S. K. Sinha, D. E. Moncton, D. C. Johnston, J. M. Newsam, C. R. Safinya, and H. E. King Jr., Phys. Rev. Lett. 58 (1987) 840.

[5] J. I. Budnick, A. Golnik, Ch. Niedermayer, E. Recknagel, M. Rossmanith, A. Weidinger, B. Chamberland, M. Filipkowski and D. P. Yang Phys. Lett. 124 (1987) 106.

[6] Y. J. Uemura, W. J. Kossler, X. H. Yu, J. R. Kempton, H. E. Schone, D. Opie, C. E. Stronach, D. C. Johnston, M. S Alvarez and D. P. Goshorn, Phys. Rev. Lett. 59 (1987) 1045.

[7] N. Nishida, H. Miyatake, D. Shimada, S. Okuma, M. Ishakawa, T. Takabatake, Y. Nakazawa, Y. Kuno, R. Keitel, J. H. Brewer, T. M. Riseman, D. L. Williams, Y. Watanabe, T. Yamazaki, K. Nishiyama, K. Nagamine, E. Ansaldo and E. Torika, Japn. J. Appl. Phys. 26 (1987) L1856.

[8] J. H. Brewer, et al., Phys. Rev. Lett. 60 (1988) 1073.

[9] J. M. Tranquada, D. E. Cox, W. Kunnmann, H. Moudden, G. Shirane, M.Suenaga, P. Zolliker, D. Vaknin, S. K. Sinha, M. S. Alvarez, A. J. Jacobson and D. C. Johnston, Phys. Rev. Lett. 60 (1988) 156.

[10] P.W. Anderson, Science 235 (1987) 1196; P.W. Anderson, G. Baskaran, Z.Zou and T. Hsu, Phys. Rev. Lett. 58 (1987) 2790.

[11] V. I. Emery, Phys.Rev. Lett. 58 (1987) 2794.

[12] M. Cyrot, Solid State Commun. 62 (1987) 821.

[13] J. R. Schrieffer, X. -G. Wen and S. -C. Zhang, Phys. Rev. Lett. 60 (1988) 944.

[14] A. Aharony, R. J. Birgeneau, A. Coniglio, M. A. Kastner Z. Phys.B 71. (1987) 57.

[15] P. Gutsmiedl, G. Wolff and K. Andres, Phys. Rev. B 36 (1987) 4043.

[16] J. I. Budnick, B Chamberland, D. P. Yang, Ch. Niedermayer, A. Golnik,E. Recknagel, M. Rossmanith and A. Weidinger, Europhys. Lett. (1988) 651.

[17] D. R. Harshman, G. Aeppli, G. P. Espinosa, A. S. Cooper, J. P. Remeika, E. J. Ansaldo, T. M. Riseman, D. L. Williams, D. R. Noakes, B. Ellman and T. F. Rosenbaum , Phys. Rev. B 38 (1988) 852.

[18] R. B. van Dover, R. J. Cava, B. Batlogg and E. A. Rietman, Phys.Rev. B 35 (1987) 5337.

[19] B. Batlogg, A. P. Ramirez, R. J. Cava, R. B. van Dover and E. A. Rietman, Phys. Rev. B 35 (1987) 5340.

[20] Y. Kitaoka, K. Ishida, S. Hiromatsu and K. Asayama, J. Phys. Soc. Japn. 57 (1988) 734.

[21] I. Watanabe, K. Kumagai, Y. Nakamura, T. Kimura, Y. Nakamichi and H. Nakajima, J. Phys. Soc. Jpn. 56 (1987) 3028.
[22] R. Birgeneau, to be published Phys. Rev. B.
[23] A. Weidinger, Ch. Niedermayer, A. Golnik, R. Simon, E. Recknagel, J. I. Budnick, B. Chamberland and C. Baines, Phys. Rev. Lett. 62 (1989) 102.
[24] A. Golnik, Ch. Niedermayer, E. Recknagel, M. Rossmanith, A. Weidinger, J. I. Budnick, B. Chamberland, M. Filipkowski, Y. Zhang, D. P. Yang, L. L. Lynds, F. A. Otter and C. Baines, Phys. Lett. A 125 (1987) 71.
[25] Dawson, K. Dibbs, S. P Weatherby, C. Boekema, K-C. B. Chan, J. Appl. Phys. (1988).
[26] P. Barlet et al. Phys. C 153-155 (1988) 1115
[27] Ch. Niedermayer, M. Rossmanith, A. Golnik, E. Recknagel , A. Weidinger, J. I. Budnick, B. Chamberland and C. Baines, Hyperfine Int. (1988), in print.
[28] H. Yasuoka et al., Talk presented at NMM 88 conference in Munich, to be published Hyp.Int.
[29] K. Asayama et al., Talk presented at Int. Conference on Magnetism, Paris 1988, to be published J. de Physique; and Y. Kitaoka et al., Talk presented at Electronic Processes in High T_c Superconductivity, Rome, Oct. 1988.
[30] K. Ishida, Y. Kitaoka, K. Asayama, H. Katayama-Yoshida, Y. Okabe and T. Takahashi, J. Phy. Soc. Japn. 57 (1988) 2897
[31] A. Weidinger, R. Schöllhorn, W. Paulus, H. Eickenbusch, et al., to be published.
[32] W. W. Warren Jr., R. E. Waldstedt, R. F. Bell,G. F. Brennert, R. J. Cava, G. P. Espinosa and J.P.Remeika, Physica C 153-155 (1988) 79.

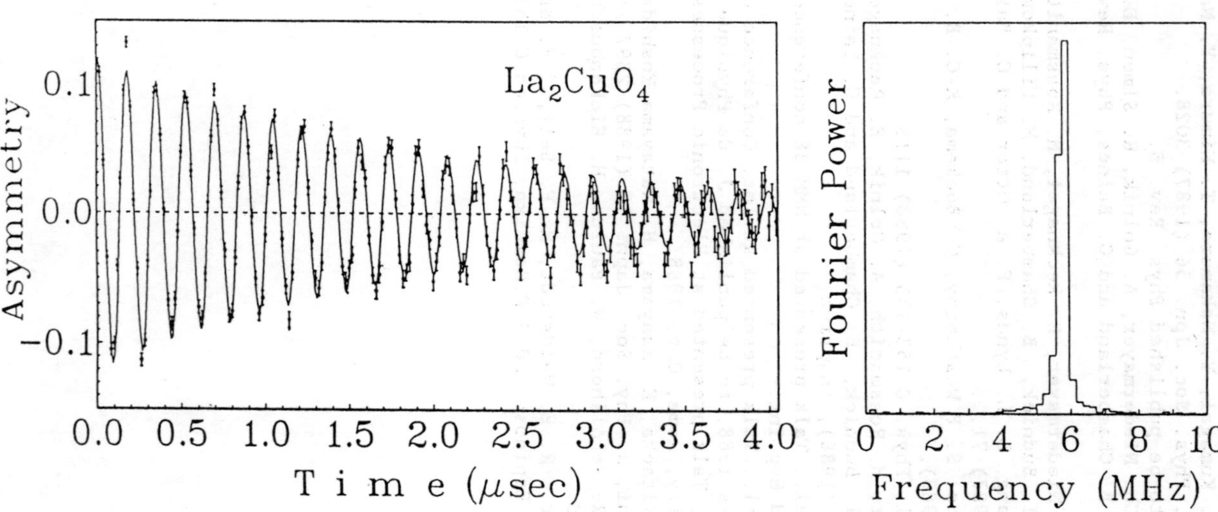

Fig. 1: μSR spectrum and its Fourier-transform in La_2CuO_4 at 4.2 K. A single sharp frequency of approximately 6 MHz is observed.

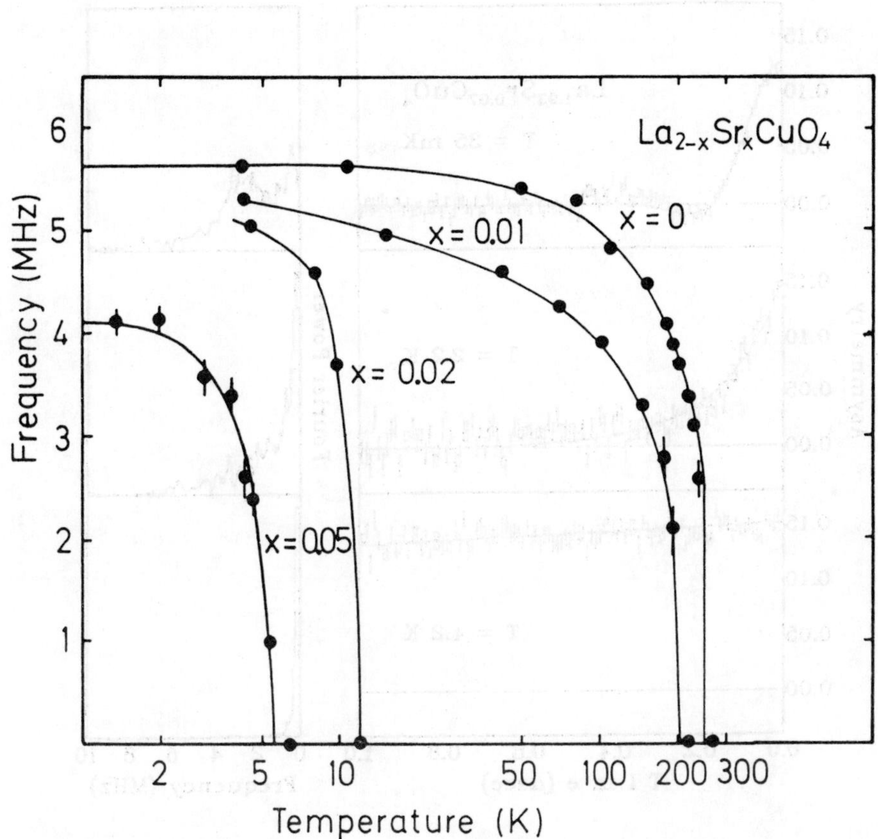

Fig. 2: Temperature dependence of the μSR frequencies observed in the undoped and slightly Sr doped samples. Lines are a guide to the eye.

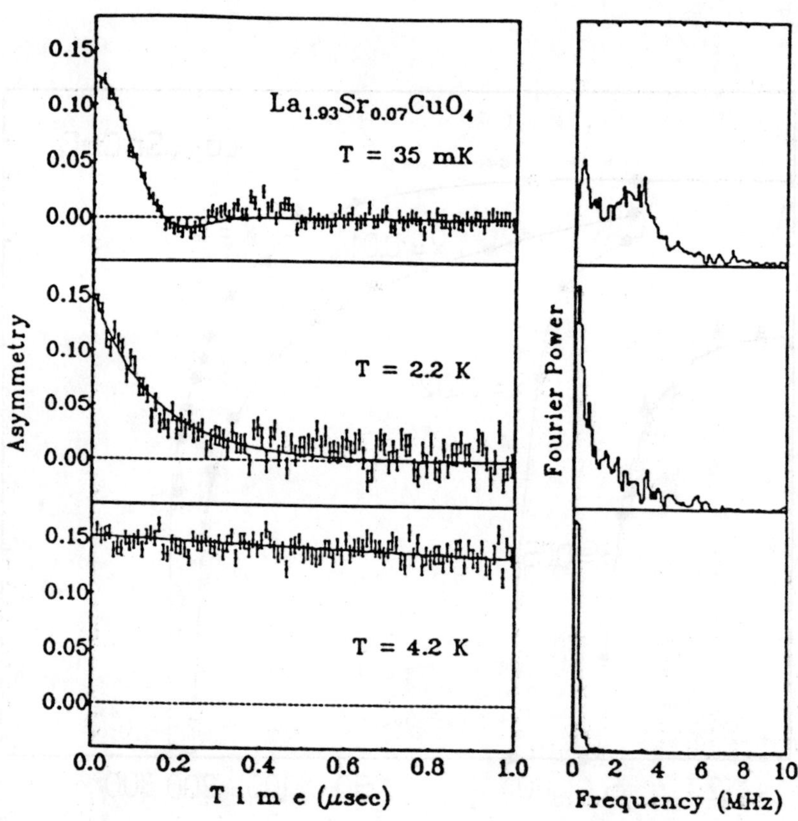

Fig. 3: μSR spectra and Fourier-transforms for $La_{1.93}Sr_{0.07}CuO_4$ at different temperatures.The solid lines are theoretical fit curves.

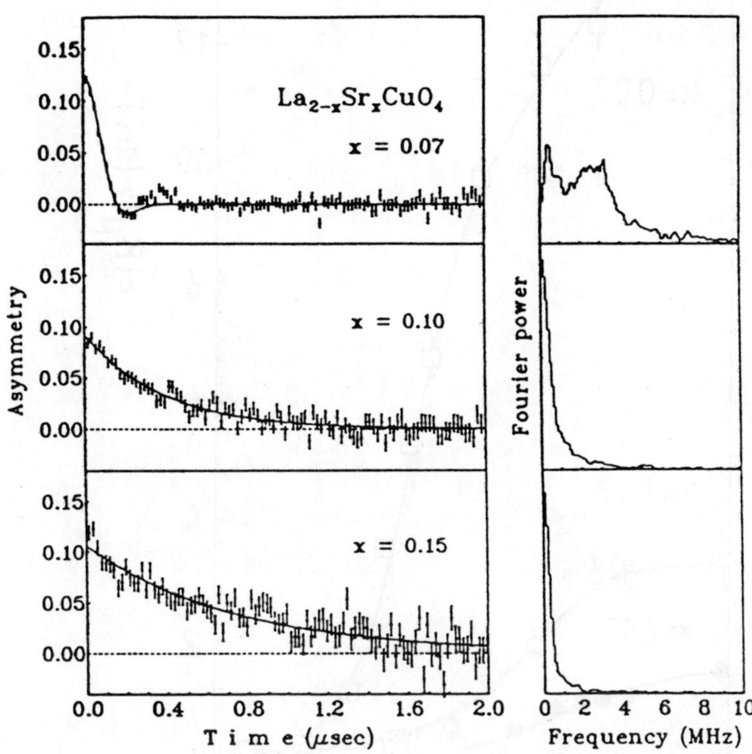

Fig. 4: μSR spectra and Fourier-transforms for $La_{2-x}Sr_xCuO_4$ with different Sr-concentrations x. The temperature of the samples was 35mK.

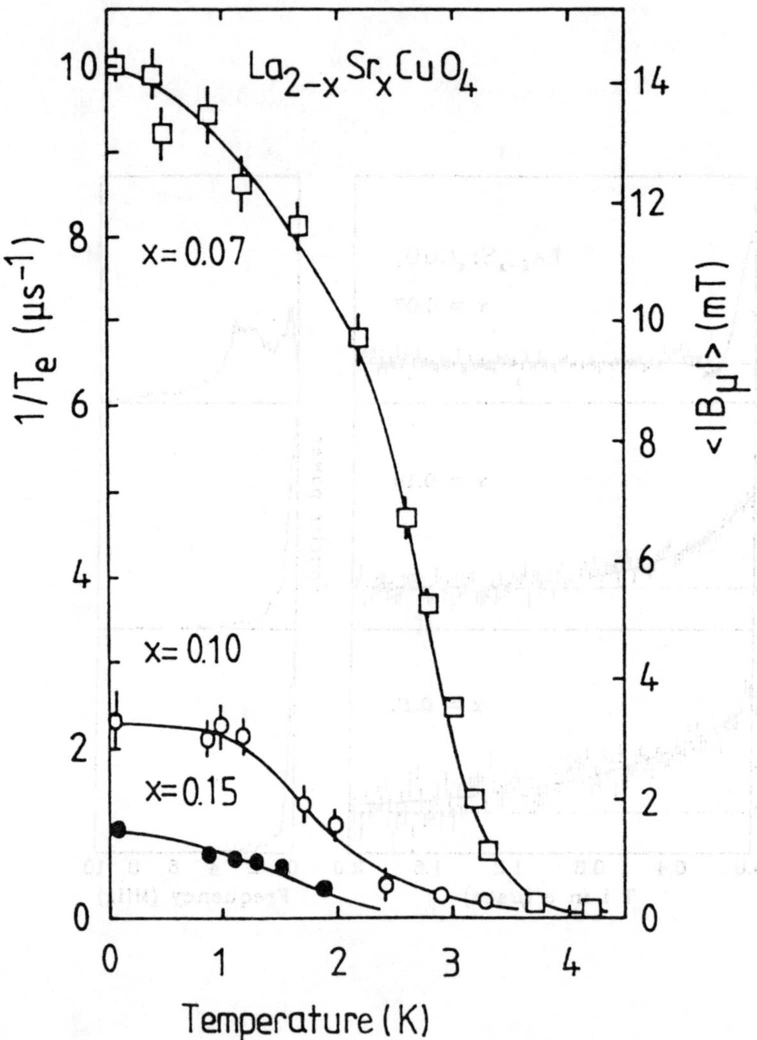

Fig. 5: The decay rate $1/T_e$ of the muon spin polarization as a function of temperature for $La_{2-x}Sr_xCuO_4$ with different Sr-concentrations. The scale on the right hand side refers to the average magnitude of the internal magnetic field.

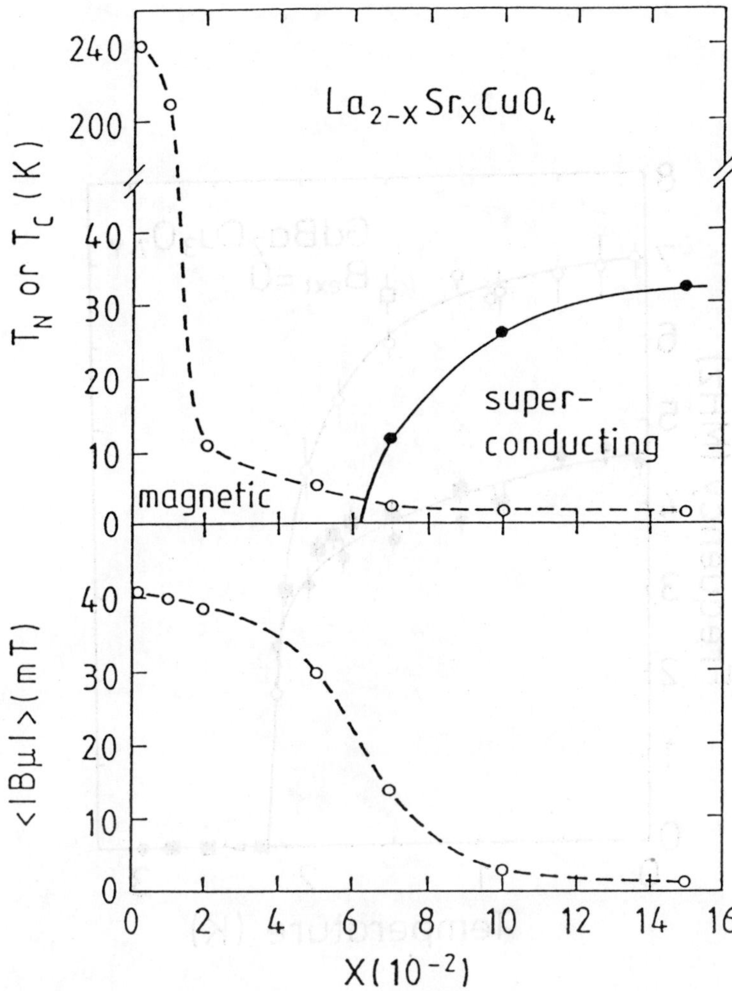

Fig. 6: Phase diagram of $La_{2-x}Sr_xCuO_4$ describing the magnetic and superconducting properties as a function of the Sr-concentration x. In the upper part, the magnetic ordering temperature T_N and the superconducting transition temperature T_C are displayed. In the lower part the average magnitude of the internal magnetic field at the muon site $< |Bm| >$ at 35mK is shown. The lines are a guide to the eye.

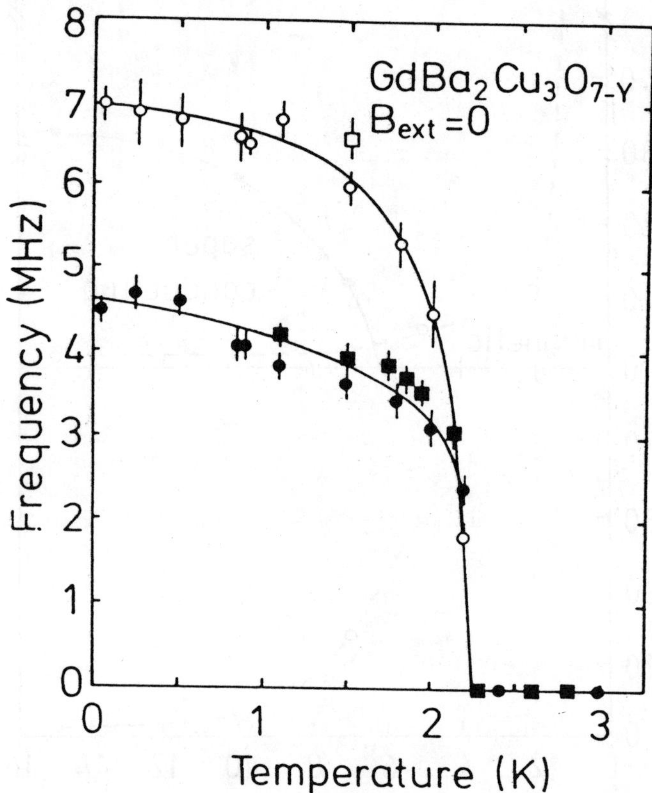

Fig. 7: Temperature dependence of the two μSR frequencies observed in $GdBa_2Cu_3O_{7-y}$. The circles are data from the 60K sample and squares from the 90K sample.

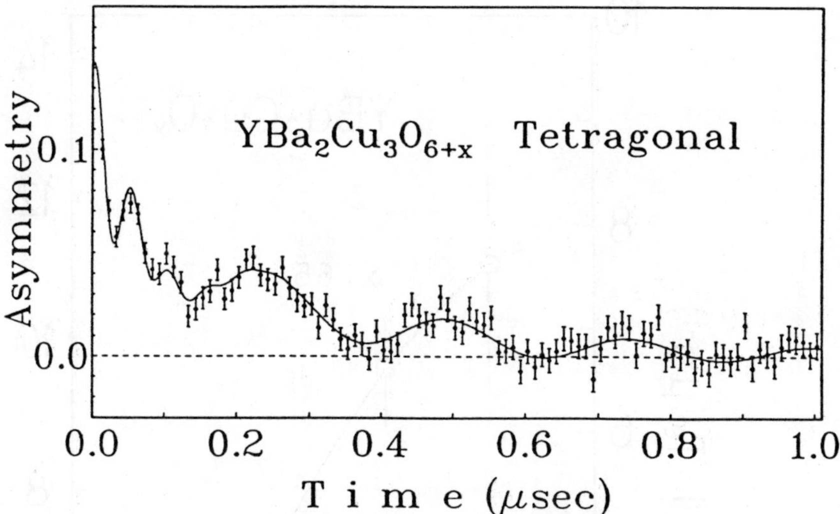

Fig. 8: μSR spectrum for tetragonal $YBa_2Cu_3O_{6+x}$. Besides a 4 MHz frequency also a higher frequency component of 18 MHz is visible.

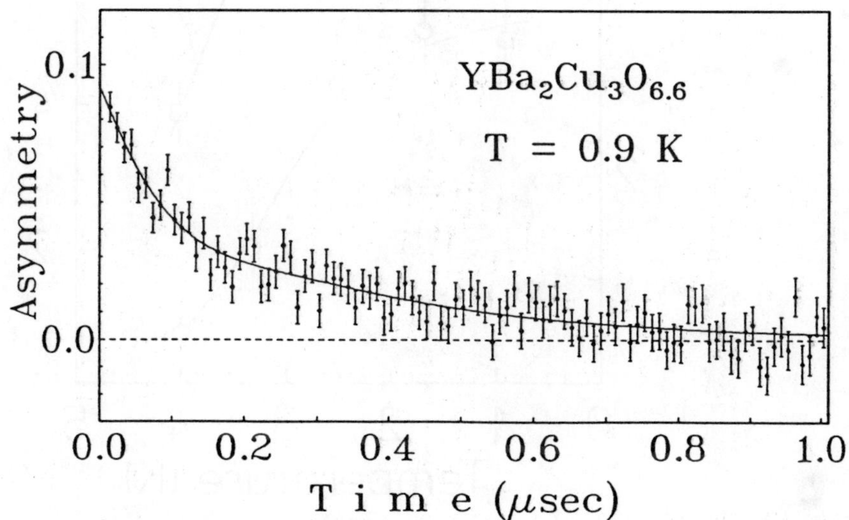

Fig. 9: μSR spectrum of $YBa_2Cu_3O_{6.6}$ at 0.9 K. The fast decay of the asymmetry definitely requires magnetic moments of electronic origin.

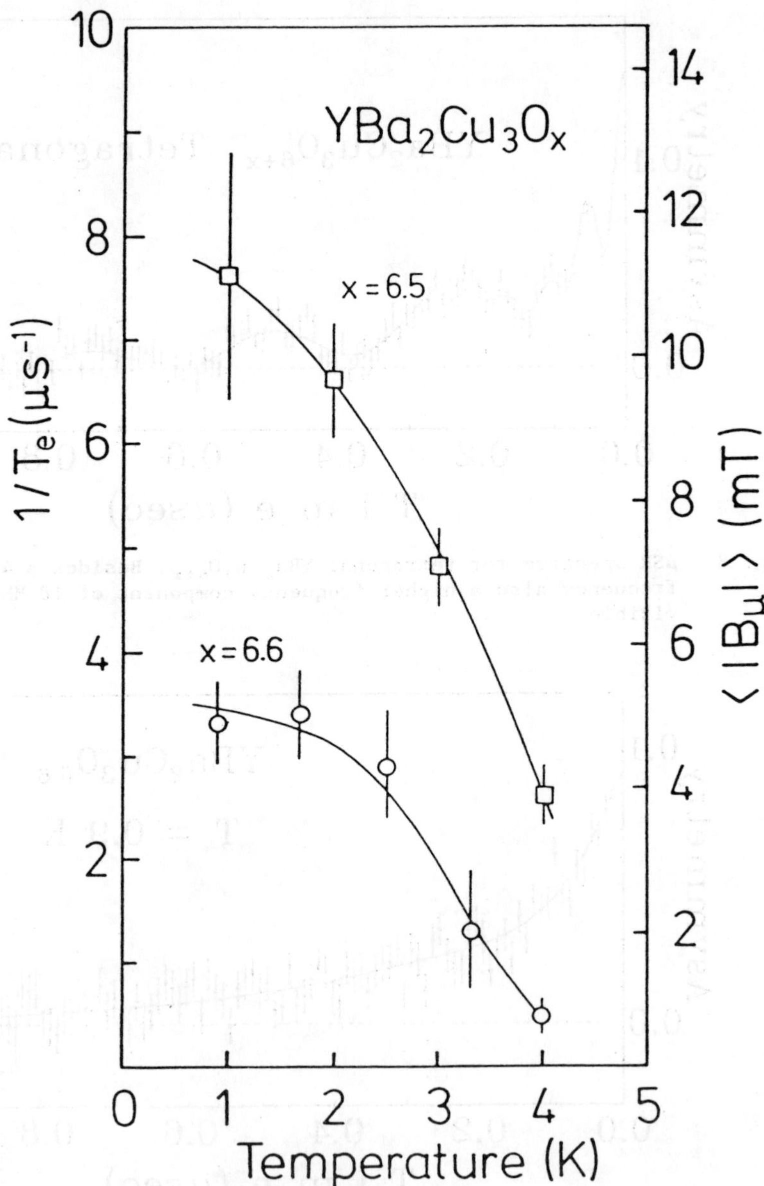

Fig. 10: The decay rate $1/T_e$ of the muon spin polarization as a function of temperature for $YBa_2Cu_3O_{6+x}$ where x = 0.5 and 0.6.

MAGNETICALLY MODULATED MICROWAVE ABSORPTION METHOD FOR STUDYING SUPERCONDUCTORS: DISTINCTION BETWEEN INTRINSIC AND EXTRINSIC RESPONSES.

K. Moorjani, B. F. Kim, F. J. Adrian and J. Bohandy
The Milton S. Eisenhower Research Center
The Johns Hopkins University Applied Physics Laboratory
Laurel, Maryland 20707

ABSTRACT

The already considerable utility of microwave techniques, which detect superconducting transitions by observing the corresponding change in microwave resistive loss, has been further enhanced by development of the novel Magnetically Modulated Microwave Absorption (MAMMA) technique, which observes only magnetic-field-dependent changes in this microwave loss. This method shares with other microwave methods their ability to detect and characterize superconducting transitions in inhomogeneous samples that lack the continuous conduction path required for dc resistance measurements, but it is spared the ambiguities of responses from metal-insulator transitions, resistance-changing phase transitions, etc., which, unlike superconducting transitions, are not magnetic field dependent. The MAMMA method is easily applied using a conventional ESR spectrometer which applies a small ac magnetic field to the sample and phase detects the microwave power reflected from the cavity at the ac modulation frequency. The method's only perceived complication is that, in addition to the response due to intrinsic sample resistivity, there can be extrinsic, magnetic-field-dependent, microwave responses due to Josephson junctions, flux-trapping, etc. at grain boundaries. Here we show that these two types of responses are readily separated and the intrinsic MAMMA signal is highly useful for detection and characterizing superconducting transitions.

1. INTRODUCTION

The microwave methods for studying superconducting transitions at high frequencies are well established[1]. Though superconducting transitions as seen in the dc resistance measurements become less abrupt with increasing frequency, the method has a number of advantages since it probes the entire sample, at least within the penetration depth, and provides a tool for investigating superconductivity even in discontinuous sample which cannot be studied by conventional resistance measurements. In this paper, we report a magnetically modulated microwave absorption (MAMMA) technique[2-4], which besides having the desirable features mentioned above, possesses a number of other advantages. Principal amongst them are: (1) high sensitivity achieved by field-modulation that allows narrow band amplification and phase detection at the ac modulation frequency, (2) the superconducting transition is recorded as a peak allowing well defined critical temperature measurement and permitting easy resolution of multiple superconducting phases, and (3) the method only records those changes in microwave resistance which are magnetic field dependent so that superconducting transitions are easily differentiated from non-field dependent transitions as, for example, insulator-metal transitions. In addition to detecting superconducting transitions, the method also is readily applied to investigating such properties as: (1) the magnetic field dependence of the superconducting transition which yields estimates of the critical field(s), (2) various anisotropies in the oxide superconductors such as the critical field along different crystallographic axes and (3) effects of Josephson junctions in these granular high temperature superconductors.

This last application of the method reflects a complication, which although essentially harmless, has caused considerable confusion. Most MAMMA applications utilize the response due to magnetic-field dependent change in the intrinsic resistance of the sample during the superconducting transition, which response is an absorption-like peak at T_c, but often this intrinsic MAMMA signal has superimposed on it other microwave responses due to extrinsic magnetic-field-dependent effects associated with grain boundaries. After a review of the MAMMA technique and some applications, the paper will focus on the distinction between the intrinsic and extrinsic MAMMA responses.

2. EXPERIMENTAL DETAILS

The measurement of microwave absorption generally involves monitoring the Q-factor of a microwave resonator. The power reflected from a waveguide cavity in a bridge is recorded and its variations provide a measure of corresponding change in the Q of the cavity due to absorption of cavity field energy by the sample. The reflected power as a function of the sample temperature behaves similarly to the dc resistance indicating the transition to the superconducting state.

The MAMMA technique described here is a refinement of this general method of microwave absorption and is closely related to that of electron spin resonance (ESR). Utilizing a conventional homodyne-type ESR spectrometer operating at X band (ca 9.1 GHz), the method is based on observing that part of the change in microwave loss which is produced by application of a small ac magnetic field ($\leq$ 5 Oe). This is readily accomplished by the ESR spectrometer's standard operating mode which phase detects the microwave power reflected from the cavity at the ac field modulation frequency (10 kHz). The ac field is superimposed on a dc magnetic field, which can be varied over a wide range and thus permits observations of dc magnetic field effects, but is always large enough so that the combined ac + dc field never changes sign since this would complicate the phase detection process. The response vs. temperature (Fig. 1) is proportional to the product of the temperature-derivative of the resistance ($\partial\rho/\partial T$) and the field-derivative of the critical temperature ($\partial T_c/\partial H$). The latter factor assures that only the field-dependent transitions, (which superconducting transitions are) are observed while the factor $\partial\rho/\partial T$ results in a peak signal.

The MAMMA response of Nb, a conventional superconductor, shown in Fig. 2, has only one feature: the expected peak at T_c = 9.3K. Also shown in Fig. 2 is the shift in the position of the peak as a function of the strength of the applied dc magnetic field which corresponds to the decrease in T_c with the increasing field, as expected for a superconductor. Fig. 3 shows the conventional microwave absorption for V_2O_3 which shows an abrupt fall at ~ 155K due to the well-known insulator-metal transition in this material and a lack of MAMMA response since the transition is not magnetic-field-dependent. Thus the method provides an exquisite tool for differentiating between superconducting and non-superconducting transitions.

3. APPLICATIONS OF THE MAMMA TECHNIQUE

A typical response from the ceramic oxide superconductor $YBa_2Cu_3O_7$ in a moderate field of 1 KG is shown for a bulk and thin film sample in Fig. 4a and 4b, respectively. The transition in the bulk sample is rather broad indicating a range in composition and/or structure of the sample with corresponding variations in T_c which cannot be revealed by dc resistivity measurements. The thin film sample deposited by the laser-assisted-processing (LAP) method[2,5] on an unheated fused silica substrate without post-deposition annealing, shows three sharp superconducting transitions indicative of three individually homogeneous, but slightly different superconducting regions in the sample. This film lacked a continuous conduction path and, thus, dc resistance measurements are useless for demonstrating the presence of superconducting regions in it.

Another example of the sensitivity of the MAMMA technique to structural/composition variations is vividly seen in the response of a sample of nominal composition $BiSrCaCu_2O_x$ (Fig. 5). Two prominent peaks indicating superconductive transitions at 72 K and 110 K, and a broad weak transition centered at ~ 100 K, in major and two minor phases of the sample respectively are observed.[6] The integrated intensities of these MAMMA peaks indicate the minor 110 K and 100 K phases constitute ~ 5% of the sample, assuming the remainder of the sample to be in the major 72 K phase. The magnetic field dependence of the transitions in this sample determined from the dependence of the MAMMA response on applied field is considerably stronger (ca. $dT_c/dH = -27$ K/Tesla for the 72 K transition) than found for $YBa_2Cu_3O_7$ (ca. -1.5 K/Tesla).[7] The adverse implications of this result for maximum critical current in the Bi-Sr-Ca-Cu-O superconductors, as compared with $YBa_2Cu_3O_7$, is consistent with another very recent report.[8]

The height of the MAMMA peak at T_c has been found to be proportional to the amount of superconducting material in an inhomogeneous sample. This has been used in several applications, one of which was to show that massive doses of γ-irradiation had no effect on a powder sample of $YBa_2Cu_3O_7$, which result is significant for possible utilization of these superconductors in space-based electronics.[9]

The fact that the MAMMA peak(s) represents the contributions from different regions of an inhomogeneous sample with a range of T_c's has many potential applications in examining the quality of superconducting materials and films prepared from them. One striking example is of LAP prepared films of Bi-Sr-Ca-Cu-O on (100) cubic zirconia.[10] Two films identically prepared except that in one case the deposition substrate was unheated during deposition, while being heated to 300°C in the other, gave identical dc resistance vs. temperature curves, but the MAMMA response showed the latter, heated-substrate film was clearly superior.

4. INTRINSIC VS. EXTRINSIC MAMMA RESPONSE

As noted earlier, application of a modulating magnetic field to a ceramic superconductor can have effects other than changing the intrinsic resistance of the sample, and some of these effects can contribute to the microwave response. The most commonly observed effect is shown in Fig. 6a for for a bulk sample of $YBa_2Cu_3O_7$ in a low (30 G) field.[11] Here, this effect exhibits itself as a strong rise in the overall MAMMA response at T_c reaching a constant plateau at roughly 60 K, which rise almost obscures the MAMMA peak at T_c. In other cases, this broad signal decreases at lower temperature and appears as a very broad peak in the MAMMA response vs. T. Another effect which almost invariably accompanies the broad peak is that the observed response becomes increasingly noisy below T_c. Here, this "noise" reaches a limit at about 60 K and remains constant down to 10 K. The first of these effects is believed to be the broad low-field microwave absorption observed by many workers[12-15] as indicated by the fact that it disappears upon application of a moderate (ca 1 KG) field as shown in Fig. 4a. This broad absorption and the "noise" observed in these granular superconductors have alternately been interpreted as due to phase slippage in a system of random superconducting regions weakly coupled by Josephson junctions[12,13] or thermally activated flux creep in these type II superconductors.[16]

We believe the Josephson junction model is more reasonable considering the dependence of this low-field absorption and microwave "noise" on particle size and temperature, as shown in Figs. 6b and 6c, Here, both effects largely disappear for particles below 10 μm in size, at which point there should be relatively few intergrain boundaries capable of forming a Josephson junction.[11] Furthermore, the temperature independence of these effects argues strongly against the thermally activated flux creep model.[11]

Irrespective of their exact explanation, however, both the broad peak or plateau and the "noise" differ from the sharp MAMMA peak observed at T_c in that the former are extrinsic effects associated with magnetic-field-dependent phenomena at grain boundaries in the granular polycrystalline samples and perhaps also twin boundaries in the individual crystallites, whereas the latter is an intrinsic effect associated with the temperature- and magnetic-field-dependent internal resistance of the sample. Further evidence for this distinction, which it is very important to make in applying the MAMMA method, is that: (1) only a single very sharp peak arising out of a flat, constant-noise baseline is observed in the conventional superconductor Nb, cf. Fig. 2, even in low field, which simple metal is unlikely to have the complex grain boundaries of the oxide superconductor, (2) as shown in Fig. 6, the height of the intrinsic MAMMA peak at T_c is independent of particle size, and (3) these effects vary considerably among the different ceramic oxide superconductors. For example, the rise in the baseline and the "noise" below T_c is much smaller relative to the intrinsic MAMMA peak in a bulk sample of Bi-Sr-Ca-Cu-O (cf. Fig. 5) than in $YBa_2Cu_3O_7$.

In summary, the MAMMA method is a sensitive method for detecting and characterizing superconducting transitions. In applying it, however, it is important to distinguish between the intrinsic MAMMA response and the extrinsic responses. Although extrinsic effects are useful for characterizing the granularity of the ceramic superconductors and can even indicate the occurrence of a superconducting transition, the intrinsic MAMMA response is better suited to the latter purpose because it is a sharp response observed only in the superconducting region and is largely independent of such parameters as external field, particle size, granularity, etc. Fortunately, the different MAMMA responses are readily distinguished because the broad peak or plateau, which can largely obscure the intrinsic MAMMA peak as shown in Fig. 6a, can be greatly reduced by application of moderate magnetic field (ca 1 KG) which has only a minimal effect on the position of the intrinsic MAMMA peak.

Work supported by the Department of the Navy, Space and Naval Warfare Systems Command under Contract No. N00039-87-C-5301.

REFERENCES

1. Pippard, A. B., Proc. Ray. Soc. A 191, 370 (1947); *ibid* A 203, 98 (1950); Maxwell, E, Marcus, P. M. and Slater, J. C., Phys. Rev. 76, 1332 (1949); Gittleman, J. I. and Bozowski, S., Phys. Rev. 161, 398 (1967).
2. Moorjani, K., Bohandy, J., Adrian, F. J., Kim, B. F., Shull, R. D., Chiang, C. K., Swartzendruber, L. J., and Bennett, L. H., Phys. Rev. B 36, 4036 (1987).
3. Kim, B. F., Bohandy, J. Moorjani, K. and Adrian, F. J., J. Appl. Phys. 63, 2029 (1988).
4. Moorjani, K., Kim, B. F., Bohandy, J., and Adrian, F. J., Reviews of Solid State Science, 2, 263 (1988).
5. Kim, B. F., Bohandy, J., Moorjani, K. and Adrian, F. J., AIP Conf. Proc., 165, 182 (1988).
6. Bohandy, J., Adrian, F. J., Kim, B. F., Moorjani, K., Shull, R. D., Swartzendruber, L. J., Bennett, L. H. and Wallace, J. S., J. Superconductivity 1, 191 (1988).
7. Adrian, F. J., Bohandy, J., Kim, B. F., Moorjani, K., Wallace, J. S., Shull, R. D., Swartzendruber, L. J., and Bennett, L. H., Physica C 156, 184 (1988).
8. Caplin, D., Nature 335, 204 (1988).
9. Bohandy, J. Suter, J., Kim, B. F., Moorjani, K. and Adrian, F. J., Appl. Phys. Lett. 51, 2161 (1987).

10. Kim, B. F., Bohandy, J., Phillips, T. E., Green, W. J., Agostinelli, E., Adrian, F. J., Moorjani, K., Swartzendruber, L. J., Shull, R. D., Bennett, L. H. and Wallace, J. S., Appl. Phys. Lett. 53, 321 (1988).

11. Bohandy, J. Kim, B. F., Adrian, F. J. and Moorjani, K., Phys. Rev. B, (in press).

12. Blazey, K. W., Müller, K. A., Bednorz, J. G., Berlinger, W., Amoretti, G., Bulluggiu, E., Vera, A. and Mattacotta, F. C., Phys. Rev. B 36, 7241 (1987).

13. Khachaturyan, K., Weber, E. R., Tejedor, P., Stacy, A. and Portis, A. W., Phys. Rev. B 36, 8309 (1987).

14. Stankowski, J., Kahol, P. K., Dalal, N. S. and Moodera, J. S., Phys. Rev. B 36, 7126 (1987).

15. Bhat, S. V., Ganguly, P., Ramakrishina, T. V., and Rao, C. N. R., J. Phys. C. Solid State Phys., 20, L559 (1987).

16. Yeshurun, Y. and Malozemoff, A. P., Phys. Rev. Lett. 60, 2202 (1988).

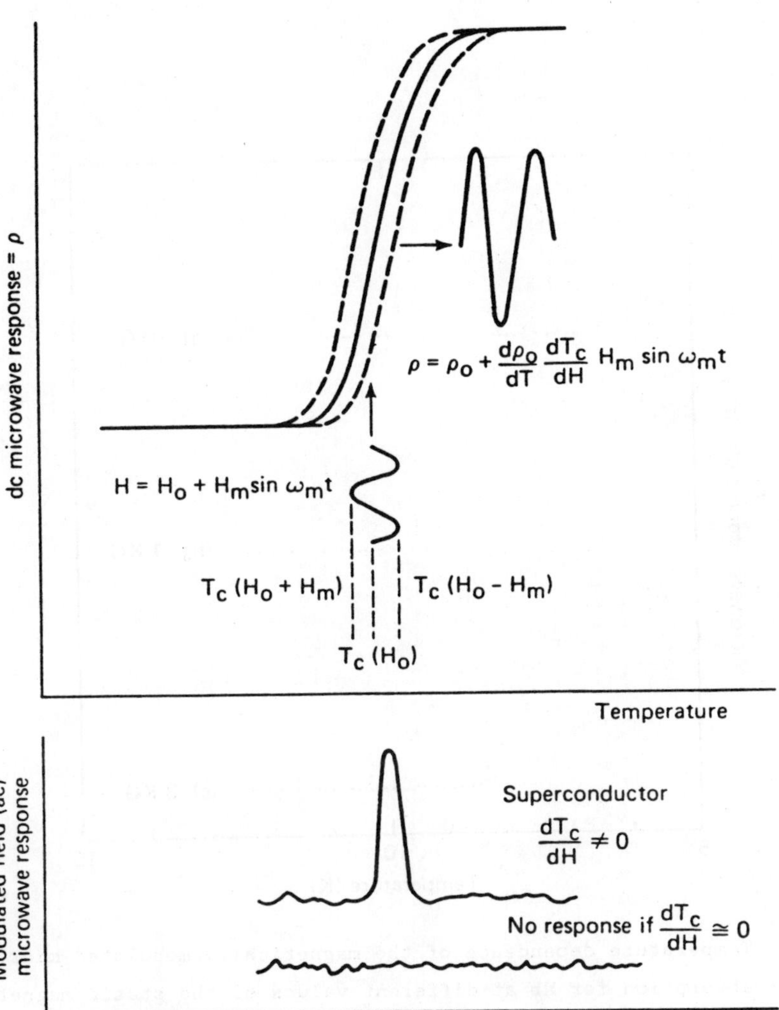

Fig. 1 Schematic description of the magnetically-modulated microwave absorption for a superconductor ($dT_c/dH \neq 0$) and a field-independent ($dT_c/dH = 0$) transition.

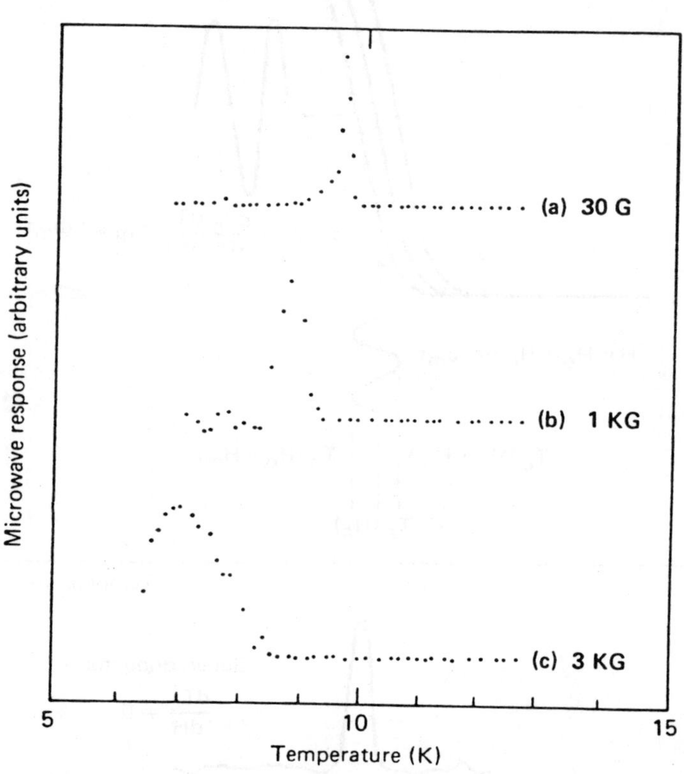

Fig. 2 Temperature dependence of the magnetically-modulated microwave absorption for Nb at different values of the static magnetic field.

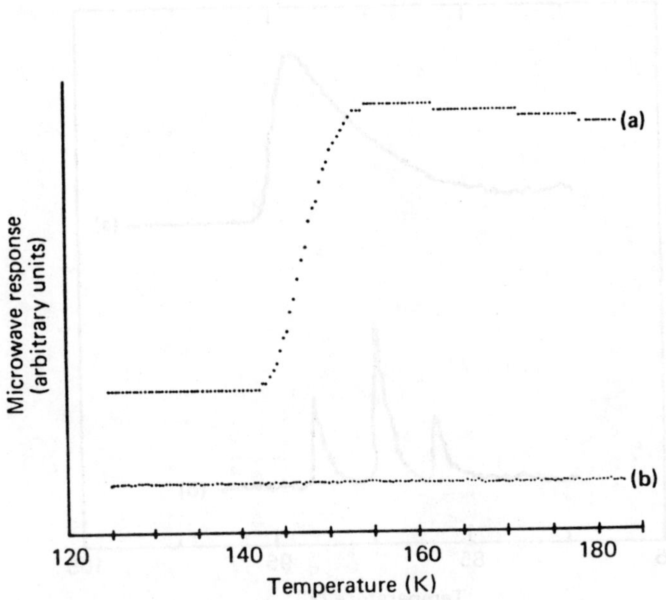

Fig. 3 Microwave response for V_2O_3 versus temperature. (a) Conventional microwave absorption and (b) magnetically-modulated microwave absorption.

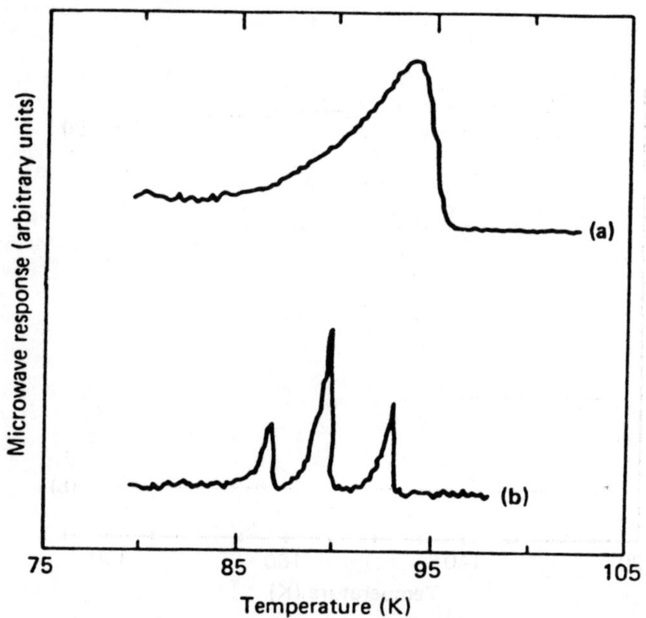

Fig. 4 Magnetically-modulated microwave absorption vs. temperature for $YBa_2Cu_3O_{7-y}$ in a 1 KG external magnetic field. (a) Bulk sample with one superconducting phase and (b) superconducting thin film sample deposited on unheated, fused silica substrate and without post-deposition annealing.

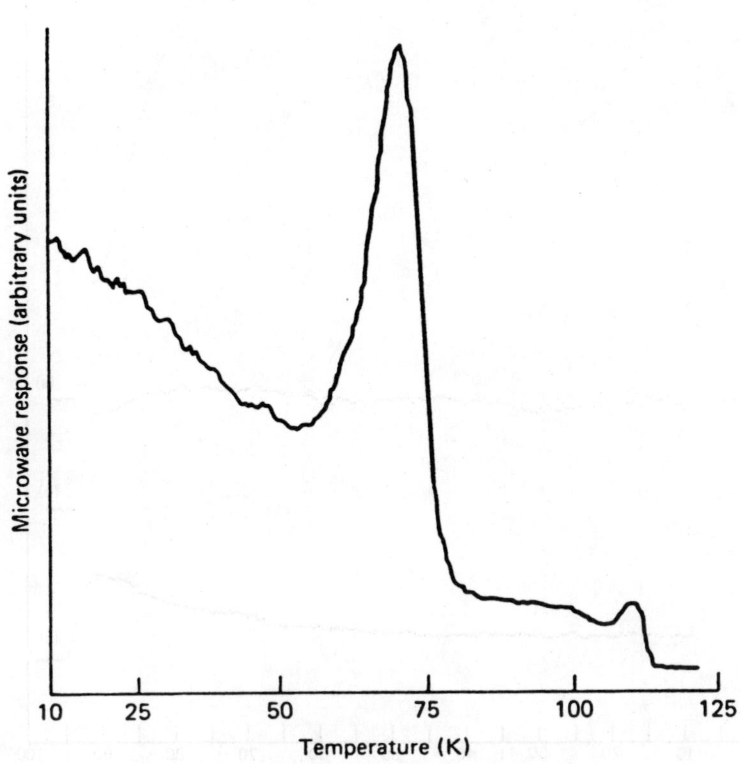

Fig. 5 Magnetically-modulated microwave absorption of $BiSrCaCu_2O_x$ as a function of temperature in a 30 G external magnetic field.

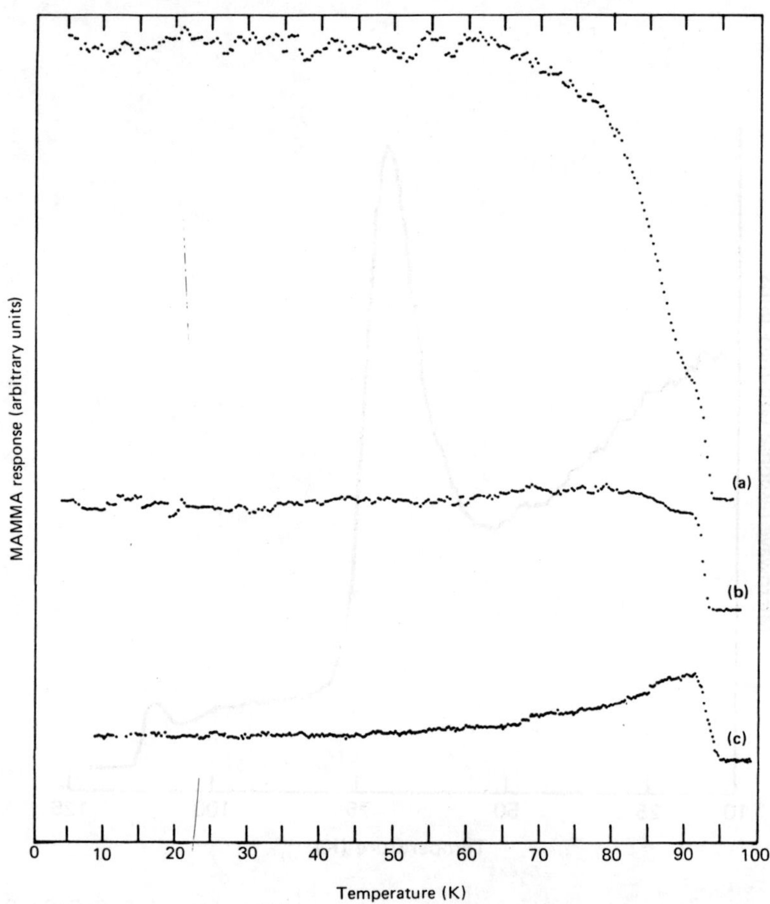

Fig. 6 MAMMA response vs. temperature for $YBa_2Cu_3O_7$ powders of various particle sizes in an external magnetic field of 30 G. The particle size, S, for each curve is: (a) S < 63 μm, (b) S < 38 μm, (c) S < 10 μm.

MAGNETICALLY MODULATED MICROWAVE REFLECTION (MMR) CHARACTERIZATION OF HIGH T_c SUPERCONDUCTORS

T. A. Mahl, J. P. DeLooze, P. K. Kahol, and N. S. Dalal,
The Department of Chemistry, West Virginia University,
Morgantown, WV 26506-6045

ABSTRACT

The Magnetically-modulated Microwave Reflection (MMR) technique, using a modified electron paramagnetic resonance (EPR) spectrometer, provides a quick, lead-less method for studying characteristics of high-T_c superconductors. The MMR signal is proportional to changes in the microwave reflection cavity's Q-factor; which, in turn, is altered by the change in the surface impedance of the sample upon entering its superconducting state. The MMR methodology appears to have promise for becoming a highly-sensitive and non-abrasive technique for determining the T_c and certain other superconductor characteristics. Applications of the technique are illustrated by measurements on variously prepared $YBa_2Cu_3O_{7-x}$ samples.

INTRODUCTION

Current research from our laboratory[1-4] and elsewhere [5-14] has led to the development of what we believe is a highly-sensitive, quick, and reliable procedure for detecting superconducting properties of the recently discovered, high T_c, ceramic superconductors. The underlying principle of the technique is that the surface impedance of these superconductors starts to decrease abruptly at the temperature of the onset of superconductivity ($\approx T_c$) and that their microwave impedance is a sensitive function of an externally applied magnetic field. It is also found that this high sensitivity of the microwave impedance is preserved when the externally applied magnetic field is modulated at audio frequencies up to 100 kHz. This makes it possible to take advantage of the high sensitivity inherent in the phase-sensitive detection and narrow band amplification techniques. Here we summarize some details of the methodology as developed in our laboratory, with illustrations of some typical results.

APPARATUS AND METHODOLOGY

A schematic diagram of the experimental arrangement is shown in figure 1. The main part of the apparatus is an electron paramagnetic resonance (EPR) spectrometer incorporating a high Q, reflection-type microwave cavity. We employ a Bruker ER-200D, X-Band (≈ 9.5 GHz) EPR spectrometer using a TE_{102} microwave cavity with Q ≈ 5,000. The Bruker ER-200D model employs 100kHz magnetic field modulation and lock-in detection. The only significant modification needed is a pair of Helmholtz coils and a precision constant-current source capable of generating a steady magnetic field of 0-7.0 Tesla. This arrangement enabled us to nullify the 2.0-3.0 Tesla remnant field of the electromagnet of the EPR spectrometer, and to obtain a magnetic field scan of -4.0 to +4.0 Tesla.

The MMR measurements are made by placing a few milligrams of the material (powder or pellet form) in a 3-4 mm diameter quartz tube. This tube is positioned at the center of the cavity such that the sample is subjected to the maximum of the microwave magnetic field in the cavity. The sample temperature is varied by using liquid helium and is controlled by a digital temperature controller, Oxford Instruments model DRC1 (precision of the order of 0.1K). The microwave output which corresponds to the MMR signalis monitored as the power reflected from the cavity. As the sample becomes superconducting, its microwave impedance decreases, thereby resulting in a concomitant increase in the amount of the microwave power reflected from the cavity . Thus, for a given sample, a plot of the MMR signal versus temperature would yield direct information on the evolution of superconductivity in the sample.

RESULTS AND DISCUSSION

A typical MMR signal from a $YBa_2Cu_3O_{7-x}$ sample in the vicinity of T_c (≈ 95.5 K) is shown in Figure2. It wasobserved that the MMR signal intensity, measured as I_{p-p} in figure 2, increases rapidly below T_c and can thus be used as a probe of the phase transition (≈ T_c). Figure 3 shows a plot of the amplitude, I_{p-p}, as a function of temperature. The temperature at which the MMR signal starts to increase rapidly (≈ 95.5 K) corresponds to the T_c of this material. The sudden change in the MMR's signal due to temperature change at fixed magnetic fields greater than ≈ 50G , as seen in figure 4, can be used as a probe for measuring T_c's precisely, as has aslo been shown independently by Moorjani et. al. [6]. The signal in figure 4 was obtained from two samples having T_c's differing by approximately 1 K being measured at the same time. The two steps observed at 90.7 K and 91.5 K correspond to the T_c's of the two preparations. This demonstrates the surprisingly high resolution of the methodology for T_c determination. The results clearly demonstrate that the method provides a quick, sensitive and non-abrasive procedure for determining the T_c without any need for attaching contacts or metal deposition. This method should be valuable for examining effects of process variables on the change in the superconducting behavior of a given sample, since the same sample can be investigated by the MMR method at successive stages. Using this method, we have been able to measure the change in T_c due to the effects of an applied magnetic field. As can be seen from figure 5, the T_c of the YBaCuO compound decreases at a rate $dT_c / dH \approx 3.0$ K/T, 1T = 10,000G.

The MMR method can also be utilized for the estimation of the superconducting grain size via the change in the width of the MMR signal [1] of figure 2. The same procedure also leads to an estimate of the lower critical field, H_{c1}, for these samples as demonstrated elsewhere [1,8,10].

REFERENCES

[1] J. Stankowski, P.K. Kahol, N.S. Dalal and J.S. Moodera, "Possible Josephson oscillations spectra and electron paramagnetic resonance of Cu^{2+} in Y-Ba-Cu-O," *Phys. Rev.* B 36, 7126 (1987).

[2] J. Stankowski, J. Pichet, C.P. Poole Jr., T. Datta, P.K. Kahol, N.S. Dalal, J.S. Moodera, "EPR and Josephson absorption in YBaCuO high temperature superconductor", *Ferroelectrics*, 78, 231 (1988).

[3] N.S. Dalal, P.K. Kahol, "Electron paramagnetic resonance and microwave response in oxide superconductors", *Progress in High Temperature Superconductivity*, 7, 196-217 (1988).

[4] N.S. Dalal, T.A. Mahl, P.K. Kahol and J.P. DeLooze, "Characterization of ceramic superconductors by magnetically modulated microwave reflectio (MMR)", *Advanced Characterization techniques for Ceramics*, in print.

[5] S.V. Bhat, P Ganguly, T.V. Ramakrishna, and C.N.R. Rao, "Absorption of electromagnetic radiation by superconducting $YBa_2Cu_3O_7$: an oxygen-induced phenomenon," *J.Phys. C.* 20, L559 (1987).

[6] K. Moorjani *et al.*, "Superconductivity in bulk and thin films of $La_{1.85}Sr_{0.15}CuO_{4-\delta}$ and $Ba_2YCu_3O_{7-\delta}$," *Phys. Rev. B* 36, 4036 (1987).

[7] R. Durny *et al.*, "Microwave absorption in the superconducting and normal phases of Y-Ba-Cu-O," *Phys Rev. B* 36, 2361 (1987).

[8] K.W. Blazey *et al.*, "Low-field microwave absorption in the superconducting copper oxides," *Phys. Rev. B* 36, 7241 (1987).

[9] K. Khachaturyan, E.R. Weber, P. Tejedor, A.M. Stacey and A.M. Portis, "Microwave observation of magnetic field penetration of high-T_c superconducting oxides," *Phys. Rev. B* 36, 8309 (1987).

[10] C. Rettori, D. Davidov, I. Belaish, and I. Felner, "Magnetism and critical fields in the high-T_c superconductors $YBa_2Cu_3O_{7-x}S_x$ (x=0,1): An ESR study," *Phys. Rev. B* 36, 4028 (1987).

[11] M. Peric, B. Rakvin, M. Prester, N. Brnicevic, "Size of Josephson junctions in Ba-Y-Cu-O compounds," *Phys. Rev. B* 37, 522 (1988).

[12] A. Dulcic, B. Leontic, M. Peric, and B. Rakvin, "Microwave Study of Josephson Junctions in Gd-Ba-Cu-O Compounds," *Europhys. Letters* 4, 1403 (1987).

[13] M.D. Sastry, A.G.I. Dalvi, Y. Babu, R.M. Kadam, Y. Yakhmi and R.M. Iyer, "Possible role of Cu^{2+} - Cu^{4+} pairs in the superconductivity of $YBa_2Cu_3O_{7-x}$ from electron spin resonance observations *Nature* (London) 330, 49 (1987).

[14] S. Tyagi, M. Barsoum, K.V. Roa, V. Skumryev, Z. Yu and J.L. Costa, "Low-field AC susceptibility and microwave absorption in YBaCuO and BiCaSrCuO superconductors", *Physic* C 156, 73-78 (1988).

FIGURE CAPTIONS

Fig. 1 Schematic diagram of the modified EPR apparatus used for the MMR measurements.

Fig. 2 Typical MMR peak for a YBaCuO sample.

Fig. 3 Temperature dependence of the MMR signal amplitude for the YBaCuO sample used for the plot in Fig 2.

Fig. 4 MMR plots from a mixture of two YBaCuO samples with different T_c's at fixed magnetic field, demonstrating the high resolution of the method for measuring T_c's of mixtures.

Fig. 5 The dependence of the T_c on the applied magnetic field (dT_c/dH).

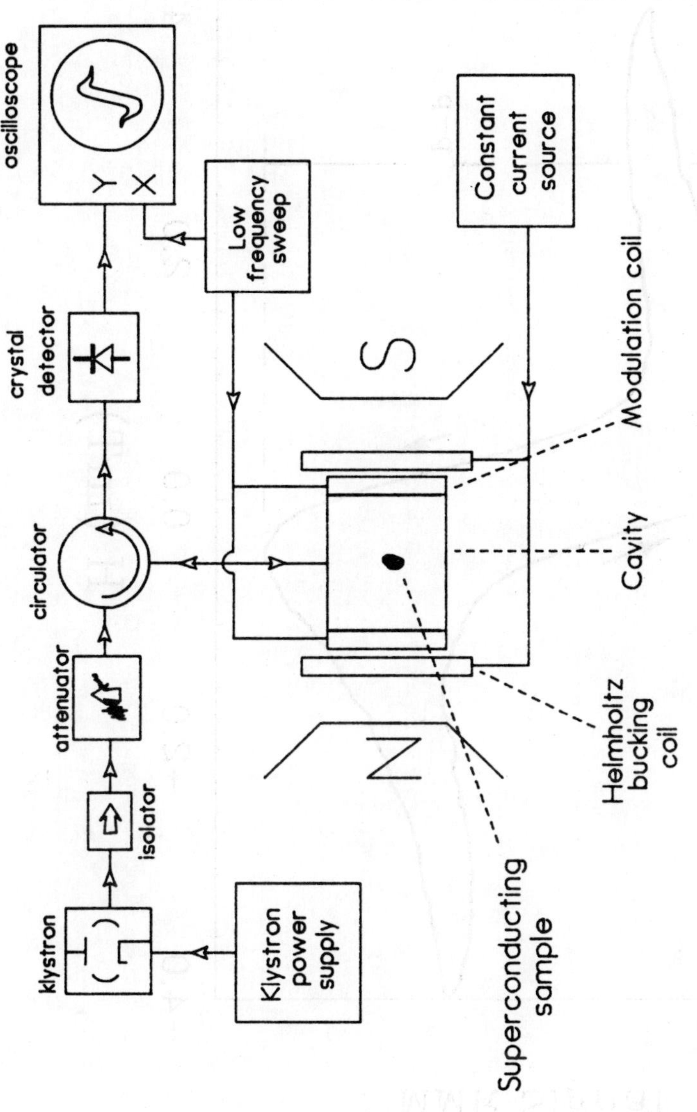
klystron
isolator
attenuator
circulator
crystal detector
oscilloscope
Y
X
Low frequency sweep
Klystron power supply
Constant current source
S
N
Modulation coil
Cavity
Helmholtz bucking coil
Superconducting sample

Fig. 1

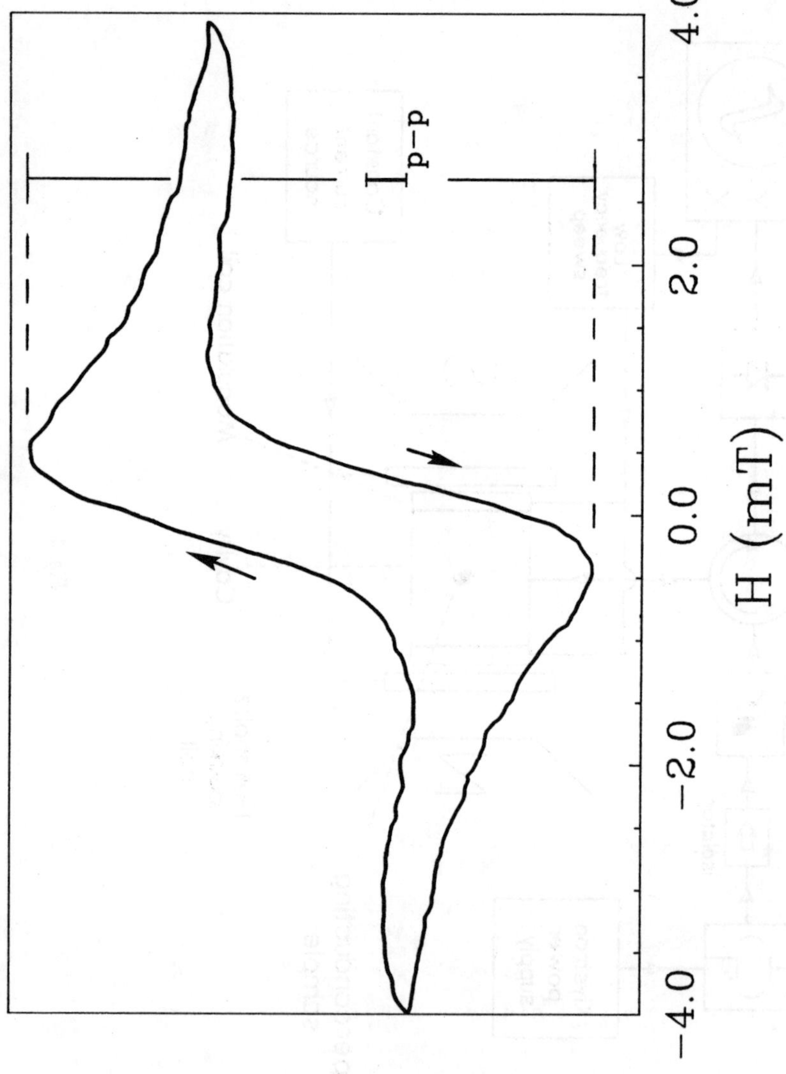

Fig. 2

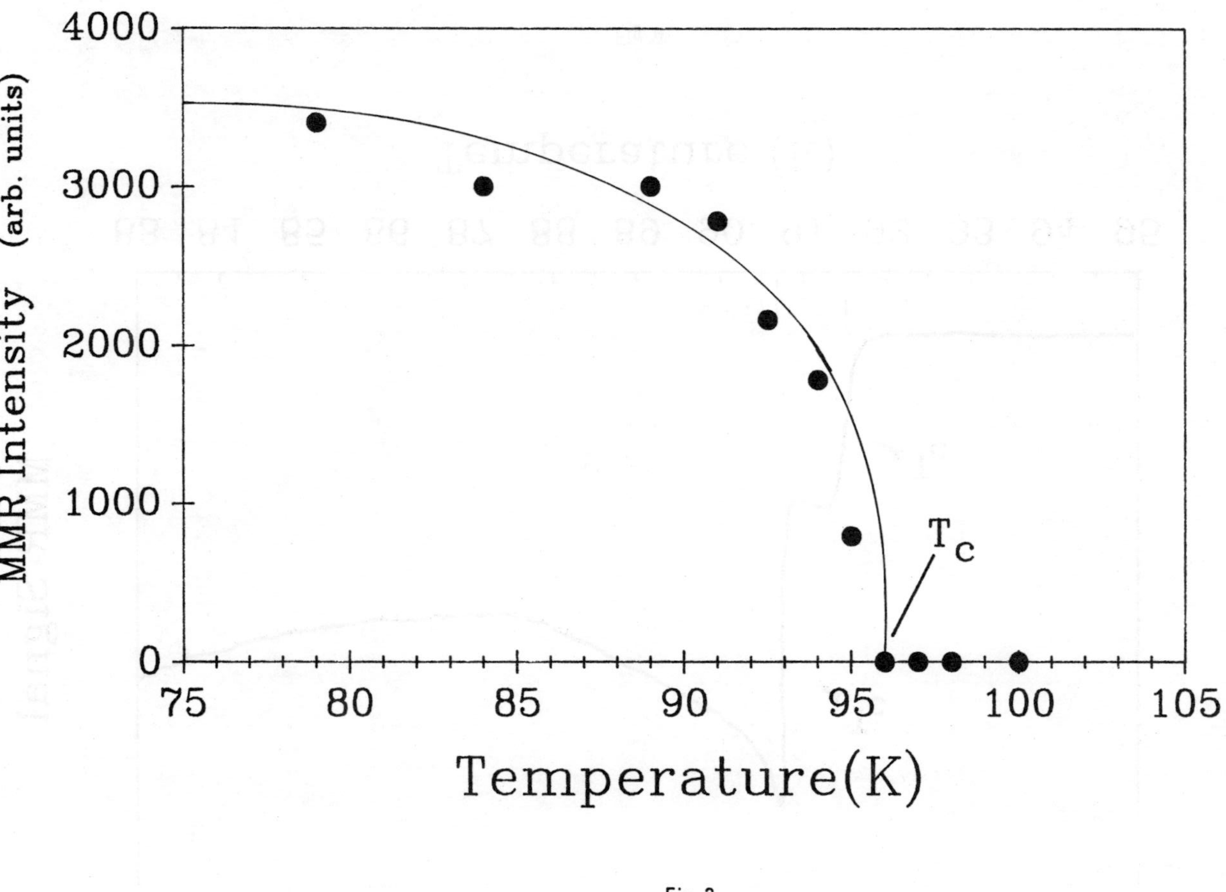
MMR Intensity (arb. units)
4000
3000
2000
1000
0
75
80
85
90
95
100
105
Temperature(K)
T_c

Fig. 3

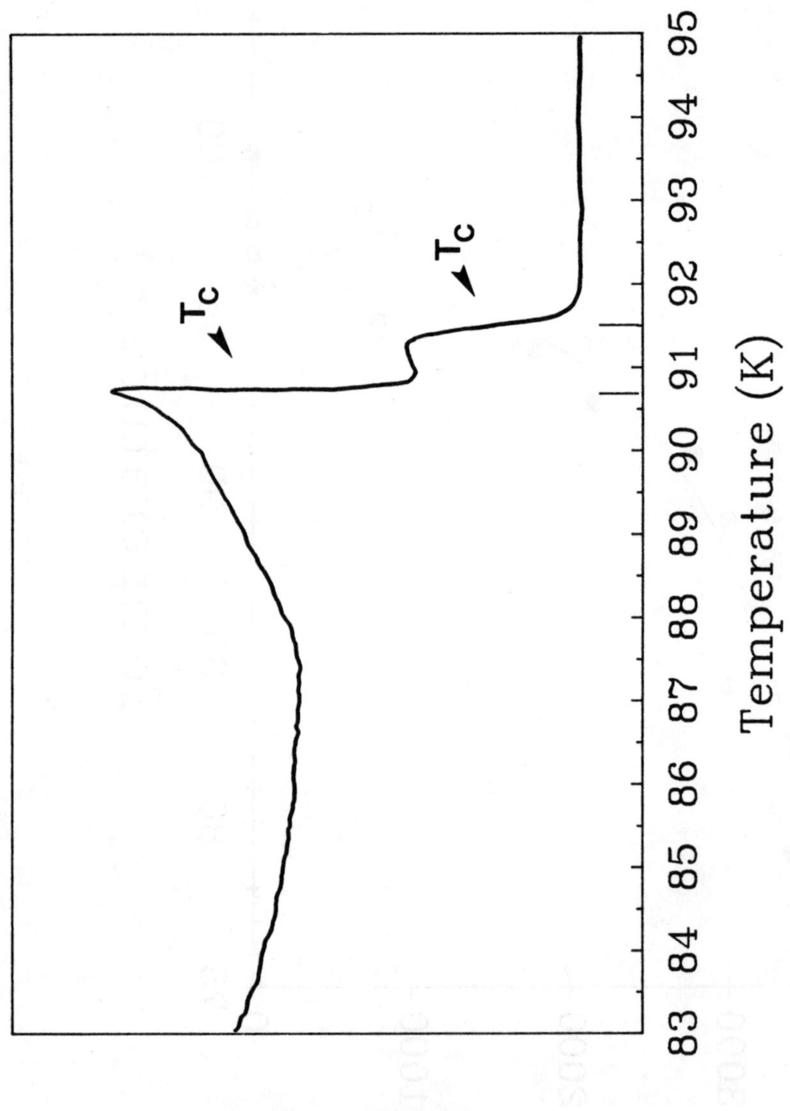

MMR Signal
Temperature (K)
83 84 85 86 87 88 89 90 91 92 93 94 95
Tc
Tc

Fig. 4

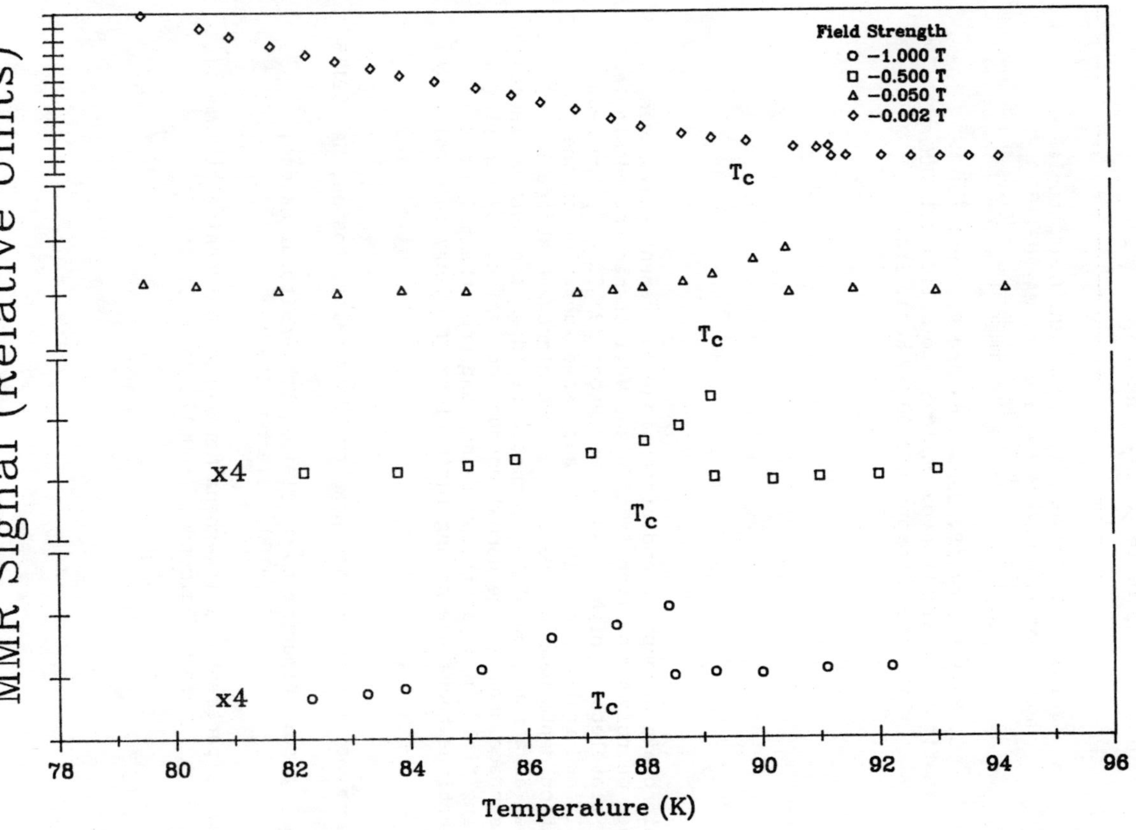

Field Strength
o −1.000 T
□ −0.500 T
△ −0.050 T
◇ −0.002 T
Tc
Tc
Tc
Tc
x4
x4
MMR Signal (Relative Units)
Temperature (K)
78
80
82
84
86
88
90
92
94
96

Fig. 5

FIELD MODULATED MICROWAVE ABSORPTION IN $YBa_2CU_3O_{7-X}$ NEAR T_C

R. Karim, S. A. Oliver, A. Widom* and C. Vittoria
Northeastern University, Boston, MA 02115

G. Balestrino†, S. Barbanera and P. Paroli‡,
Institute di Elettronica dello Stato Solido del CNR,
Via Cineto Romano 42, 00156 Roma, Italy

Novel microwave absorption and dispersion measurements have been performed on polycrystalline samples and well characterized single crystal platelets of high T_c superconductors $YBa_2Cu_3O_{7-x}$.The results are explained in terms of fluxoids and rapid variation of the penetration depth near and below T_c. The microwave absorption increases as a function of the magnetic field as a result of increase in the surface area of the normal conducting region. By applying a modulating field in a particular manner and from thermodynamic considerations there are strong indications of energy absorption by fluxoids.

*Department of Physics, Northeastern University, Boston, MA 02115

†also: Dipartmento di Fisica, Universita di Saleno, I-84100, Salerno, Italia

‡also: Dipartmento di Ingegnena Meccanica, Universita di Roma II, Via O. Raimondo, I-000173 Roma, Italia

I. Introduction

Non resonant microwave absorption experiments provide a valuable technique for investigating high T_c superconductors. Novel effects associated with granular superconductivity at low fields (H<200 Oe) have been reported[3-9]. Our experiment was carried out to measure microwave absorption and dispersion near the superconducting transition temperature T_c on bulk material and single crystal platelets of $YBa_2Cu_3O_{7-x}$ with emphasis on their angular dependence on the applied magnetic fields.

The experimental technique involves the use of a standard ESR spectrometer at sample temperatures near T_c[1]. In typical ESR measurements only microwave absorption is measured as the static field (H) is swept through magnetic resonance[2]. Dispersion measurements are rarely done using this technique, although most ESR systems are capable of performing this measurement. The microwave dispersion is proportional to the shift in cavity frequency which is measured by monitoring the DC voltage output of the automatic frequency control (AFC) circuit in the microwave bridge. The AFC dc voltage is proportional to the change in frequency required to "lock" the klystron source frequency to the resonant frequency of the microwave cavity.

Initially, two sets of experiments were performed on the single crystals with different field configurations. In the first set of measurements the DC field (H) is swept while the microwave field (h_{rf}) direction was perpendicular to it. The modulating field (h_m) was applied parallel to H. This is similar to the low field non-resonant experiment performed by others[3-7]. From now on this will be referred to as the parallel($\|$) configuration. In the other Magnetically Modulated Microwave Absorption (MMMA) experiment h_m was applied perpendicular to H, but parallel to h_{rf}. We will refer to this experiment as perpendicular($\perp$) configuration. In all of the above experiments the temperature was kept fixed below T_c. In the second set of experiments H was fixed but T was varied near T_c. The r.f magnetic field source was applied in the platelet plane, the 'c' axis being normal to the platelet plane.

We found that in the ($\|$) configuration d/dH (power absorbed) as a function of H had a maximum at 10 Oe and the signal was symmetrical about H=0. This indicates minimum absorption occurs at H=0 and it increases with $|H|$. This contrasts with the reported[8] maximum absorption at H~2 Oe. Hysteresis in the absorption near H=0 could possibly shift the extremum in absorption away from H=0.

Obviously, for this configuration one should see some sort of change in the signal, d/dH (Power Absorbed), as T is lowered through T_c. This was indeed observed and an almost monotonically decreasing signal, flattening out approximately 6-8 K below T_c was seen. The flat portion of the curve showed some fine structure. The dispersion flat portion of the curve showed some fine structure. The dispersion which is proportional to the change in cavity frequency was observed to show some marked extremums.

The ($\perp$) configuration experiment exhibited an entirely different behavior as compared to the ($\|$) configuration. Firstly, d/dH (Power Absorbed) versus H showed no change in absorption with respect to H. This remarkable feature can be completely explained by thermodynamic coupling properties of fluxoids.

Secondly, the temperature experiment in this configuration showed some marked extremums in d/dH (Power Absorbed) near T_c. However, the absorption peaks occurs at temperatures shifted from the ($\|$) configuration.

We interpret our data in terms of existence of fluxoids in these superconductors. In the ($\|$) configuration the absorption as a function of H has a minimum at H=0 and increases with $|H|$. This has been explained in terms of weak links[9] assumed to exist in these high T_c superconductors. As the field is increased, the number of flux lines or fluxoids also increases (proportionally) the area of normal conduction region relative to the superconductive regions. This leads to an increase in the microwave absorption. The ($\perp$) configuration shows no dependence of the absorption on H. From thermodynamic considerations of the energy coupling of a fluxoid we conclude the dependence of $Z(\omega+io^+, H)$ on the modulating field, h_m takes the form $Z = Z(\omega+io^+, H+h_m \cos\Theta)$, where $\cos\Theta$ is the angle between H and h_m. Later experiments carried out on bulk samples with various angles between H and h_m also fit the predicted angular dependence quite well.

II. Sample Characterization

Single crystals of the high temperature superconductor ($YBa_2Cu_3O_{7-x}$) have been grown by a melt flux method, using as a solvent the eutectic of two of the constituent oxides, namely CuO and BaO, and adding to it a few mole % of Y_2O_3.[10] The initial composition of the melt, in mole %, was Y_2O_3 (3%), BaO (27%), and CuO (70%). Crystal growth was carried out in air, oxygen annealed at 600°C overnight and slowly cooled. Single crystal platelets of YBCO were obtained with surface areas up to several square millimeters, and a typical thickness of 0.1mm. A crystal of sample dimensions (1 x 1 x .05mm^3) was used in the microwave absorption experiments.

Typical crystals from the melt were characterized by both X-ray diffraction and ac susceptibility measurements. Laue photographs show the surface of the platelet to be normal to the c crystallographic axis. The value of the lattice parameter c was then measured by a single axis dif-fractometer using Co K_α radiation, from the (0 1 11) symmetric reflection (2 Θ Bragg = 115^0).

The a.c. magnetic susceptibility was measured using a SQUID magnetometer as a null detector, at an operating frequency of 80 Hz, and an oscillating field intensity of about 10^{-6}T.[11] The magnetic susceptibility variation was recorded as a function of temperature from 125K down to 4.2K. Onset of the observed diamagnetic transition was considered to be the onset of superconductivity in the samples.

For the as grown crystals the value to the lattice parameter along the crystallographic c-axis was found equal to 11.724 ± 0.003A with a transi-tion temperature of T_c = 70 ± 3K. In addition, the diffraction peak were observed to be quite broad, indicating a spread of c-axis values over the sample. The average lattice value for the c-axis was also found to be larger than expected for a 92K superconducting sample (for example 11.677A in Ref. 12). This behavior indicates a large oxygen deficiency (x ≈ 0.4) in the sample.[13] Oxygen was reintroduced into the sample by annealing in 99.99 pure oxygen gas for 48 hours at 450°C, and then furnace cooling to room temperature. The effects of this annealing process was a narrowing of the diffraction peaks (indicating an increased and more homogenous distribution of the oxygen ions over the sample), a decrease in the lattice parameter c to 11.694A and an increase in the transition temperature to 85 ± 3K. The sample composition after annealing was found to be $YBa_2Cu_3O_{7-x}$. Due to twinning, which microscopically mixes 'a' and 'b' axes, the crystals (even to X-rays) are macroscopically tetragonal i.e. the twins are so small that the X-rays cannot distinguish that the sample is orthorhombic .

The sample magnetization (M) was measured using a SQUID magnetometer, with magnetic field (H) applied parallel to the c-axis at temperatures above and below T_c (7, 80, and 120K). Above 85K the sample exhibited typical paramagnetic behavior, as M aligns parallel to, and scales with H . Below 85K the sample becomes diamagnetic, with dM/dH being zero at 300 Oe and 1 kOe at 80K and 7K, respectively We were unable to obtain H_{c2} at any temperature, due to inavailability of sufficiently high fields. The Meissner effect was found to be of order 11-12% at 7K. In deducing this value the demagnetizing field in the platelet was taken into account.[14]

III. Microwave Results

The microwave absorption and dispersion of a rectangular cavity excited in a TE_{102} rectangular mode tuned at 9.15 GHz was measured. The modulating field h_m was set at 4 G p-p throughout the experiment and the r.f power in was held at 2 mW in the microwave cavity.

The superconducting state is affected by the application of external magnetic fields. A fairly linear region of the microwave response with respect to the modulating field amplitude and microwave field amplitude was used for the experiments. In all the temperature experiments the samples were zero field cooled whereas in the field swept experiments the samples were cooled at H≈-20 Oe. Figures 2 to 4 refer to experiments carried out on the single crystal. In figures (2a) and (2b) the derivative of the microwave absorption as a continous function of the swept field, H is displayed. Figure (2a) corresponds to the $_\perp$(∥) configuration experiment and Figure (2b) corresponds to the ($\perp$) configuration experiment under same settings of gain, power and field strength.

Clearly Figure (2a) indicates that the derivative of the sample absorption has a maximum near H=10±2 Oe. and is antisymmetrical about H=0. Actually, there is a phase change of π in the detection process about H=0. Thus, the absorption derivative is symmetrical about H=0 and the absorption itself has a minimum at H=0 and then increases, asymptotically tending to saturation. On the other hand Figure (2b) indicates little change in absorption as a function of H.

The temperature dependence of the microwave absorption and dispertion shown in Fig 3. and Fig 4. were taken at H=0 and h_m=4 G p-p. Variations in the AFC error voltage due to temperature variations of the cavity have been subtracted from the sudden changes of AFC voltages near T_c.

Figure (3a) displays the absorption derivative as a function of temperature in the (‖) configuration. The monotonic decrease in this signal is spread over a range of approximately 6-8 K. Some fine structure is also evident in addition to the main absorption curve. Temperature measurements are accurate to +/- 5K.

Figure (3b) displays the absorption derivative as a function of temper-ature in the (⊥) configuration with H=0.

Figure (4a) and (4b) display the dispersion signals as a function of temperature in the (⊥) and (‖) configurations, respectively. The AFC error voltage has been converted to shift in cavity frequency after caliberation.

Note that the vertical axes in both plots have been normalized, to their values at 120 K and the data has been smoothed by a standard algorithm.

Figure 5 refers to experiments on the bulk sample (approx. 0.3 mm radius). The calculated curve for the absorption derivative and the experimental data points for various angles between H and h_m keeping both $|H|$ and $|h_m|$ constant are compared .

IV. Discussion and Conclusion

(a) Field Dependent Experiments

Various models have been proposed to explain the low field absorption in the high T_c cuprates[3-9]. Viscous vibration of fluxoids driven by microwave currents causing the Lorentz force per unit length of a flux line to be $(1/c)(j \times \emptyset_o)$ was found to be the cause of microwave absorption by type II superconductors[15]. A network of internal Josephson junction current each reducing the screening of normal electrons by supercurrent according to the " diffraction" expression $I = I_o \sin(n\pi)/n\pi$, where $n = \emptyset/\emptyset_o$ and $\emptyset_o$ is the flux quantum are among the other interesting theories that have been suggested[16].

We have used the phenomenalogical model of Portis et.al[17] in order to interpret the experimental data. The salient points of the model are :(i) The magnetic field penetration into the bulk superconductor is assumed to have the form of fluxoids , each having the quantum of magnetic flux $\emptyset_o$ and cross sectional area A_1 .(ii) When these fluxoids thread through the superconductor , they create regions with a normal conductiviuty σ, inducing a macroscopically averaged resistivity proportional to the number of fluxoids, or equivalently proportional to the magnetic field B,

$$\rho_n = (A_1 B/\emptyset_o \sigma) \quad (1)$$

Equation (1) along with the usual London equation (penetration depth λ) yields the constitutive equation

$$\vec{E} = \rho_n \vec{J} + (4\pi\ \lambda^2/c^2)(\partial\vec{J}/\partial t) \quad (2)$$

Equation(2) implies that the electromagnetic distrurbances within the superconductor propagate with a complex wave vector k,

$$\frac{1}{k} = -i \left[\lambda^2 + i \left(\frac{\rho_n c^2}{4\pi\omega} \right) \right]^{1/2} \quad (3)$$

Equation(3) translates into a surface impedence of

$$Z_s = (4\pi\omega/c^2 k), \quad (4)$$

or equivalently (for a platelet of thickness d)

$$Z_s(\text{platelet}) \cong \left(\frac{4\pi\omega}{c^2 k} \right) \left[\frac{1-\cos(kd)}{\sin(kd)} \right] \quad (5)$$

Thus the surface impedence depends on the magnetic field in a complex manner.

To test the above phenomenological model (via the angular dependence of the impedance on the modulating field direction) one notes that the energy of a single fluxoid in the presence of the modulating field intensity $\vec{h}$ obeys the thermodynamic law

$$\delta U = \frac{1}{4\pi} \int \vec{H} \cdot \vec{\delta B}\, d^3r \qquad (6)$$

which for a single fluxoid thread implies

$$\delta U_{fluxoid} = \frac{\phi_o}{4\pi} \int \vec{h}_m \cdot \vec{dl} = \frac{\phi_o}{4\pi} \int \cos\Theta\, h_m\, dl \qquad (7)$$

where $\cos\Theta$ is the angle between the fluxoid thread direction (length element) $\vec{dl}$ and the modulating field $\vec{h}_m$.

The poynting theorem relates the energy of the fluxoids to the surface impedence and in terms of the coarse grained average (over many fluxoids) the surface impedance obeys (from the coupling in equation(7)),

$$Z_s(\vec{B},\vec{h}_m) = Z_s(B,h_m\cos\Theta) \qquad (8)$$

where Θ is now the angle between $\vec{B}$ and $\vec{h}$. Equation(8) is the predicted angular dependance of the fluxoid model , at least in the regime $h \lesssim B$. In figures 1 and 4 , Equation (8) is indeed verified for the angles $\Theta=0$ and $\Theta=90^\circ$ for the single crystal and for various angles using the bulk sample of YBCO .

It seems strange that $Hc_1 \approx$ 300 Oe at 80 K , and yet fluxoids are assumed to penetrate by applying a field ~ 10 Oe. However, the latter agrees with Gammel et al.(18), who see fluxoids in YBCO cooled in a 13 Oe field.

(b) Temperature Dependent Experiments

The difference in the temperature dependence of d/dH (Power Absorbed) in (||) and ($\perp$) configuration can be explained on the crystal anisotropy and platelet dimensions which will considerably affect the geometrical shape of the fluxoids formed. The rapid variation of the penetration depth near T_c is also a factor to be considered .As can be seen from equation (3) the propagation constant k can easily have some contribution from λ at temperatures very close to T_c.

The dispersion curves show some similarity in structure in the (||) and (⊥) configuration, but the absorption peaks are slightly shifted from each other with respect to temperature.

The absorption derivative may be interpreted as a measure of the number of fluxoids trapped as a function of temperature. This implies hysteritic effects as a function of field as observed by us and in ref. 17.

For the (||) configuration the nature of absorption below T_c is simply understood from the fact that above T_c there are no fluxoids so there is maximum absorption. Below T_c the absorption decreases. Thus, as the temperature is varied from above to below T_c one would indeed expect a change in the MMMA signal. For the (||)configuration this differential in absorption depends on the anisotropy of fluxoid formations (as a function of the direction of H) or simply that there is absorption in the superconducting region as well below T_c. We find that MMMA signal strengths for H parallel and perpendicular to the (c) axis to be within 15% , eliminating the anisotropy concept. We conclude that in the (⊥) configuration microwave absorption occurs in the superconducting regions rather than in the fluxoid region .

Aknowledgement

This work was partially supported by PROGETTO STRATEGICO MATERIALI SUPERCONDUCTTORI AD ALTA TEMPERATURA CRITICA OF C.N.R.

REFERENCES

1. R. Karim, S.A. Oliver and C. Vittoria, in press on J. Superconductivity.

2. See for example C. Kittel, Introduction to Solid State Physics, 6th ed, NY, Wiley (1986)

3. C. Rettori, D. Davidov I. Belaish and I. Felner, Phys. Rev. B, 36, 4028 (1987).

4. S.V. Bhatt, P. Ganguly, T.V. Ramakrishnan and C.N.R Rao, J. Phys. C, 20, L559-L563 (1987).

5. D. Shaltiel, J. Genossar, A. Grayevsky, Z.H. Kalman, B. Fisher and N. Kaplan, Solid State Cmm., 63, 987 (1987)

6. R. Durny et-al Phys. Rev. B, 36, 2361 (1987)

7. M.D. Sastry, A.G.I. Dalvi, Y. Babu, R.M. Kadam, J.V. Yakmi and R.M. Iver, Nature, 350,49 (1987).

8. K. W. Blazey, K.A. Muller, J.G. Bednorz, and W.Berlinger Phys. Rev. B, Vol 36, 7241, (1987).

9. K.A. Muller, M. Takashige, and J.G. Bednorz, Phs. Rev. Lett. 58 1143 (1987).

10. G. Balestrino, S.Barbanera, P.Paroli, in press on J. Cryst. Growth (Special issue devoted to the high-Tc superconductors)

11. S. Barbanera, M.G. Castellano, V. Foglietti (to be published)

12. R.J. Cava, B.Batlogg, R.B. VanDover, D.W. Murphy, S.Sunshine, T. Siegrist, J.P. Remeika, E.A. Rietman, S. Zahrusk and G.P. Espinosa , Phys. Rev. Let. 58, 1676 (1987).

13. A.Ono, Jpn. J. Appl. Phys. 26, L1223 (1987).

14. L.D. Landau, E.M. Lifshitz, and L.P. Pitaevskii, Electrodynamics of Continuous Media, Second Edition, Pergamon Press, Oxford, U.K., 182 (1984).

15. For a review , see C.G. Kuper in 'The theory of superconductivity' (Oxford , London 1968) Chap. 7.

16. A.Dulcic et al., Phys. Rev. B38, 5002 (1988)

17. A.M.Portis, K.W.Blazey, K.A.Muller and J.G.Bednorz, Europhysics letters, 5, p467-72, Mar 1,1988.

18. Gammel et al., Phys Rev. Lett. 59 2592 (1987).

Figure 1 (a) Magnetization of a single crystal high T_c superconductor is plotted as a function of the applied field at T=100 k.

(b) Same plot below T_c (T=85 k).

Figure 2 (a) Field Derivative absorption (P) of a single crystal high T_c superconductor is plotted as a function of the swept field, H_{dc} in the parallel configuration experiment.

(b) Same sample in the perpendicular field configuration.

Figure 3 (a) Field derivative absorption (P) of the same sample as above is plotted as a function of the temperature in the parallel configuration. The magnetic field H was fixed at 0 Oe.

(b) Same sample in the perpendicular field configuration. Again H was fixed at 0 Oe.

Figure 4 (a) Shift in the cavity frequency of the same sample as above is plotted as a function of the temperature in the parallel configuration. Again H was fixed at 0 Oe.

(b) Same sample in the perpendicular configuration at H=0 Oe.

Figure 5 Absorption derivative signal level of the bulk sample YBCO is plotted as a function of 'Vertical Field' The solid line represents the calculated fit. The value of H and h_m were held constant.

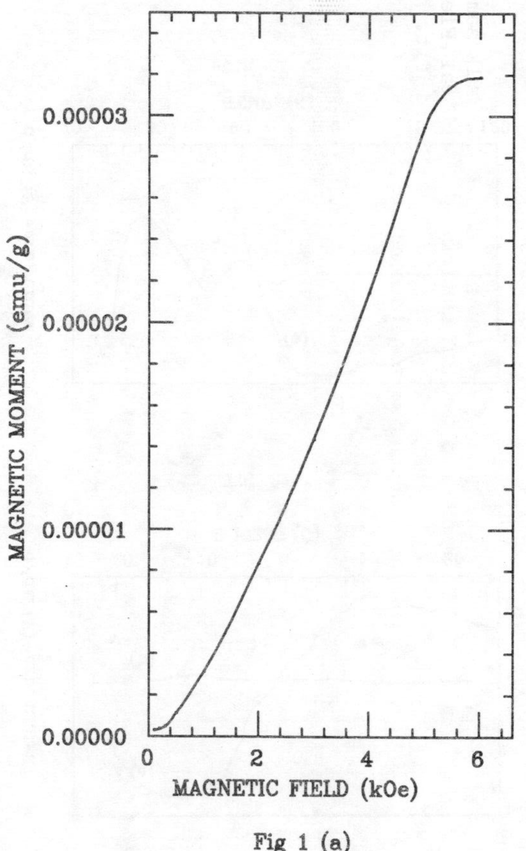
MAGNETIC MOMENT (emu/g)
0.00003
0.00002
0.00001
0.00000
0
2
4
6
MAGNETIC FIELD (kOe)

Fig 1 (a)

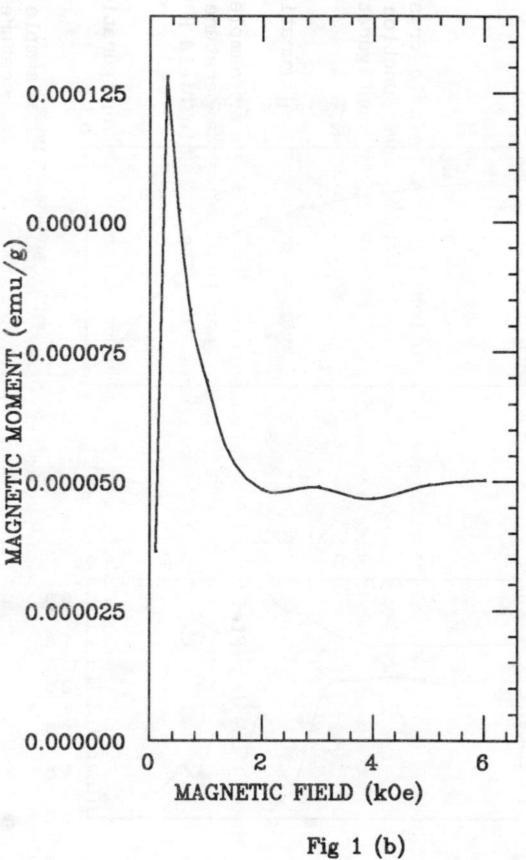
MAGNETIC MOMENT (emu/g)
0.000125
0.000100
0.000075
0.000050
0.000025
0.000000
0
2
4
6
MAGNETIC FIELD (kOe)

Fig 1 (b)

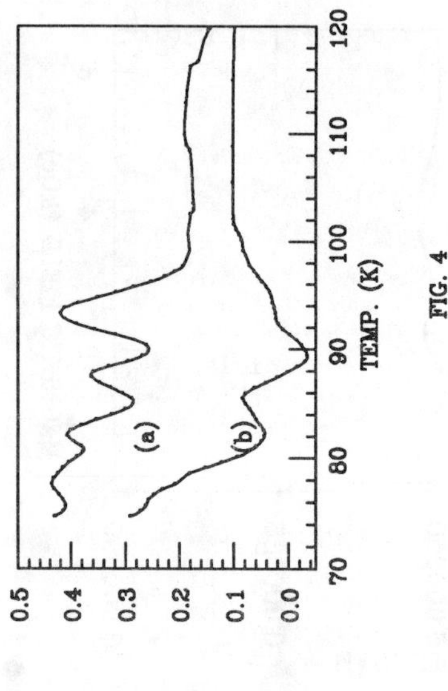

FIG. 2

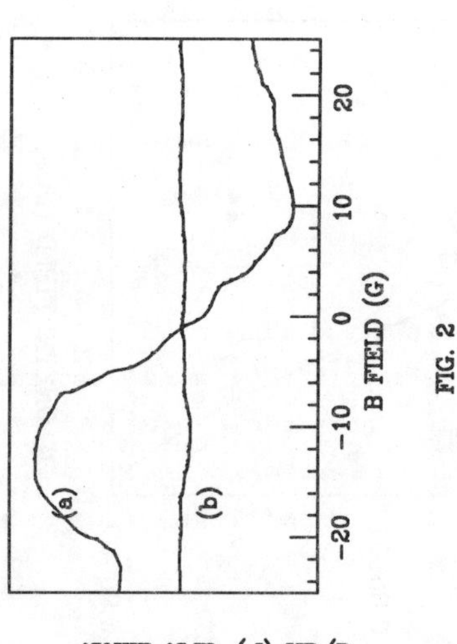

FIG. 3

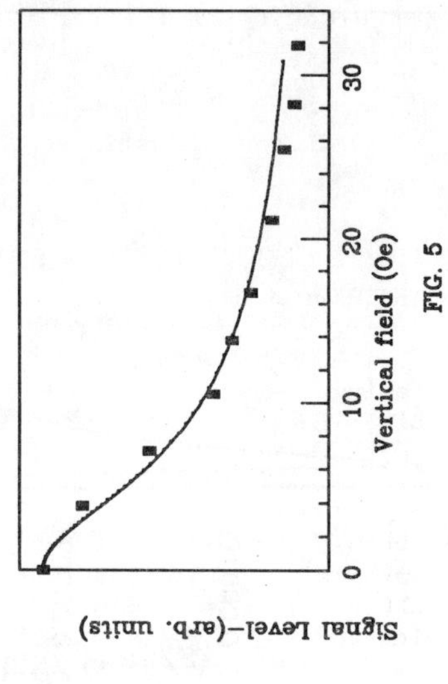

FIG. 4

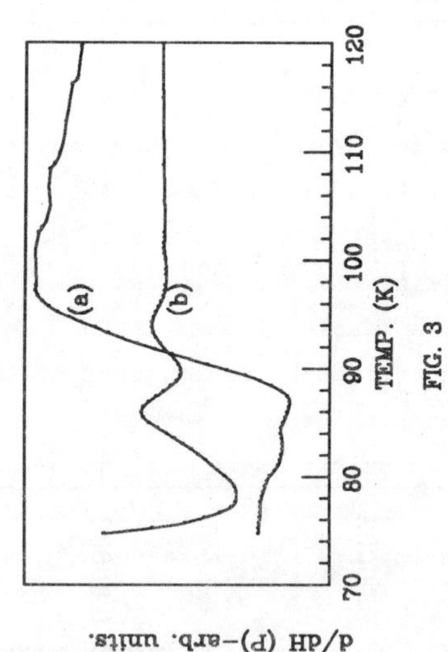

FIG. 5

MAGNETIC PROPERTIES OF A CHEMICALLY SYNTHESIZED Bi(Pb)SrCaCuO SUPERCONDUCTOR

D.R. Lundy, J.J. Ritter, L.J. Swartzendruber,
R.D. Shull, and L.H. Bennett
National Institute of Standards and Technology
Gaithersburg, MD 20899

ABSTRACT

It has been reported that the presence of lead serves to increase the fraction of high temperature phase in the Bi-Sr-Ca-Cu-O system prepared by a solid state reaction. To test if this is also the case for material prepared by a chemical method, a lead containing bismuth superconductor ($Bi_{1.5}Pb_{0.5}Sr_{1.5}Ca_{1.75}Cu_2O_x$) was chemically synthesized and its magnetic properties measured. The material obtained contained a large fraction (≈50%) of a phase with a superconducting onset temperature near 110 K. Inspection of the AC susceptibility with a small applied transverse magnetic field indicated the presence of additional superconducting phases with lower onset temperatures. Flux depinning was found to occur at relatively low fields.

INTRODUCTION

The superconductor BiSrCaCuO (BSCO) is of interest because of the presence[1,2] of a high temperature phase with a superconducting onset temperature of 110 K, ≈15 K higher than the highest found for the BaYCuO (BYCO) system. The presence of lead appears to encourage or to stabilize the formation of the high temperature BSCO phase[3,4]. In this work, a lead-containing BSCO was synthesized and magnetically characterized by measurement of the AC susceptibility and by obtaining hysteresis loops at several temperatures. The material was found to consist primarily of the phase with a 110 K onset temperature. Only a small number of flux pinning sites were observed at 95 K. Flux depinning was also observed.

SAMPLE PREPARATION

A Bi(Pb)SrCaCuO sample, nominally $Bi_{1.5}Pb_{0.5}Sr_{1.25}Ca_{1.75}Cu_2O_x$, was prepared by a chemical method[5)] as follows: fifteen millimoles (mmol) of bismuth hydroxide, 5 mmol $PbCO_3$, 12.5 mmol $SrCO_3$ and 20 mmol Cu from the appropriate weight of basic copper carbonate are dissolved in 200-250 ml of 85% lactic acid at 150 °C with continuous stirring. After several hours a clear blue solution free of suspended solids was obtained. Volatile materials were distilled off by raising the temperature to 250 °C, leaving behind a homogeneous dark viscous mass. The resultant plastic mixture was cast into a cylindrical mold at 200 °C. After cooling to room temperature, the solidified ingots were calcined at 450 °C in air. Final sintering of the molded shape was conducted at 850 °C in air. From this treatment a sample which readily crumbled into a powder was obtained. Energy-dispersive x-ray analysis in a scanning electron microscope revealed two major types of particles: one close to the target composition and the other lead rich. The exact role of the lead is not clear but attempts to produce the material without it resulted in much smaller fractions of the phase with a 110 K onset temperature.

MAGNETIC MEASUREMENTS

AC susceptibility was measured in a Hartshorn-type bridge with an applied ac field of about 0.5 Oe at a frequency of 1.68 kHz.

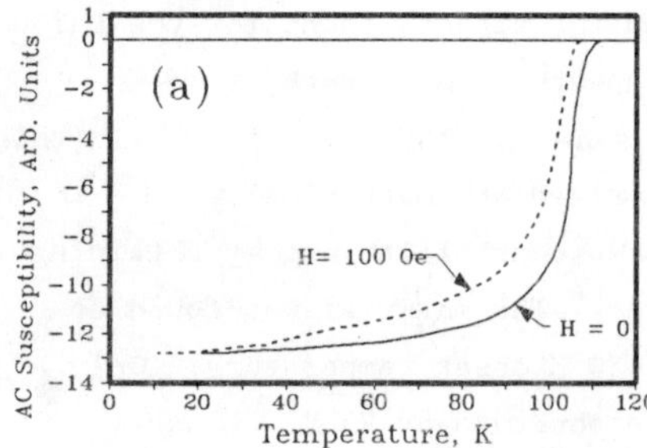

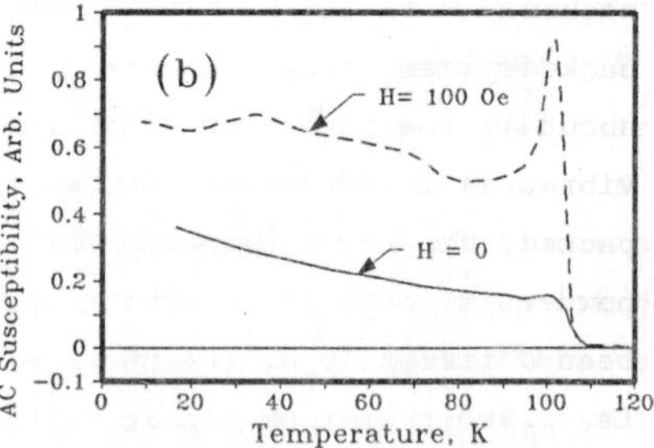

Fig. 1. Real part (a) and imaginary part (b) of the AC susceptibility as a function of temperature for a chemically synthesized sample of Bi(Pb)SrCaCuO with zero applied field (solid line) and with an applied DC field of 100 Oe (dotted line). The DC field was oriented at right angles to the AC field.

Measurements were made with and without the application of a 100 Oe field. The real part of the susceptibility (Figure 1a) shows an onset temperature of 110 K, shifted to 105 K with application of the field. The imaginary part of the AC susceptibility (Figure 1b) also shows an onset temperature of about 110 K without the applied field. When the 100 Oe field is imposed, additional "phases" are indicated with onset temperatures of about 80 K and 40 K. This means that the interpretation of measurements below 80 K will be complicated by the presence of these additional phases.

A series of hysteresis curves were obtained between 95 K and 10 K in a vibrating sample magnetometer (VSM). A loop obtained at 95 K showed very little hysteresis, indicating that the effective number of flux pinning sites is quite small at this temperature. Estimates of the Meissner effect fraction and of H_{c1} may be obtained from such a loop. The Meissner effect fraction, obtained from using the initial slope after cooling to 95 K in zero field was approximately 50%, in rough agreement with that obtained from the AC susceptibility. The field at which the virgin curve of the loop first deviated from linearity gave a rough estimate of 20 Oe for H_{c1} at 95 K. A high value for H_{c2} at 95 K was indicated by a nearly flat magnetization vs. applied field curve at 10 kOe.

The material studied here was in the form of a powder and the results obtained can be somewhat influenced by how tightly the powder is packed. For example a 'dimpling' in the hysteresis loop for loosely packed powder can be seen in Figure 2. This effect appeared only below about 40 K and may arise, at least in part, from an artifact due to vibration of the sample in the VSM. As the material is more tightly packed, the loops approach that seen in Figure 3 for tightly packed powder. A similar dimpling in a superconducting hysteresis loop has been observed by Shaplygin et al.[6)] in a tightly packed sample of $La_{1.85}Sr_{0.15}CuO_4$ sample at 20 K.

FLUX DEPINNING

Hysteresis loops obtained from superconductors are the result of

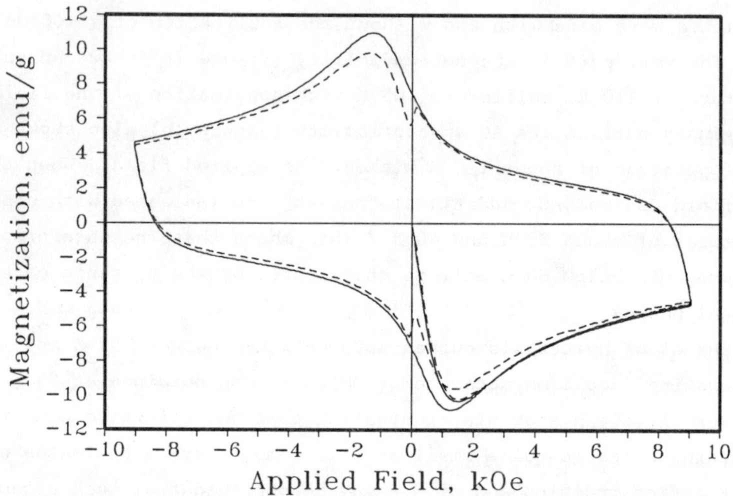

Fig. 2. Hysteresis loop for a chemically synthesized sample of Bi(Pb)SrCaCuO at 10 K. The dotted line is for a loosely packed powder and the solid line is for the same powder after tight packing.

flux pinning. For fields below H_{c1} no flux penetrates the sample and no loop is seen, i.e. the magnetization process is reversible. For fields greater than H_{c1}, flux penetrates the sample and, in a type-II superconductor, the mixed state is formed. Some of this flux can be pinned by, for example, defects in the crystalline lattice, in which case the magnetization process will no longer be reversible and a magnetization loop with hysteresis will be obtained. For sufficiently high fields, though still less than H_{c2}, the flux pinning may no longer be effective and the magnetization process again becomes reversible. Nelson[7)] has referred to this depinning as a phase transformation from a "braided flux" into a liquid of disentangled rods. Figure 4 shows hysteresis loops of our Bi(Pb)SrCaCuO sample at two temperatures which illustrate such high field flux depinning. There is a field, illustrated by the arrows in Figure 3a, below which flux pinning occurs, and above which the flux is depinned. We find that this depinning field, H_{cp}, can be expressed as a function of temperature by

$$H_{cp} = H_{co}(T_c - T)^\alpha$$
$$= H_o(1-t)^\alpha$$

where T is the temperature, T_c is the superconducting onset temperature, $t=T/T_c$, and $H_o=H_{co}T_c^\alpha$. A plot of $\ln(H_{cp})$ vs. $\ln(1-t)$ for the temperature range of 50-95 K is displayed in Figure 3b. The plot is a straight line with slope $\alpha=2.3$, very close to the value 5/2. Below 50 K the values of H_{cp} deviate from the straight line fit of Figure 3b due, at least in part, to the presence of other superconducting phases in the material with onset temperatures near or below 50 K. The deviation is such that the measured H_{cp} becomes much larger than that predicted by extrapolation of the line in Figure 3b. Yeshurun and Malozemoff[8)], studied depinning of flux in BYCO by a different technique wherein the sample temperature was changed in fixed field. They found that the temperature at which flux pinning occurred was could be described by a power law in the field with an exponent of 3/2. The reason for this discrepancy in exponent value is currently under investigation.

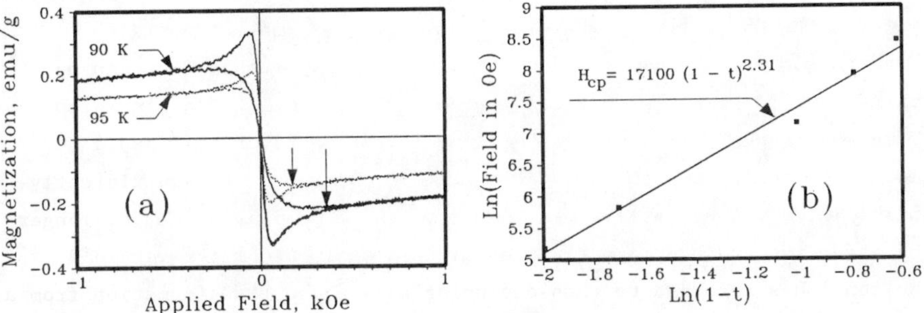

Fig. 3. (a) Hysteresis loops for a chemically synthesized sample of Bi(Pb)SrCaCuO obtained at 90 K (solid line) and 95 K (dotted line). The arrows indicate the field above which the magnetization is reversible, i.e. has very little or no hysteresis. (b) Log-log plot of the depinning field, H_{cp}, vs. one minus the reduced temperature. The temperature range is 95 K to 50 K. The line is a least squares fit to the data points shown.

ACKNOWLEDGEMENTS

We thank R. Drew and D. Mathews for technical assistance.

REFERENCES

1 H. Maeda, Y. Tanaka, M. Fukutomi, and T. Asano, Jap. J. Appl. Phys. Lett. (in press) (1988).

2. C. W. Chu, L. Bechtold, L. Gao, P. H. Hor, Z. J. Huang, et al., Phys. Rev. Lett. 60, 941 (1988).

3. R. Cava, preprint, (1988).

4. R. M. Hazen, C. T. Prewitt, R. J. Angel, N. L. Ross, L. W. Finger, et. al., Phys. Rev. Lett. 60, 1174 (1988).

5. J.S. Wallace, J.J. Ritter, E. Fuller, L.H. Bennett, R.D. Shull, and L.J. Swartzendruber, Phys. Rev. in press.

6. I.S. Shaplygin, I.A. Konovalova, V.B. Lazarev, E.A. Tishchenko, A.I. Bazhan, and A.B. Sushov, JETP Lett. 46, S126 (1987).

7. D.R. Nelson, Phys. Rev. Lett. 60, 1973 (1988).

8. Y.Yeshurun and A.P. Malozemoff, Phys. Rev. Lett. 60, 2202 (1988).

Critical Current vs Temperature in the High-Tc $Y_1Ba_2Cu_3O_{7-\delta}$ Superconductor

G. C. Vezzoli

U.S. Army Materials Technology Laboratory

Emerging Materials Division

Watertown, MA 02172

Terence Burke

U.S. Army Electronics Technology and Devices Laboratory, LABCOM

Fort Monmouth, New Jersey 07703-5000

M. Moon, B. Lalevic, A. Safari & A. Sundar

Rutgers University, Department of Electrical Engineering

New Brunswick, New Jersey

ABSTRACT

We have studied the polycrystalline ceramic oxide superconductor $Y_1Ba_2Cu_3O_{7-\delta}$ with respect to critical current by use of single and repetitive pulses of 2 us - 1 s duration and magnitude of 20 to 2200V ($J=10^2 - 10^4$ A/cm^2). These pulses have been employed to interrogate the superconducting state of the high-T_c material and to study the recovery to the normal state. The response of the magnitude of the voltage across the sample to constant-magnitude applied pulses is a non-monotonic function of the ambient temperature upon rewarming and not a function of the steady state resistance. This J_c-equivalent response curve shows second order structure and several regions of recovery to the normal state. The time required for voltage recovery is <u>decreased</u> with increasing temperature and shows a <u>ms</u> time-constant. It is very doubtful that these results can be a consequence of I^2R heating via the applied pulses because contact resistance is extremely low and because the circuit is current-limited.

INTRODUCTION

In this work we address the high-T_c ceramic oxide superconductor[1-5] $Y_1Ba_2Cu_3O_{7-\delta}$ in polycrystalline form. One of the major research objectives in this new class high-T_c superconductors is to build an understanding of critical current (J_c) characteristics and to enhance J_c particularly in polycrystalline and thin film materials. Because $Y_1Ba_2Cu_3O_{7-\delta}$ is highly anisotropic in structure[6] (Fig. 1), the critical current properties are extremely complex as are the magnetic properties. From our previous study[7] we believe that the material is best described as a sustitutive defect A_2BX_4 structure in the form $(Y_1{}^{3+}Cu_1{}^{3+})$ 1 $(Ba_2{}^{2+}Cu_2{}^{2+})$ $1O_{8-w}$ where for superconductivity W = 1.1; (the Cu^{3+} sites are associated with chains in basal plane). Because in polycrystalline materials of this superconductor the individual grains do not line up relative to common structural sub-units, the measured value of J_c must include contributions from more than a single sub-structural unit. Distinguishing between these individual contributions does not seem possible in a polycrystalline sample in the superconducting state. However, such distinction may indeed be feasible during the recovery to the normal state especially if transitory (metastable) states are involved during recovery. The mechanism giving rise to critical current density in a superconductor must involve the kinetic energy of the Cooper-paired electrons. For small values of the momenta of a Copper pair, the current density is proportional to the momenta (P). In general, the current density (J) is proportional to the rate at which the free energy changes as a function of the momenta of a Copper pair[8,9]. As the free energy of the superconducting state approaches that of the normal state, the current density as a function of momenta falls to zero and the transition to the normal state occurs. Conceptually, when the kinetic energy associated with a Cooper pair exceeds the Cooper-pairing energy, then J_c is achieved and the transition to the normal state occurs. In low T_c superconductors this energy is the forbidden gap. In high-T_c materials the energy gap may refer to a different phenomenon.[10] The coherence length then is the distance over which electrons are Cooper-paired. This is a relatively low number in high-T_c materials (~15A) and suggests local atomic behavior.

EXPERIMENTAL PROCEDURES

Polycrystalline pellets of $Y_1Ba_2Cu_3O_7$ were synthesized at the U.S. Army Materials Technology Laboratory by Benfer[11] and at Rutgers University by Saffari et al[11] employing established methodologies[1-3]. At the Materials Technology Laboratory undoped as well as 1.2% tin- and 1.2% fluorine-added material was prepared. The MTL sample that was undoped yielded an onset temperature of 90°K. The sample to which 1.2% tin was added showed an onset between 91 and 92°K. Both samples gave zero resistance at 84°K and a center of the transition region at 88°K. The Rutgers sample showed an onset temperature of 83-82°K.

The experimental circuits and electrode configurations employed in this study are given in Fig. 2. In experiments that were designed to measure decay times every precaution was taken to minimize or eliminate any parasitic effects of capacitance and inductance.

EXPERIMENTAL RESULTS AND DISCUSSION

A. Studies of Pulsed Current vs Temperature

In Fig. 3 we present the electrical resistance (4-terminal) versus decreasing temperature data (middle) obtained upon rapidly cooling the superconducting material to liquid nitrogen ambient. The termocouple in these experiments was placed within a mm of the sample, hence reflects the temperature of the near-environment of the sample. The first order decrease in electrical resistance in Fig. 3 identifies the transition to the superconducting state.

In the rewarming direction (toward right at lower region of the graph), we show the response of the sample to 35 applied current pulses (of magnitude 2.0A, duration 1 sec, and supplying a current density of about $60A/cm^2$). These pulses are superimposed upon the existing dc to produce the resistance versus temperature curve on the X-Y recorder in Fig. 3. In this experiment the electrode configuration given as #1 in Fig. 2 was utilized. In the re-warming direction the temperature scale is expanded: The current-voltage behavior is observed to be hysteretic. Clearly, there is no response observed on the X-Y recorder nor from the digital micro-voltmeter over the temperature range of the first sixteen of these pulses. We believe that in the bulk of the sample this re-warming range does not exceed T_c. The electrical resistance still measures as zero on both instruments, with the digital voltmeter oscillating from plus to minus with values of a fraction of a microvolt. Following the above region is a zone of constant response (17th to 22nd pulse).

Above a given temperature range, however, the instrumentation shows a voltage response to the applied pulses that increases with temperature. The voltage (and hence resistance) returns to the zero measured level at the end of the superimposed applied pulse. The region of small increasing response to the pulses extends from the 23rd to the 26th pulse in Fig. 3. A region of much greater response magnitude then dominates the behavior from the 27th to the 35th pulse. In the upper inset (of the figure) oscillograms are given to show the response of the voltage across the sample to the applied superimposed pulse as a function of time on a ms scale. It can be seen from these oscillograms that as the ambient temperature increases, the slope of the response also increases. The pulse height of response is plotted against re-warming temperature in Fig. 4 for two separate experiments. Figs. 3 and 4 are the equivalent of a critical current vs temperature curve (J_c vs T) because at temperatures above 80K (the X data points) the material will support an electric field thereby, showing electrical resistance, this occurring because J_c has been exceeded by the current which accompanies the applied pulse. In both experiments of Fig. 4 a region of no-response is followed by a region of constant amplitude response, and then followed by two separate regions of increasing magnitude response labeled C and D. The inception of these regions of response are not associated with any change which we could measure departing from zero resistance in the steady state. The differing J_c properties of the coordinated species of the superconductor structure (such as the Cu^{3+} chain regions and the Cu^{2+} inverted pyramidal with planar base regions) may possibly give rise to the different response regions such as A and B. Not until after the 35th pulse does the dc steady-state resistance begin to measurably rise as temperature increases. This then refers to still another recovery region. The experiment was repeated using electrode configuration #2 from Fig. 2 with the voltage electrodes across the bulk of the sample, yielding similar results in that the region of real response to the applied pulse was <u>not</u> associated with a measured re-gaining of dc resistance. Oscillograms showing the eventual response to the superimposed voltage pulse as a function of re-warming are shown sequentially in Fig. 5a and 5b. An actual positive response is not observed until the re-warming condition (still zero measured dc resistance) corresponding to the sixth oscillogram (lowermost right). In Fig. 6 the slope of this recovery is plotted (for different regions of interest) from the data of Fig. 5. It is clearly shown that this slope increases with increasing temperature. Substituting copper metal for the sample and then re-doing the entire set of above experiments indicated the static response of a typical metal undergoing no transformation and hence <u>no</u> unusual dependence on the temperature of the environment.

The above data indicates a ms response-time to the recovery of resistance or more precisely the acceptance of a bonafide field when J_c is exceeded with increasing temperature for superimposed current-limited pulses. This ms response time may have both non-thermal and thermal mechanisms responsible for its origin. Any thermal contribution would have to be due to a small I^2R effect arising from the small contact resistance and later arising after the initial re-assumption of resistance upon exceeding J_c.

It is clear from Figs. 3 and 4 that J_c is an inverse function of temperature. The magnitude of the current response increases with increasing temperature because the <u>difference</u> between the voltage established by the constant current pulse at a given re-warming temperature ($J_{(I,T)}$) and the critical current for that temperature ($J_c(T)$) is increasing with increasing temperature. Hence $V=(E)(d)= E(d)=\frac{(d)}{\sigma(T)}(J_{(I,T)} - J_{c(T)})$. The greater the difference term, then the greater the amount of resistance which is devleoped. This is because the kinetic energy exceeds the Cooper-pairing energy and the resulting difference is directly related to electrical resistance.

We can develop the relationship to describe the measured voltage of the pulse in Figs. 3 and 4 by employing the analysis given in the appendix to this paper.

Further indications of regions of recovery from the superconducting state to the normal state, and the possibility of transitory phases, are given in Fig. 7. It is shown therein that the initial region of recovery, employing ms pulses, is accompanied by transient voltage collapses. These are shown in the left-hand area of Fig. 7a, b, and d, and in the uppermost trace (110V) in Fig. 7c. The traces in Fig. 7a, b, and d are translated manually to the right with increasing temperature for clarity to the observer. In Fig. 7c we observe that the transient collapse in voltage is not observed for applied voltages of 20 to 50V, but only for 100V (and presumably higher). In Fig. 7d the uppermost right-hand traces refer to the zone of the final stages of recovery. Intermediate between the region of transients and the region of final recovery is a region (designated as #2) where a flat on-state exists and a transient is not clearly observable. This is in consonance with the data of Figs. 3 and 4. In the lower right of Fig. 7d we give the normal state pulsed behavior as well as the low-temperature state at 50 and 100V applied potential respectively.

Fig. 8 gives pulsed response and recovery data employing still a third sample with electrode configuration #3 in Fig. 2. This experiment was conducted in order to determine whether placing the voltage electrodes on the edge of the same side of the sample caused a change in the results. This was performed in order to compare results from a configuration with a greater measurement contribution from surface currents to configurations with greater measurement contributions from bulk currents. The results from the experiments described in Fig. 8 are very similar to those from Figs. 3-7 which had electrodes across the sample, namely the temperatures corresponding to the onsets of the voltage response to the superimposed pulse was not associated with an increase in dc resistance. As temperature continued to increase, an envelope type response is observed.

We must address the question whether the data of Figs. 3-8 is simply or partially representative of a pulsed Joule-heating effect. If I^2R heating were <u>exclusively</u> causing the material to be pulsed out of the superconducting state because of additional heat over and above ambient, then we should expect a continuous monotonic response curve rather than the discontinuous recovery behavior observed in Fig. 4. Furthermore, we would also expect that pulses of widely varying width (and duty cycle) such as those employed in Fig. 3 and those employed in Fig. 7 should yield very different recovery and response behavior. However, both sets of these data indicate at least three real recovery regions. Additionally, since the material is a superconductor, the Joule-heating could only arise from contact resistance while in the superconducting state. The contact resistance in the present experiments was about a maximum of 1.0 mΩ as determined from two-terminal measurements. After initial resistance is reassumed, then additiona I^2R effects can of course arise. Hence it is possible that thermal components are involved (or superimposed) in these data but it is very doubtful that Joule-heating could exclusively explain the above data.

AC susceptibility studies by Goldfarb et al[12] have shown that in oxygen-annealed polycrystalline superconducting samples of this material there are two regions of T_c behavior versus field amplitude. These are derived by studying the peak loss characteristics as well as the Meissner transition (which itself shows two regions of behavior). Because of lack of single crystal data it is not clear whether the origin of these two types of behavior is due to fundamental properties of the superconductor such as structural polyhedra, or due to coupling effects arising from crossing grain boundaries. It is shown that the data giving rise to the above regions of behavior changes dramatically when the sample is pulverized into a fine power, thus changing grain-to-grain mismatch. Hence it is likely that grain boundaries heavily influence at least one form of the above behavior. Recent Hall data on a similar material is interpreted by grain boundary phenomena[13]. This description of the detailed work of Goldfarb et al[12] is included herein because it is the only other indication to our knowledge of two types of superconducting behavior in a single phase material, and in this sense may be related to our own J_c data.

B. Studies of Voltage Decay at 77^{0}K Using Short Pulses

In Fig. 9 we show the superconductor response to a voltage pulse of about 4-20 us duration and manitude varying from 50V to 2200V. Fig. 9a gives an oscillogram corresponding to a small magnitude and a large magnitude double-pulse and single pulse The double pulse is shown at 77K and reflects decay to zero resistance in about 1.4 μs. The larger magnitude pulse is also shown at room temperature on a different scale. It is clear that in the superconducting state there are two regions of voltage decay: 1) a region of "overshoot" or inductive emf; and 2) a region of what appears to be resistive decay. These two regions have time constants of about 100 ns and about 1-1.4 μs respectively. Fig. 9f and g gives multiple traces showing the effect of re-warming using an applied pulse of 2200V giving a current density $3x10^3 A/cm^2$ and not indicating any tendency to break out of the superconducting density by exceeding J_c. (There was no dc applied field in these studies, but merely a short applied pulse not superimposed on any other signal). Figs. 9b and 9c show decay of the voltage on 200 and 500 ns/div scales. Fig. 9c also includes an oscillogram of the net result when a parallel copper short is placed across the superconductor at low temperature (lower trace). Figs. 9d and 9e give the resulting traces when copper wire is substituted for the sample at room and low temperature respectively. Fig. 9f gives, from bottom to top, the voltage decay across the superconductor at 77^oK, the trace upon considerable re-warming, and the trace corresponding to source voltage prior to reaching the superconductor. The source voltage shows only an inductive type decay and does not include the longer resistive decay observed in the superconductor. Fig. 9g gives the re-warming data until the normal state is recovered at room temperature. The inductive contributions to these data suggest an inductive coupling emf which appears inescapable. The resistive decay seems to be considerably longer than what would be expected from a non-equilibrium viewpoint. The possibility must be considered that this resistive decay of about a µsec may be at least in part due to the capacitance of the sample arising from its polycrystalline nature. Capacitance effects may certainly arise from grain boundaries in the sample. Secondly, this time period could be associated with the time required to establish the appropriate Cooper-paired current distribution through the sample and along the surface. Regardless of such effects, we would tend to expect some finite time period for the superconductor to expel the transient magnetic field associated with the current transmitted through the material at a given temperature (contrasting the longer logarithmic dependence of flux properties with temperature). Furthermore, there must be a finite time associated with the capability of the material to become resistanceless in a phase transition of the Ginzburg-Landau type. This may simply be the time for the Cooper-pairing of injected electrons with each other and with itenerant electrons. Our value of 1 us would then set an upper limit on this time parameter.

When in the superconducting state, but prior to the application of a voltage, the material must exhibit some ratio of intenerant Cooper-paired electrons to normal electrons. This is because the delocalized electrons under no electric field undergo motion randomly dictated by kT. Pairing will then depend upon the probability of interaction between an itenerant conduction electron and the mediating boson, as well as the mediating boson and another conduction electron with opposite spin and momentum. This probability depends upon the spheres of influence of these particles, i.e. the coherence length of the electron pair and the proximity of the boson (or its degree of localization or delocalization). When an electric field is applied (with $J < J_c$), then electron motion is no longer random but directed. The region of influence then of the subject particles changes, and additional Cooper-pairing takes place, this changes the ratio of paired to normal electrons. This additional Cooper-pairing must require a finite time prior to reaching a state of resistancelessness. We suspect on a local level that this time interval will be much less than 1 us. However, the time for this pairing (in the sense of distribution) to influence the entire bulk interelectrode gap (and hence be measurable in our experiment) may be much longer and attain at least 1-2 us. Along these lines it seems appealing to conduct time-dependence thermal equilibrium experiments at the onset temperature and during the narrow temperature region where the resistance is collapsing toward the superconducting state. Equally important would be to perform time-dependent studies in the thermal region between the departure from essentially linear R vs T behavior and the achievement of the onset condition. This latter region seems to us sufficiently important to merit far more study than has been heretofore conducted. We have observed that this temperature signifying deviation from linear R vs T behavior also identifies the beginning of a sharply increasing $+R_H$ Hall coefficient (in R_H vs T studies)[14].

FIGURE LEGEND CAPTIONS

1. The crystalline unit cell structure of $Y_1Ba_2Cu_3O_{7-}$.

2. The circuit schematic for measuring J_c.
3. Superconducting transition and re-acceptance of pulsed ($\sim$ 1s) voltage with re-warming at $J > J_c$ for $T < T_c$.
4. Magnitude of reaccepted superimposed pulse vs temperature. Inset is from ref. 14

5a,b Applied pulse and recovery of resistance vs time.

6. Slope of resistance-recovery (volts/time) vs increasing temperature.
7. Oscillograms showing recovery of resistance for 100 ms pulses.
8. Long superimposed pulse study (1s).
9. Short pulse study (μs).

Appendix 1: Derivation of Tc in terms of p, S, v_g

$$J_c(T) = \alpha \frac{dG}{dP}$$

J_c = Critical current density

G = Gibbs Free Energy

p = crystal momenta of Cooper-pair

α = constant

G = U - TS + PV

$\Delta G = \Delta U - T\Delta S + P\Delta \bar{V}$

E = J/σ

U = lattice energy

T = Temperature

P = pressure

$\bar{V}$ = volume

S = Entropy

V/d = E = electric field

σ = conductivity

V = voltage

d = interelectrode spacing

From Figs. 3 and 4 the voltage accepted by the superconductor upon rewarming from 77K at a given T is :

$$V(T) = \frac{(d)}{\sigma(T)} \left(J(V,T) - \alpha \frac{dG}{dp}\right)$$
$$= \frac{(d)}{\sigma(T)} \left(J(V,T) - \alpha \frac{\Delta G}{\Delta p}\right)$$

$$= (d)\rho(T)\left(J(\nu,T) - \alpha\frac{(\Delta U - T\Delta S)}{\Delta p}\right)$$
$$= (d)\rho(T)\left[J(\nu,T) - \frac{\alpha}{\hbar}\left(\frac{\Delta U - T\Delta S}{\Delta k}\right)\right]$$
$$= (d)\rho(T)\left[J(\nu,T) - \frac{\alpha}{\hbar}\left(\frac{\hbar\Delta\omega}{\Delta k} - \frac{T\Delta S}{\Delta k}\right)\right]$$
$$V(T) = (d)\rho(T)\left[J(\nu,T) - \frac{\alpha}{\hbar}\left(\hbar v_g - T\frac{\Delta S}{\Delta k}\right)\right];\quad v_g = \text{group velocity}$$
$$\Delta u = h\Delta\nu = \hbar\Delta\omega$$
$$\frac{\Delta u}{\Delta k} = \frac{\hbar\Delta\omega}{\Delta k} = \hbar v_g$$

Thus: 1. The lower the change in entropy relative to k the higher $J_c(T)$. 2. The higher the group velocity, the greater $J_c(T)$. At $T = Tc$; $J_c \doteq 0$.

$\therefore\ \hbar v_g = T_c\Delta S/\Delta k$ or $Tc = \frac{\hbar\Delta k v_g}{\Delta S} = \frac{\Delta p}{\Delta S}v_g = \left[\frac{dp}{dS}\right]v_g$; Thus $\frac{dp}{dS}$ must be very high for exciton mediated superconductivity.

$$T_c \approx \frac{dp}{dS}v_g$$

References

1. Chu, C.W., Hor, P.H., Meng, R.L., Gao, L., Huang, Z.J. and Wang, Y.Q., Phys. Rev.Lett., 58(4), 405 (1987).

2. Wu, M.K., Ashburn, J.E., Torng, C.J., Hor, P.H., Meng, R.L., Gao, L., Huang, Z.J., Wang, Y.Q., Chu, C.W., Phys. Rev. Lett., 58(9, 908 (1987).

3. Hor, P.H., Meng, R.L., Wang, Q.Y., Gao, L., Huang, Z.J., Bechtold, J., Foster, R., Chu, C.W., Phys. Rev. Lett., 58(18), 1891 (1987).

4. Bednorz, J.G., and Mueller, R.A., Z. Phys., B64, 189 (1986).

5. Michel, C., Rakko, L er, and Raveau, B., Mat. Res. Bull, 20, 667 (1985).

6. Beno, M.A., Soderholm, L. II Capane, D.W., Hinks, D.G., Jorgensen, J.D., Schuller, I.K. Segre, C.U., Zhang, K., and Grace, J.D., Appl. Phys. Lett.

7. Vezzoli, G.C., Benfer, R. and Spurgeon, W., in Novel Mechanisms of Superconductivity, Proc. from Conf. at Berkeley California, June 1987, edited by V. Kresin and S. Wolf, Plenum Publishers, New York, 1987.

8. Bardeen, J. in Superconductivity in Science and Technology, ed. by Cohen, M.H., Chicago, University of Chicago press, 1968.

9. Superconductivity by Lynton, E.A., London, Chapman and Hall, 1969.

10. Iafrate, G., US Army Materials Technology Laboratory, Ft. Monmouth, NJ 07703, private communication.

11. Benfer, R., U.S. Army MTL Watertown, MA, Safari, B., Rutgers University, Piscataway, NJ.

12. Goldfarb, L. and Chen, D.X., J. Appl. Phys. 63 (Feb. 1, 1988).

13. Zhao, Y., Xia, J., He, Z., Sun, S., Zhang, Q., Qian, Y., Chen, Z., Pan, G., Chinese Phys. Lett., 5(5), 221(1988). See also Hall experiments given by Vezzoli et al in the present volume, p.

14. Chudhari, P., IBM Yorktown Heights, New York. Note in inset deviation from linearity in J_c vs T near 87-88K similar to C-D slope change in main graph in Fig. 4.

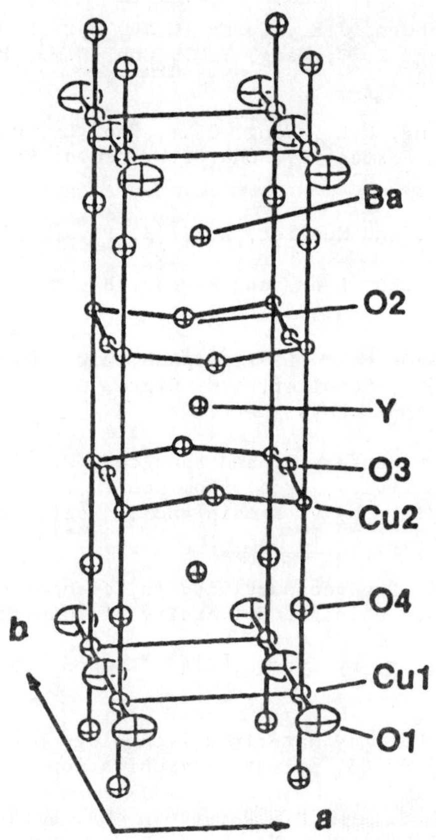

Fig. 1

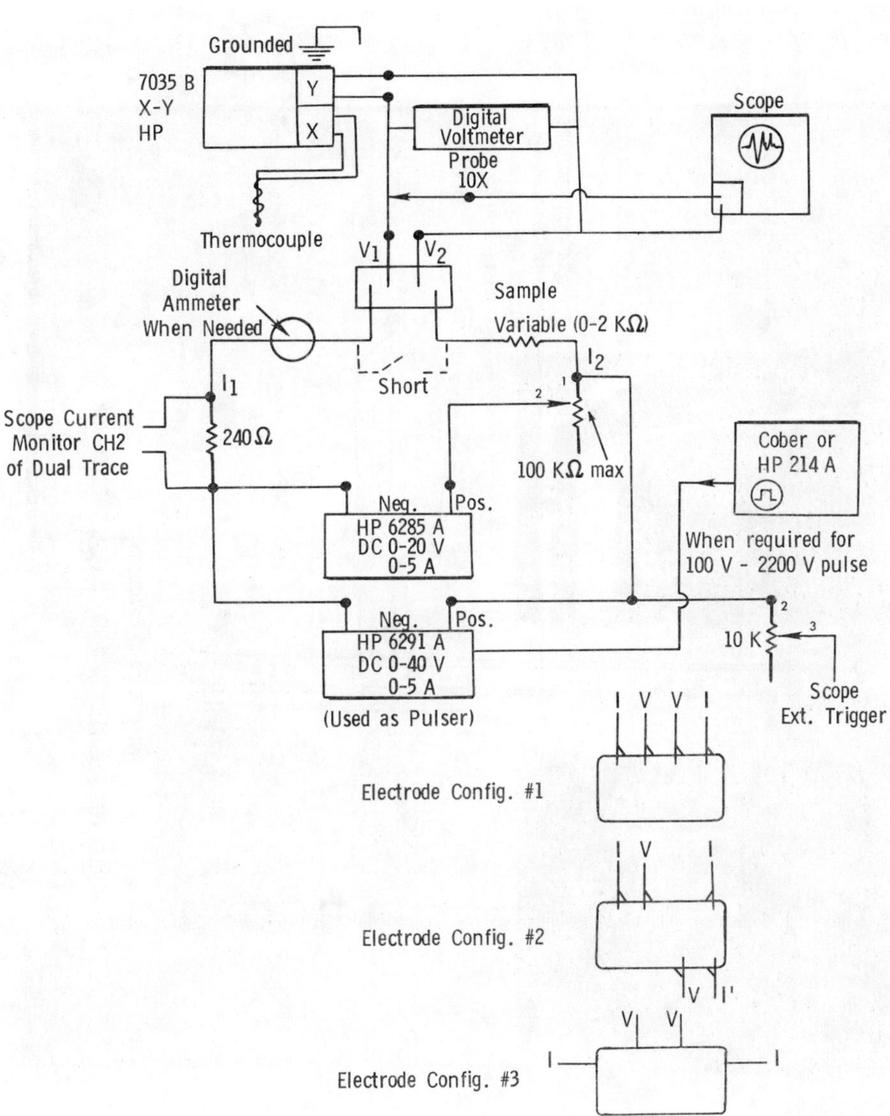

Fig. 2. J_c CIRCUIT SCHEMATIC

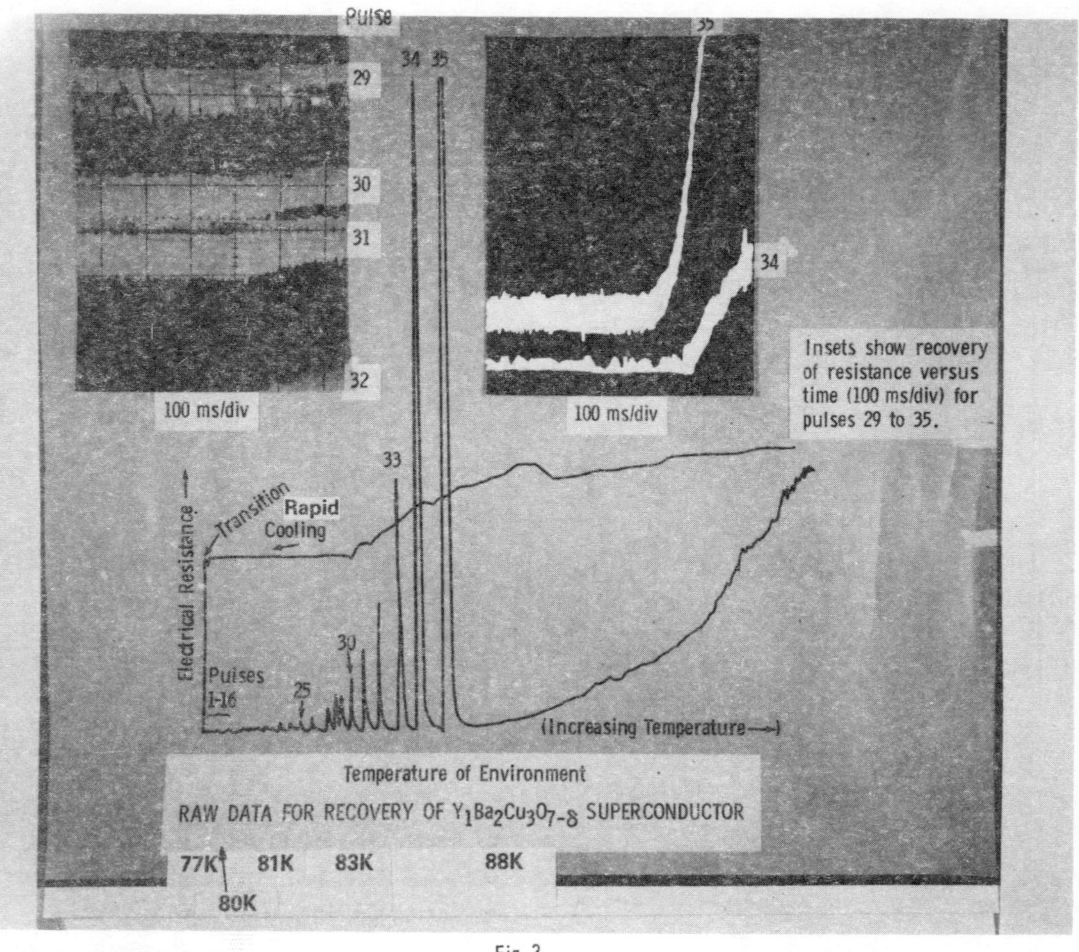

Pulse
29
30
31
32
100 ms/div
34 35
35
34
100 ms/div
Insets show recovery of resistance versus time (100 ms/div) for pulses 29 to 35.
33
Electrical Resistance →
Transition
Rapid Cooling
Pulses 1-16
25
30
(Increasing Temperature—→)
Temperature of Environment
RAW DATA FOR RECOVERY OF $Y_1Ba_2Cu_3O_{7-\delta}$ SUPERCONDUCTOR
77K
80K
81K
83K
88K

Fig. 3

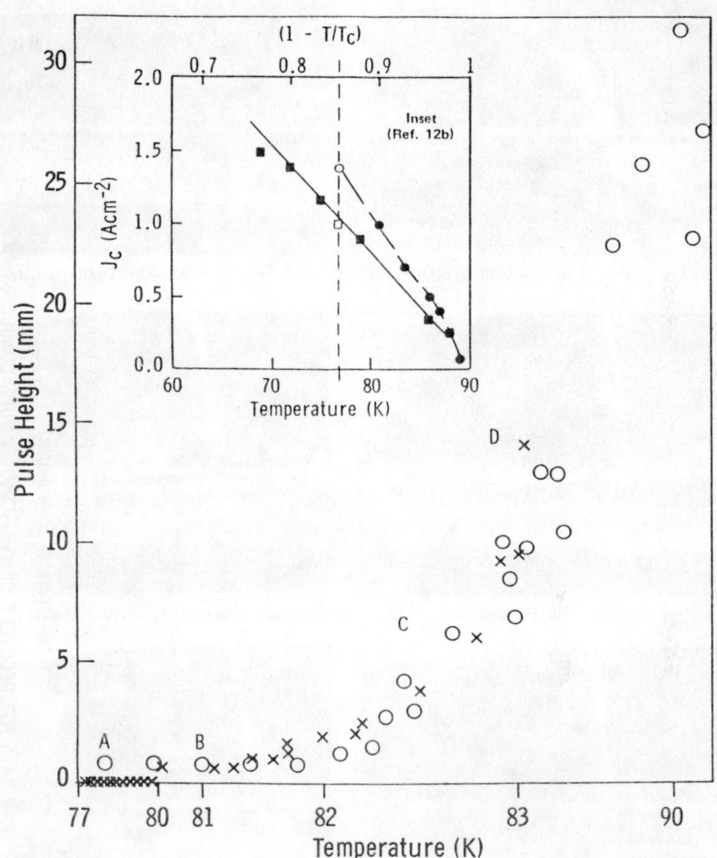

Magnitude of superimposed pulse versus temperature for $Y_1Ba_2Cu_3O_{7-\delta}$ superconductors.

Fig. 4

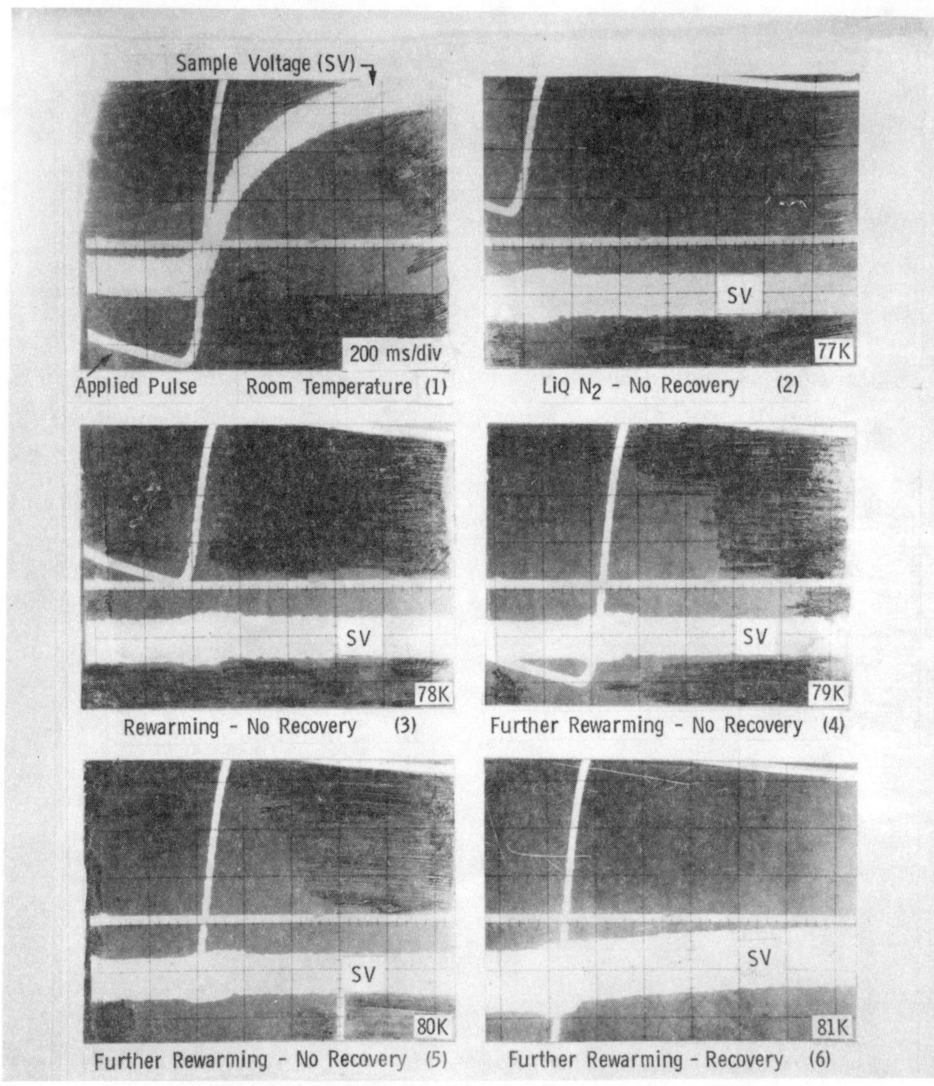

APPLIED CURRENT PULSE AND RECOVERY OF RESISTANCE
WITH INCREASING TEMPERATURE

Fig. 5a

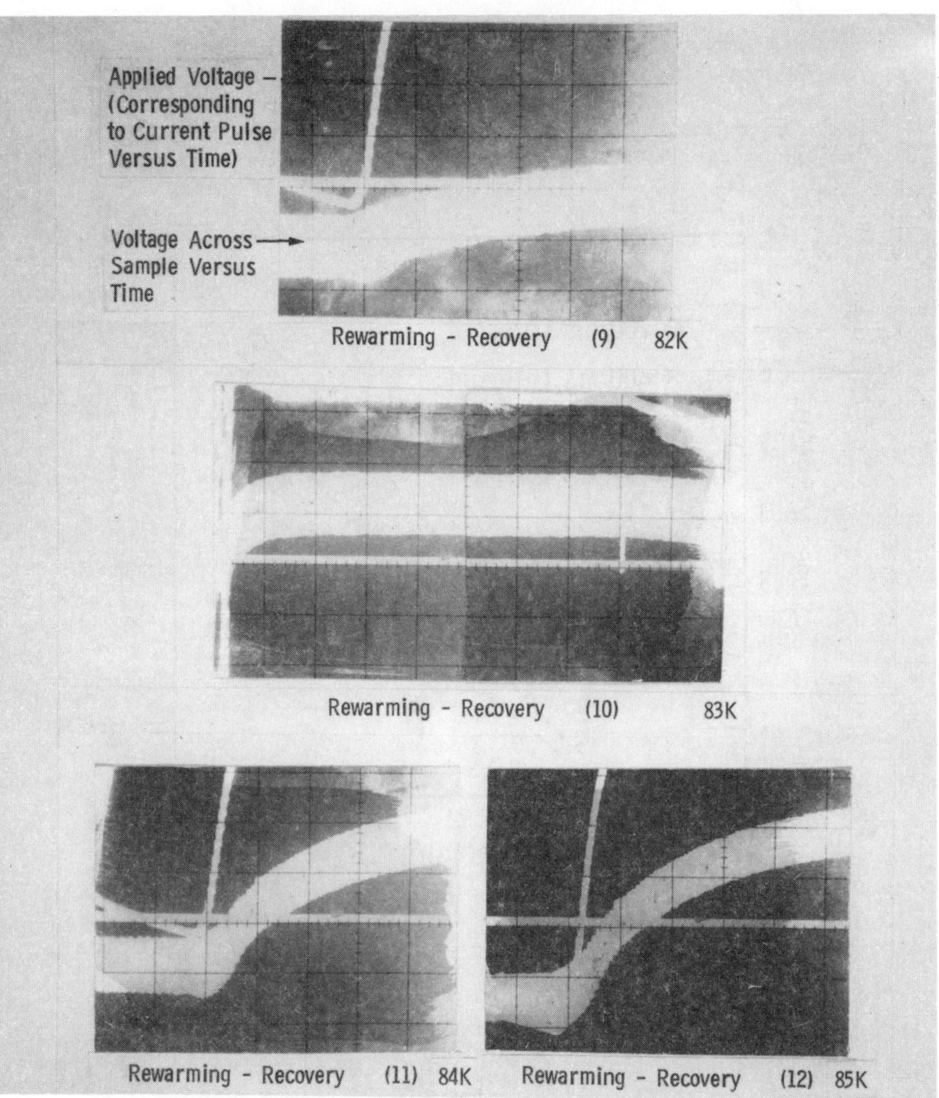

Rewarming - Recovery (9) 82K

Rewarming - Recovery (10) 83K

Rewarming - Recovery (11) 84K

Rewarming - Recovery (12) 85K

APPLIED CURRENT PULSE AND RECOVERY OF RESISTANCE
WITH INCREASING TEMPERATURE

Fig. 5b

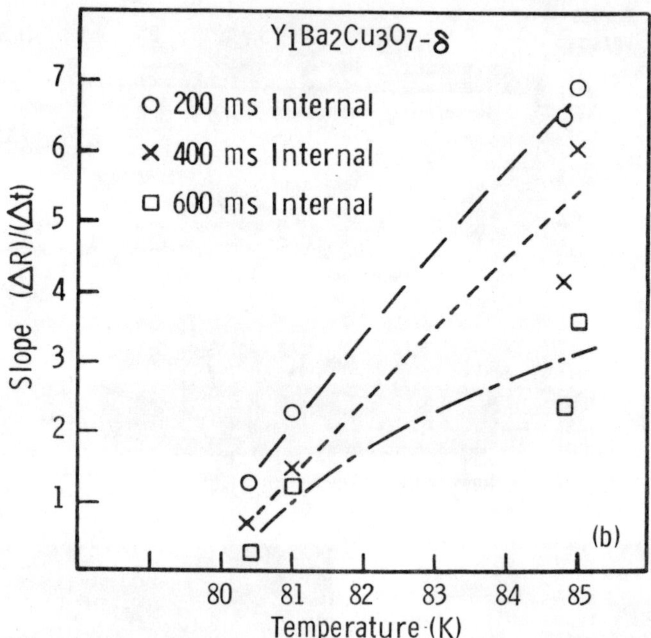

Recovery-of-resistance-slope versus temperature for $Y_1Ba_2Cu_3O_{7-\delta}$ superconductor.

Fig. 6

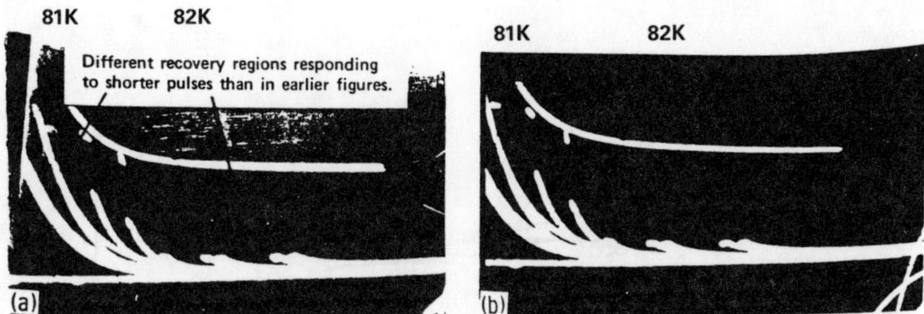

Oscillograms showing superconductor's recovery of resistance response to voltage pulse. The individual pulses are displaced to the right with increasing temperature for clarity. The pulses to the left show a voltage collapse or breakdown transient, whereas the pulses in the center show a low resistance flat region. The uppermost trace across the oscillogram refers to the normal state behavior.

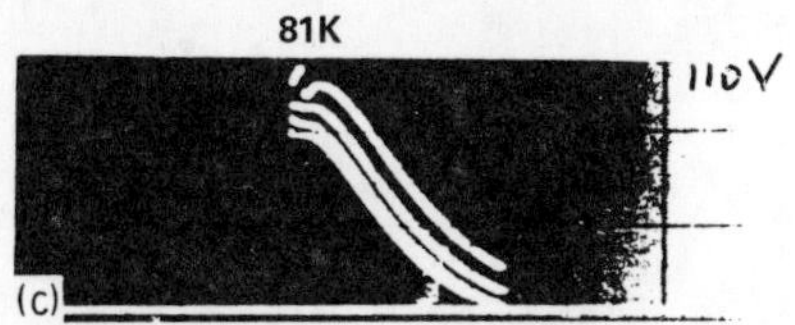

Voltage versus time (200 μs/div) in superconductor during rewarming at (from bottom to top) 20, 50, and 100 V applied amplitude. Note voltage collapse or breakdown transient in uppermost oscillogram.

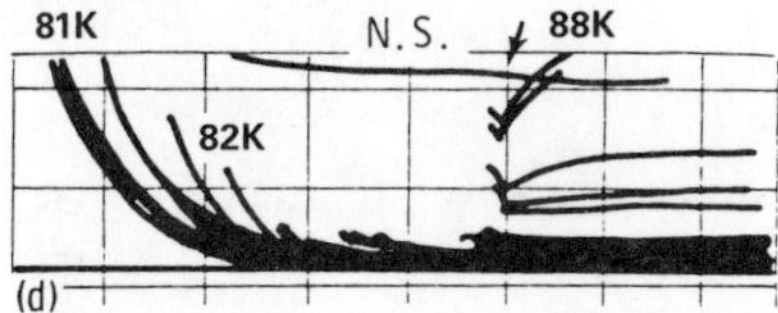

Upper left and center same as (a) and (b). Upper right shows region of recovery of normal state, lowermost traces are superconductor near T_C. Almost horizontal trace are normal state before cooling (designated N.S.).

Oscillograms showing recovery characteristics for 100 ms pulse.

Fig. 7

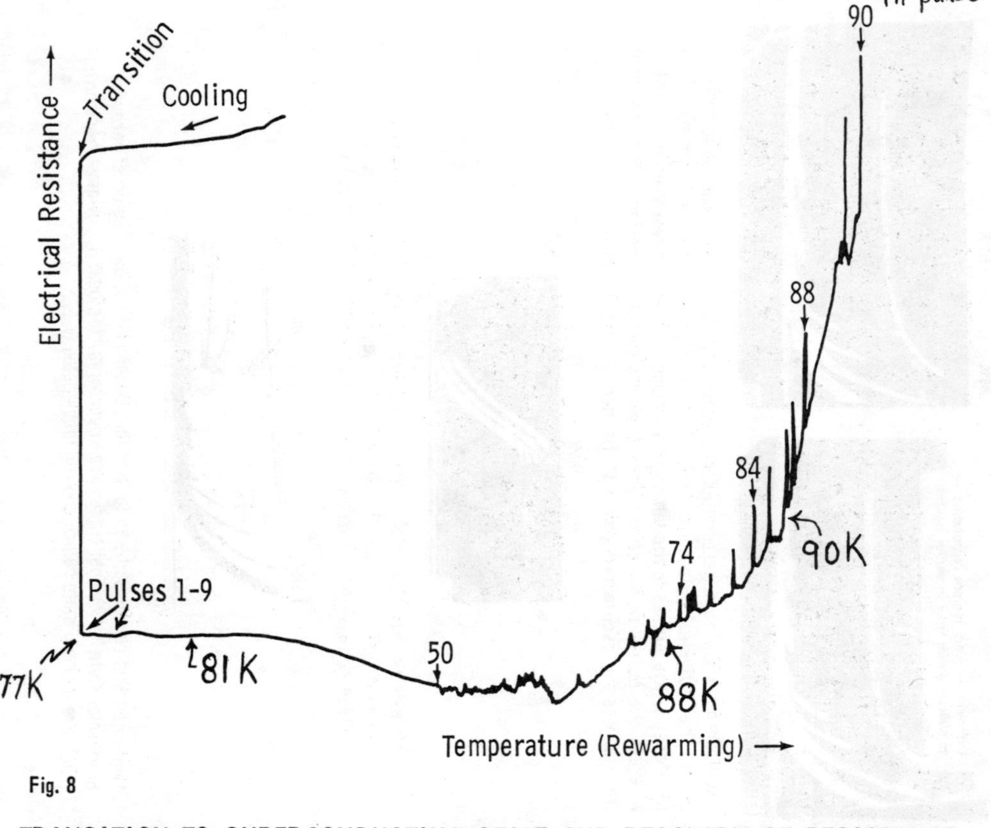

Fig. 8

TRANSITION TO SUPERCONDUCTING STATE AND RECOVERY OF RESISTANCE WITH REWARMING USING SUPERIMPOSED PULSES ($J > J_c(T)$)

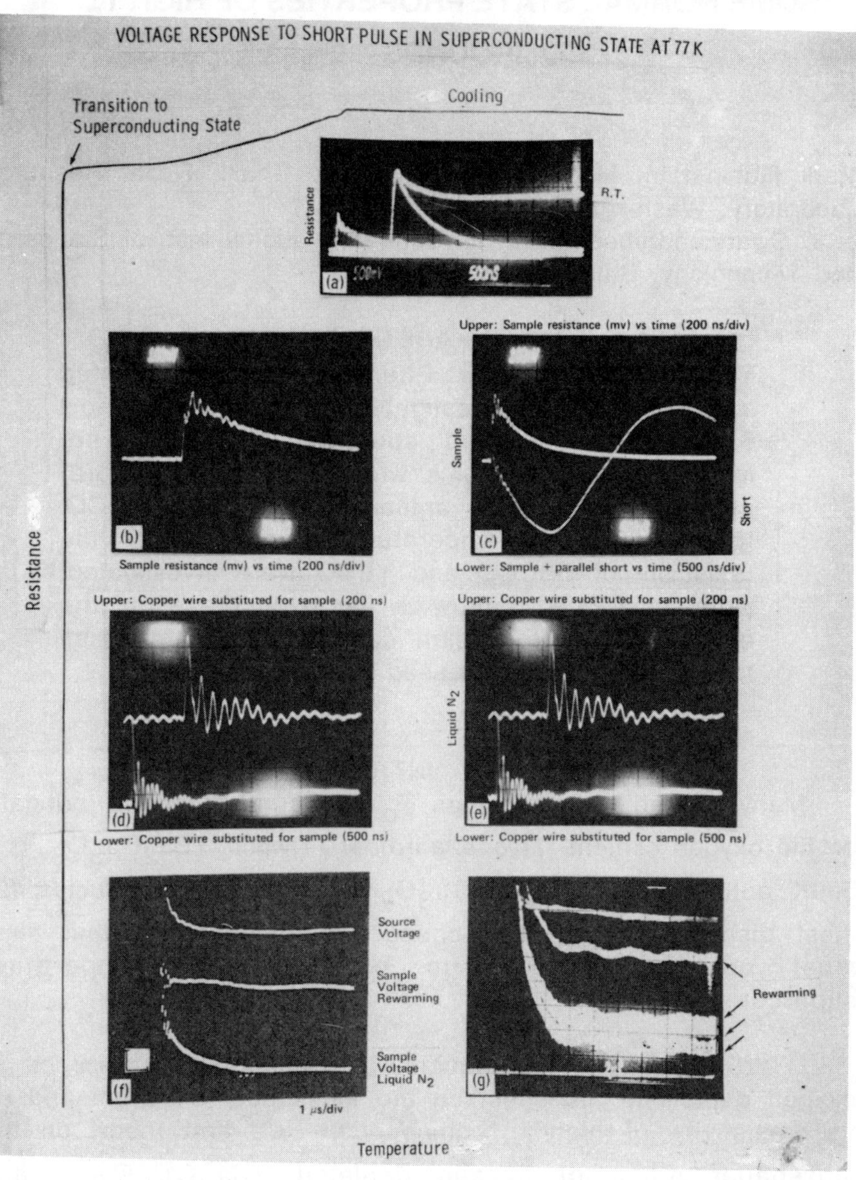
VOLTAGE RESPONSE TO SHORT PULSE IN SUPERCONDUCTING STATE AT 77 K
Transition to Superconducting State
Cooling
Resistance
R.T.
(a)
Upper: Sample resistance (mv) vs time (200 ns/div)
(b)
Sample resistance (mv) vs time (200 ns/div)
Sample
Short
(c)
Lower: Sample + parallel short vs time (500 ns/div)
Upper: Copper wire substituted for sample (200 ns)
Upper: Copper wire substituted for sample (200 ns)
Liquid N2
(d)
(e)
Lower: Copper wire substituted for sample (500 ns)
Lower: Copper wire substituted for sample (500 ns)
Source Voltage
Sample Voltage Rewarming
Sample Voltage Liquid N2
(f)
1 μs/div
Rewarming
(g)
Resistance
Temperature

Fig. 9

SOME NORMAL STATE PROPERTIES OF HIGH T_c COMPOUNDS

Mark Rubinstein, M.Z. Harford and T.K. Chaki, Naval Research Laboratory, Washington, D.C. 20375.
L. J. Swartzendruber and L. H. Bennett, National Inst. of Standards and Technology. Gaithersburg, MD 20899.

ABSTRACT

The ^{57}Fe Mossbauer Effect of oxygen-depeleted YBCO has been investigated. A magnetically split 6-line hyperfine field spectrum was observed, indicating that Fe atoms which substitute for Cu(2) atoms participates in antiferromagnetism of YBCO. In addition, the temperature dependence of the resistivity of BSCCO and YBCO were investigated. Both compounds show a large increase in the resistivity as the oxygen content is depleted from the samples with increased temperature.

INTRODUCTION

Many properties of the high T_c compounds depend critically on the oxygen content. ***Two examples:*** (1)while $YBa_2Cu_3O_6$ is a 500K antiferromagnet, $YBa_2Cu_3O_7$ is a 90K superconductor; (2) most bismuth- and thallium-based superconducting thin films must be post-annealed to produce high temperature superconductivity.

The present paper is primarily concerned with two aspects of oxygen depletion: its effect on the Mossbauer spectrum and on the resistivity of high-T_c compounds. We first report on the Mossbauer effect of oxygen depleted $YBa_2Cu_3O_{7-x}$ and $PrBa_2Cu_3O_{7-x}$ doped with ^{57}Fe . In addition to several quadupole-split spectra[1], the Mossbauer spectra display a magnetically-split six-line hyperfine field spectrum arising from those Fe atoms which substitute for Cu(2) sites in the most oxygen-deficient Y samples, and for all the Pr samples. The similarity between oxygen atom depletion and the incorporation of Pr^{4+} in place of Y^{3+}, in which both processes may lead to added electrons in the conduction band is evident.

Secondly, we have investigated the temperature dependence (at elvated temperatures) of the electrical resistivity[2] in $YBa_2Cu_3O_{7-x}$ and in $Bi_2Sr_2Ca_1Cu_2O_{8-x}$, where the oxygen content ,x, varies with temperature. Although the details are somewhat different, we find that in both compounds a large change in the resistivity occurs at high temperatures (compared to room temperature) which is correlated with the oxygen content of the sample.

MAGNETIC ORDER AND THE MOSSBAUER EFFECT

Room temperature ^{57}Fe Mossbauer spectra[3] of $RBa_2(Cu_{0.97}Fe_{0.03})O_{7-x}$[R=Y,Pr,Er] were obtained with varying x. The samples were quenched from several elevated temperatures into liquid nitrogen in order to change the oxygen content, x. In addition to a triplet of quadrupole-split pairs which we associate with the Fe atoms in Cu(1) sites (the "chains"), we observe a simple six line hyperfine splitting for the oxygen-depleted Y sample and all the Pr samples at room temperature, (but for none of the Er samples.) We interpret these hyperfine spectra as arising from those iron atoms in the Cu(2) sites(the "planar sites) which participate the antiferromagnetism of this plane. This antiferromagnetism is most likely to occur[4] when $x \approx 6.0$, or when R=Pr.

Fig I displays the Mossbauer spectra of BYCO doped with 3% ^{57}Fe quenched at a variety of temperatures. The "outer" quadrupole doublet, the "inner" quadrupole doublet and "singlet" features are apparent for samples quenched between room temperature and 800°C. Only when the sample is quenched at 950°C, and $x \approx 0.8$, does one discern a hyperfine Mossbauer spectrum at room temperature. Unlike BYCO, $PrBa_2Cu_3O_{7-x}$ doped with 3% Fe^{57} displays a room-temperature hyperfine field for all values of x. In **Fig. 2** we display the spectrum of both the

Y and Pr compound quenched at 950°C. A magnetically split six line hyperfine field with H_{eff} = 295kG, accounting for ≈ 25% of the spectral area, was found at 300K for the Y sample.

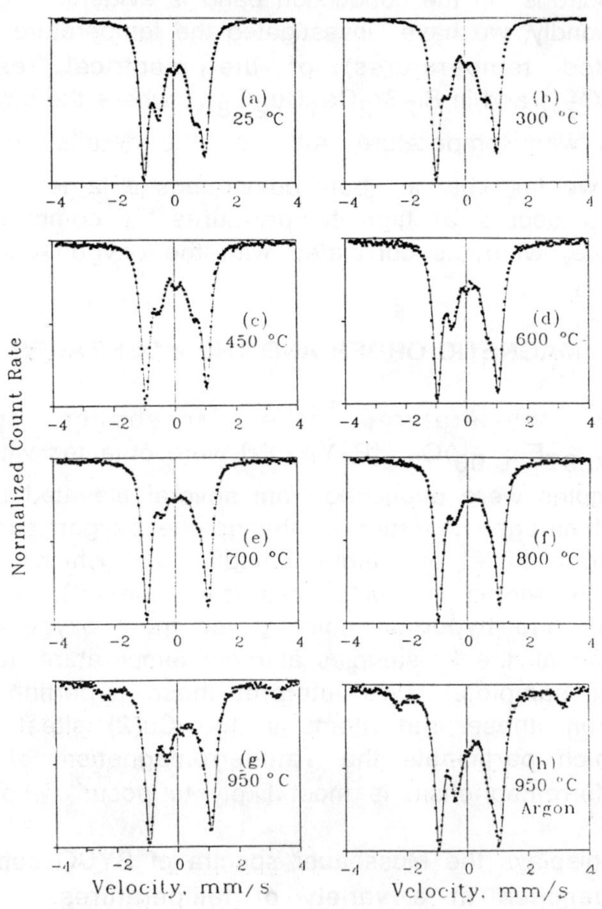

<u>FIG 1:</u> MOSSBAUER SPECTRA OF BYCO DOPED WITH 3% ^{57}Fe quenched at a variety of temperatures. The"outer doublet" and "inner doublet" are apparent. The central lines of a six-line hyperfine spectrum can be discerned in Fig 1(g) and 1(h).

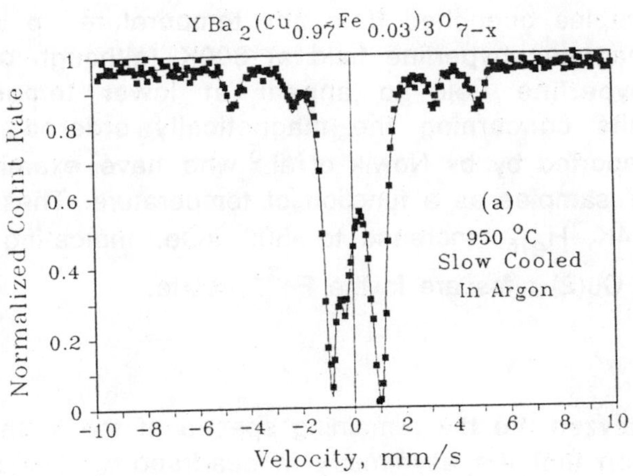

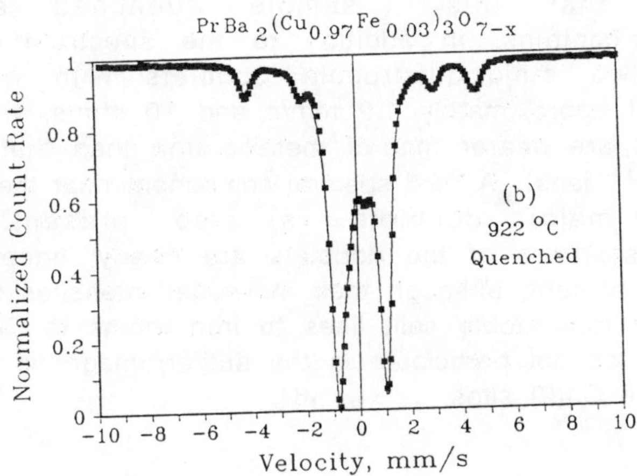

FIG 2: The room temperature Mossbauer spectra of oxygen-depleted $YBa_2Cu_3O_{7-x}$ and $PrBa_2Cu_3O_{7-x}$ that display both the six-line hyperfine-field and the manifold of quadrupole-split doublets.

For the Pr samples quenched from temperatures varying from 920°C to 30°C (x = 0.9 to x=0.1) hyperfine fields with H_{eff} = 270 kG were found, accounting of 20% of the spectral area. None of the Er samples quenched from any temperature up to 900°C showed a magnetic hyperfine field at 300K (although one might expect a hyperfine field to appear at lower temperatures.) Similar results concerning the magnetically ordered spectrum have been reported by by Nowik et al.[5] who have examined both the Pr and Y samples as a function of temperature. These authors find that at 4K, H_{eff} increase to ≈500 kOe, indicating that Fe atoms in the Cu(2) sites are in the Fe^{3+} state.

We have analyzed the the remaining spectra of the Y samples on the assumption that the asymmetry in quadrupolar line shapes is simply due to an induced preferential alignment of crystallites. We find that this sample quenched at room temperature contains, in addition to the spectrum discussed above, two major quadrupole doublets with quadrupole splitttings of approximately 1.9 mm/s and 10 mm/s, with isomer shifts which are nearer that of metallic iron than that of either Fe^{2+} or Fe^{3+} ions. A third spectral component near the center of the two major doublets is also present. The quadrupole splittings of the doublets are nearly independent of the oxygen content, although their individual intensites vary. We attribute the quadrupole split lines to iron atoms in Cu(1) sites. These sites do not participate in the antiferromagnetic order that exists on the Cu(2) sites.

HIGH TEMPERATURE PROPERTIES OF BYCO AND BSCCO.

In the preceeding discussion , we have demonstrated how the Mossbauer Effect can probe the effects of oxygen depletion on the high-T_c compounds, which we have done by examining the room-temperature spectra in samples quenched from high temperatures. This procedure allows the effects of oxygen desorption at high temperatures to be examined at ambient temperature.

There are other techniques, such as measurements of electron transport properties, which allow the sample to be probed directly at elevated temperatures where oxygen depletion is an important contribution to the physics. Below we report on our studies of high temperature resistivity and TGA (thermo-gravimetric analysis) in YBa_2CuO_{7-x} (a nominal 90K superconductor) and $Bi_2Sr_2Ca_1Cu_2O_{8-x}$ (a nominal 85K superconductor). Despite the fact that the Bi compound contains no "chains" we shall atempt to convince the reader that the resistivity of both compounds is probably dominated at high temperatures by a mechanism involving oxygen depletion - on the Cu-O chains for BYCO and on the Bi-O planes for BSCCO.

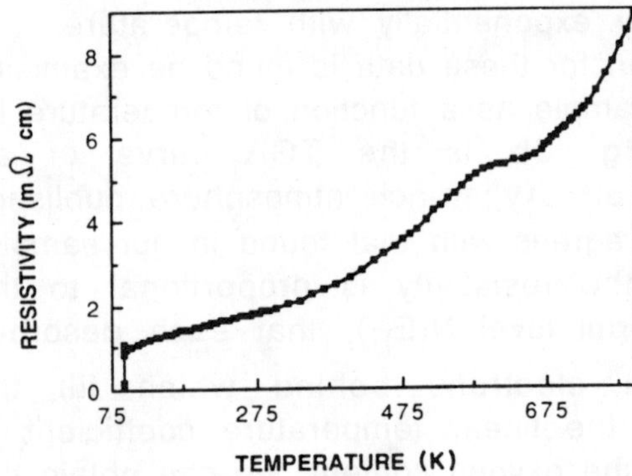

FIG. 3a. Resistivity vs. temperature curve for BYCO sample during heating.

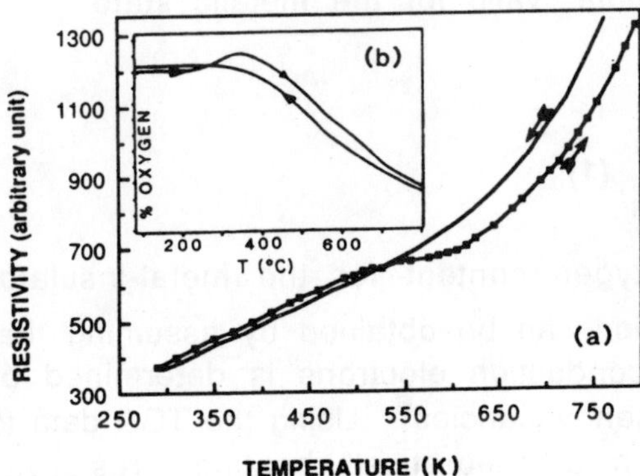

FIG. 3b. Graph of Eq. I, which can be compared to Fig. 3a. INSERT: Oxygen content determined by TGA. Heating and cooling curves are shown. Data taken from Ref. 7.

Fig. 3a displays a plot of the resistivity of a BYCO sample (which turned out to be somewhat oxygen deficient to start out) as a function of temperature in an oxygen-rich atmosphere. The abrupt transition at 90K is the superconducting-normal state transition. The well-known linear temperature dependence of the resitivity is observed from 90K to 300K. Above 300K the slope begins to increase with increasing temperature. An anomaly is noted around 600K where the resistivity becomes practically independent of of temperature. Above 700K the resistivity rapidly increases with increasing temperature, an effect which we attribute to the fact that our sample was initially oxygen deficient. The experiment was terminated at 800K, but other workers[6] have determined that the resistivity continues to increase nearly exponentially with temperature.

A reasonable explanation for these data is found be examining the oxygen content of the sample as a function of temperature by TGA. The insert of **Fig 3b** is the TGA curve of an oxygen-deficient sample in an oxygen-rich atmosphere published by Steinfink et al.[7], which agrees with that found in our sample. If we assume that the the resistivity is proportional to the density of states at the Fermi level $N(E_F)$, that each desorbed oxygen atom leaves two electrons behind which fill the conduction band, and that the linear temperature coefficient of resistance is not altered by the oxygen content, we can obtain the following approximate formula[2] valid for the metalic state

$$\rho = AT/[\delta_c - \delta(T)] \qquad (1)$$

where δ_c is the critical oxygen content for the metal-insulator transition. A similar expression can be obtained by assuming that the mean-free-path of the conduction electrons is determined by the distance betweeen oxygen vacancies[6] Using the TGA data to determine $\delta(T)$, we obtain a "theoretical" heating resistivity curves shown in the insert of Fig. 3b, which agree well with the exerimental resistivity curve shown in Fig. 3a - even with respect to their temperature hysteresis. The plateau of Fig 3a and 3b is an artifact, caused by an uptake of oxygen into the oxygen-depleted sample, which occurs at approximately $400^{o}C$; the TGA of the sample is in a pure oxygen atmosphere is shown in the insert of Fig. 3b

The high temperature properties of $B_2Sr_2Ca_1Cu_2O_{8-x}$ possesses high-temperature properties which are strikingly different from the BYCO type compounds. For example, BSCCO melts at approximately $900^{o}C$, and the molten substance can be poured into a mold to serve ,e.g., as targets for thin film fabrication.

We have investigated the temperature dependence of the electrical transport and the thermogravimetric properties, from $-200^{o}C$ to $+1000^{o}C$ of the BSCCO high-temperature superconductor. We conclude that this compound has a cooperative, simultaneous, melting - oxygen desorption - metal-insulator transition which occurs near $900^{o}C$.

Fig. 4 displays the temperature dependence of the electrical resistivity for BSCCO from ambient temperature to $950^{o}C$ in air. The inset in Fig. 5 includes the same data plotted on a logarithmic scale, plus low temperature data on the same sample,revealing the superconducting transition. The relatively small hysteresis of $\approx 50^{o}C$ at the $900^{o}C$ transition is not shown in the figure, but the existence of hysteresis demonstrates that the transition is first order.

The thermogravimetric data for BSCCO in air, heated at a rate of $3^{o}C/min$ is displayed in **Fig. 5.** At the end of the run ($1000^{o}C$), the sample is completely melted. Approximately 1 oxygen atom per formula unit is desorbed from ambient temperature to $1000^{o}C$. The inset in Fig. 5 displays the "theoretical" temperature-dependent resisistivity using Eq. l and the experimental TGA data also shown in Fig. 5. The agreement between this "theory" curve (Fig. 5 insert) and the measured resistivity (Fig. 4) demonstrates that the basic physics presumed by Eq. 1 can constitute a valid approximation to a complex situatation.

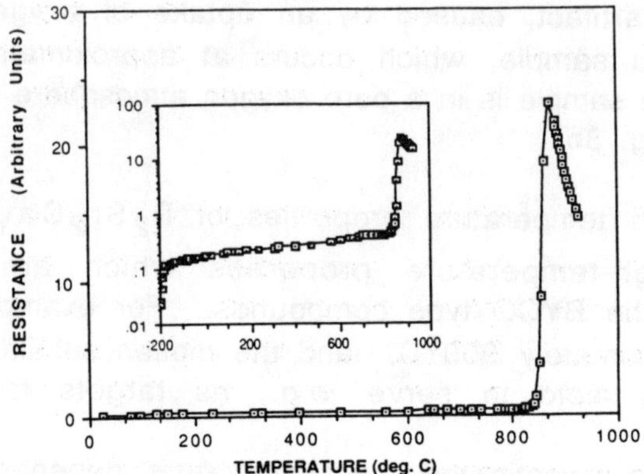

FIG. 4: Temperature dependence of the electrical resistivity for BSCCO from ambient temperature to 950°C. Inset includes the same data on a logarithmic scale, plus low temperature data on the same sample.

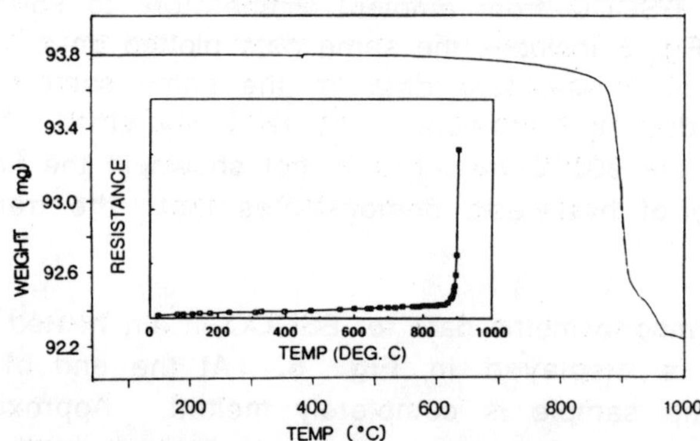

FIG. 5: Thermogravimetric data for BSSCO heating in air at 3°C/min. At end of run, sample had completely melted. Insert displays the theoretical temperature dependence of the electrical resistivity, derived using Eq. I with the TGA data as a measure of the temperature -dependent oxygen desorption parameter δ.

The temperature in the resistivity curve of Fig. 4 at which the resistivity abruptly increases two orders of magnitude coincides with the temperature in the TGA curve of Fig 5 where the weight change of the BSCCO sample abruptly decreases by an amount corresponding to 1 oxygen atom per formula weight. The abrupt decrease in resistivity is seen to be accompanied by the equally abrupt decrease in the oxygen content of the sample.

The metal-insulator/melting transition may occur due to the loss of oxygen in the Bi-O layers. Band structure calculations indicate that the Bi-O bands contribute to the density of states at the Fermi surface. Loss of oxygen from these bands will cause a charge redistribuion between the BiO planes and the CuO planes, removing carriers from the Cu-O bands, which are presumably responsible for the superconducting transition. The roles of the Cu-O chains and the Bi-O planes may be quite similar in BYCO and BSCCO, repectively, as far as their roles in donating carrriers to the conduction bands are concerned.

Above $\approx 875^{o}C$, the resistivity decreases with increasing temperature, as is observed in Fig. 4. From the high-temperature slope, assuming that the high-temperature resistivity decreases with temperature as $\rho \propto e^{(Eg/kT)}$, we deduce that an (pseudo) energy gap $E_g \approx 0.5$ e,v. develops in the liquid state of BSCCO.

REFERENCES

(1) X.Z. Zhou, M.Raudsepp, Q.A. Pankhurst, A.H. Morrish. Y.L. Luo, and I. Maartense, Phys Rev. 36, 7230 (1987); P. Boolchand. C. Blue, K. Elgaid, I. Zitovsky D. McDaniel, W. Lemar, D Farrell and B.S. Chandresekhar, Phys. Rev.,to be published , and references contained therein.

(2) T. K. Chaki and Mark Rubinstein, Phys Rev. B 36, 7230 (1987)

(3) L. J. Swartzendruber, L.H. Bennett, M. Z. Harford and Mark Rubinstein, Journal of Superconductivity 1, 219 (1988)

(4) J.M. Tranquada, D.E. Cox, W. Kunnmann, H. Moudden, G. Shirane, M Suenaga, P. Zolliker, D. Vanknin, S. Sinha, M. Alvarez, A. Jacobsen and D. Johnston, Phys. Rev. Lett. 60, 156 (1988)

(5) I. Nowik, M Kowitt, I. Felner, E.R. Bauminger, Phys. Rev. B38, 6677 (1988)

(6) A. T. Flory, M Gurvitch, R.J. Cava, G.P. Espinosa, Phys. Rev. B36 , 7262 (1987)

(7) H. Steinfink, J.S. Swinnea, A Manthiman, Z.T.Sui, and J.B. Goodenough in Proceeding of the International Workshop on Novel Mechanisms of Superconductivity, Berkeley 1985, edited by S.A. Wolf and V. Z. Kresin(Plenum, New York, 1987) p. 1067

MEASUREMENT OF MAGNETIC PROPERTIES IN A MELT CAST Bi-Ca-Sr-Cu-O SUPERCONDUCTOR

L.J. Swartzendruber and L.H. Bennett
National Institute of Standards and Technology
Gaithersburg, MD 20899

and

C.F. Gallo, Superconix, Inc.
St. Paul, MN 55144

ABSTRACT

Magnetic measurements were made to examine the superconducting properties of a crystalline chunk, consisting of many small crystals, prepared by casting from the oxide melt with a starting composition of $Bi_3Ca_2Sr_2Cu_3O_x$. AC susceptibility revealed an onset temperature of 82 K and a transition width of ≈10 K. Hysteresis loops at 70 K showed a very small hysteresis, indicating a low density of effective flux pinning sites at this temperature, similar to the behavior observed for some superconducting samples of the Bi-Ca-Sr-Cu-O system prepared by other techniques. However, the magnetization at 10 K showed no hysteresis for fields greater than ≈3.5 kOe, a much smaller value than we have previously observed for other samples of this material regardless of preparation method, or for any other material with a T_c greater than 70 K. This critical field for flux depinning, H_{cp}, varies with temperature as $H_{cp} = 4980(1-t)^{2.5}$, where t is T/T_c.

INTRODUCTION

The recently discovered Bi-Sr-Ca-Cu-O superconducting oxides appear to have certain advantages over the 1-2-3 materials of which $YBa_2Cu_3O_7$ is the prototype. Among these advantages are a phase with a higher superconducting transition temperature (T_c), the lower cost of bismuth as compared to yttrium, ease of processing particularly with respect to annealing, and indications of a higher stability with respect to air and moisture. Another important characteristic of these materials is the magnetic field dependence of their superconducting properties, which we examine here and compare with that of $YBa_2Cu_3O_7$.

Many applications of the high T_c superconductors pertain to their ability to carry large currents. The critical current density arises from the equilibrium between the Lorentz force exerted by the transport current and the pinning forces operating on the magnetic flux lines. The flux lines often form a flux lattice so that a small number of pinning centers will pin many flux lines. The observation of a hysteresis loop conforming to the Bean critical-state model suggests the presence of conventional flux pinning[1]. In this model, the magnetic critical current is derivable from the loop parameters.

In conventional superconductors, the movement of flux lines occurs with flux "creep" when pinning governs[2] and flux "diffusion" when the transport current causes energy dissipation[3]. Both of these dissipative phenomena were observed[1] in the same sample of $YBa_2Cu_3O_7$. In addition to these dynamic effects, a new behavior has been observed in the high T_c superconductors[4-8], namely the occurrence of flux depinning at a critical temperature and field. The critical field for flux depinning found in the present sample is uncommonly low.

MATERIAL

Using a technique of casting from the oxide melt, various superconducting crystals of Bi-Ca-Sr-Cu-O were obtained. The degree of crystallinity, density and superconducting properties vary according to the starting composition and process variables. One of the resultant "crystals" was actually a crystalline chunk, consisting of many small crystals, that had been prepared from the melt with a starting composition of $Bi_3Ca_2Sr_2Cu_3O_x$. The sample had an irregular shape with the longest dimension ≃3 mm and weighed approximately 9 mg. We have made magnetic measurements on this sample as part of its characterization. The results are sufficiently interesting to warrant reporting on them at this time. In particular, the effective flux pinning weaker in this sample than in any other we have observed so far.

MEASUREMENTS

The real and imaginary parts of the ac susceptibility were measured in a Hartshorn-type bridge with an applied ac field of about 40 A/m (0.5 Oe) at a frequency of 1.68 kHz. The hysteresis loops were obtained in a vibrating sample magnetometer.

The real part of the ac susceptibility shows a fairly sharp diamagnetic transition at 82 K, a 10% to 90% width of ≈15 K, and achievement of ≈80% Meissner effect at low temperature (see Fig. 1). The imaginary part of the ac susceptibility shows a weak peak at about the center of the 10%/90% transition. These observations indicate that the sample was predominately single phase, but with a distribution of transition temperatures, most likely due to compositional variations within that phase.

Fig. 2 shows a hysteresis loop at 55 K for the same material used for the measurements in Fig. 1. This loop is narrow, indicating the presence of only a small amount of flux trapping. In addition, the loop is constricted in the center, indicating the probable existence of dynamic effects similar to those found[6] in $YBa_2Cu_3O_7$.

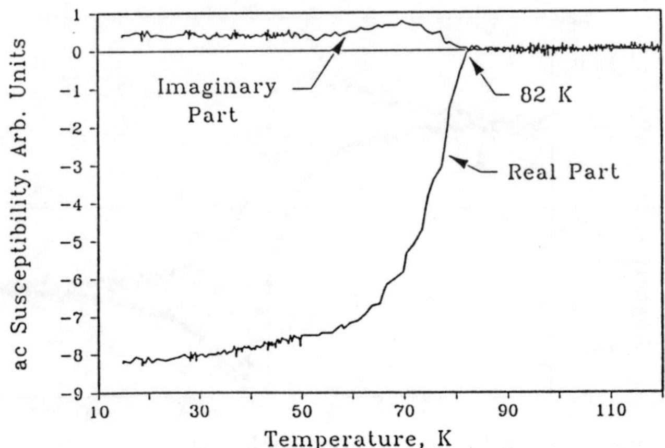

Fig. 1. AC susceptibility as a function of temperature for a crystalline chunk of Bi-Sr-Ca-Cu oxide prepared by melt casting. The data above zero represent the imaginary part of the susceptibility, while those below represent the real part. The onset temperature for superconductivity is 82 K.

The hysteresis disappears in this loop for applied fields above ≈400 to 500 Oe. This effect was observed previously (at somewhat different fields) in the hysteresis loops of Bi-Ca-Sr-Cu-O system prepared by other techniques[9,10]. Earlier, Müller, Takashige, and Bednorz[4] had noted a reversible-irreversible transition in temperature sweeps in La_2CuO_{4-y}, and this type of effect was also measured by a number of different kind of experiments in[6,8] Y-Ba-Cu-O and in[7,8] Bi-Sr-Ca-Cu-O high temperature superconductors.

Examination of the plots shown in Figs. 2 and 3 suggests the existence of a critical field, H_{cp}, above which flux pinning vanishes. The exact definition of such a critical field in our experiment has some uncertainty since the transition is gradual. If we define the critical field as the field at which the separation between the magnetization for increasing and decreasing fields is less than one-tenth the magnetization (see Fig. 3), then a log-log plot of the critical field versus the temperature interval below the critical temperature is a straight line (see Fig. 4). The exponent is very similar to what Lundy et al observed[9] in a different sample, but the prefactor is considerably different in the two samples.

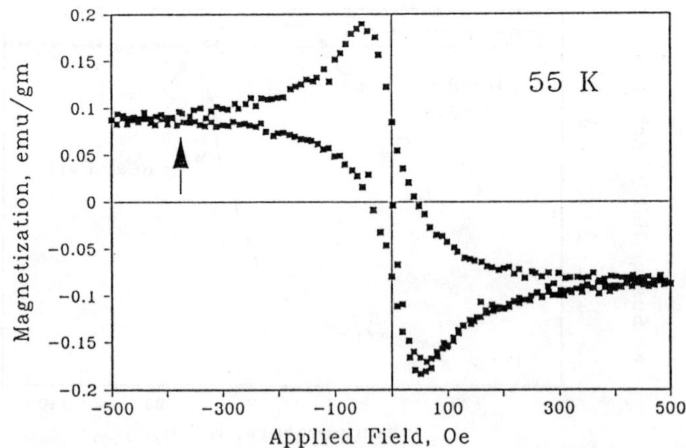

Fig. 2. Magnetization versus applied field at 55 K for the same sample as in Fig. 1. The sample is cooled from above the superconducting transition temperature in zero field, and the measurements started by increasing the field, displaying the initial susceptibility, before traversing the loop.

The value of H_{cp} was found to be orientation dependent. Hence in the crystalline chunk we have measured, its value depends critically on the positioning of the sample in the magnetometer. Thus, the results presented here represent an average over several crystalline directions. Details of the orientation dependence will have to await further experimentation with single crystals. Furthermore, if it is assumed that the sample consists of more than one phase, with more than one T_C, another interpretation of the data shown in Fig. 4 is posssible. Instead of fitting the data to one line, the H_{cp} values in Fig. 4 could be subdivided into several parallel lines with a smaller slope than that shown.

ACKNOWLEDGEMENTS

We thank R. Drew and D. Mathews for technical assistance.

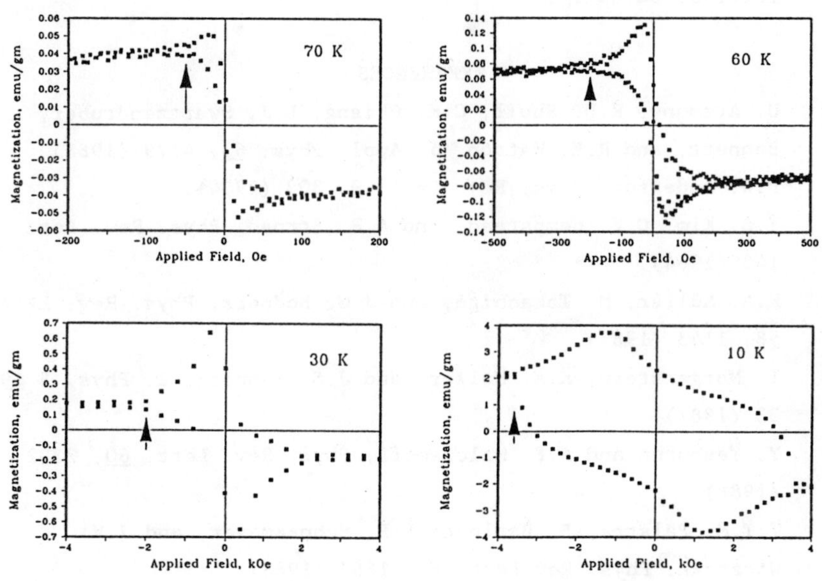

Fig. 3. Magnetization versus applied field at several temperatures, showing (arrow) the critical field for flux depinning, as defined in the text.

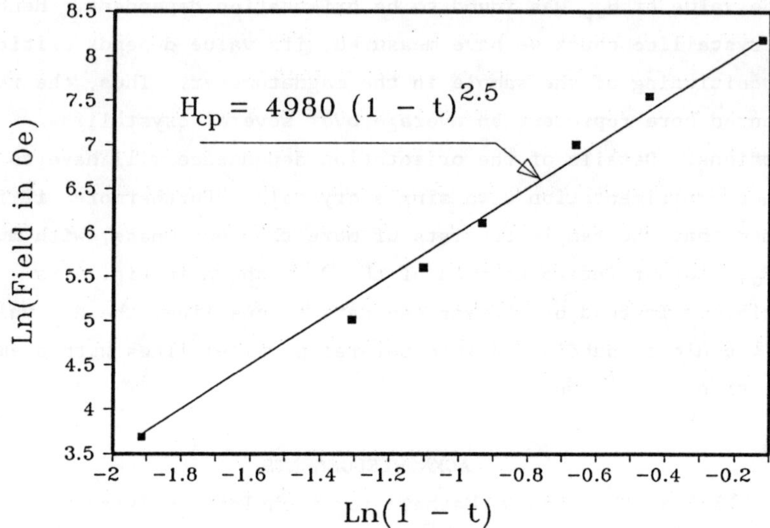

Fig. 4. Natural logarithm of the critical fields, as defined in text (see Fig. 3), versus the natural logarithm of the temperature interval below T_c.

REFERENCES

1. U. Atzmony, R.D. Shull, C.K. Chiang, L.J. Swartzendruber, L.H. Bennett, and R.E. Watson, J. Appl. Phys. 63, 4179 (1988).
2. P.W. Anderson, Phys. Rev. Lett. 9, 309 (1964).
3. Y.B. Kim, C.F. Hempstead, and A.R. Strnad, Phys. Rev. Lett. 12, 145 (1964).
4. K.A. Müller, M. Takashige, and J.G. Bednorz, Phys. Rev. Lett. 58, 1143 (1987).
5. I. Morgenstern, K.A. Müller, and J.G. Bednorz, Z. Phys. B 69, 33 (1987).
6. Y. Yeshurun and A.P. Malozemoff, Phys. Rev. Lett. 60, 2202 (1988).
7. T.T.M. Palstra, B. Batlogg, L.F. Schneemeyer, and J.V. Waszczak, Phys. Rev Lett. 61, 1662 (1988).
8. P.L. Gammel, L.F. Schneemeyer, J. V. Waszczak, and D. J. Bishop, Phys. Rev. Lett. 61, 1666 (1988).
9. D. Lundy, this workshop.
10. F.J. Adrian, J. Bohandy, B.F. Kim, K. Moorjani, J.S. Wallace,

ANALYSIS OF MICROWAVE SURFACE RESISTANCE IN HIGH Tc SUPERCONDUCTORS

Peter J. Walsh, Physics Department

Fairleigh Dickinson University, Teaneck, NJ 07660

ABSTRACT

A high frequency electromagnetic wave couples to a superconductor surface through the spatially damped surface supercurrents which penetrate the superconductor a finite distance. At any temperature below the transition temperature T_c, normal electrons are excited across the superconductor gap while some superconducting electrons lose phase coherence. Both processes remove energy from the penetrating wave producing a measurable microwave surface resistance. This surface resistance can serve as a very useful diagnostic tool for analyzing the new high T_c superconducting materials in the regime below T_c where, of course, the dc resistance is zero. An example of such data and its analysis are reviewed for YBCO where the common 1-2-3-7 90 K phase is present with the 60 K oxygen--deficient bulk phase and, apparently, the 80 K 2-4-8-20 phase. Analysis of the microwave characteristics of the high T_c superconductors should take account of their large penetration depths but small coherence lengths and moderate to small mean free paths. Traditional Mattis--Bardeen theory is reviewed along with "dirty" superconductor modification. A single integral local-limit form is presented applicable to any ratio of mean free path to coherence length.

1. INTRODUCTION

The new classes of high temperature superconductors hold the promise of important applications as low loss, low noise and wide bandwidth elements in microwave circuits.[1] In addition, investigation of the microwave surface resistance of the new superconductors offers unique insight into the structure and properties of these novel materials.

Fig. 1 presents measurements of the surface resistance of polycrystalline YBCO at 16.5 GHz obtained by Hanscom AFB scientists and analyzed with the author's assistance.[2] In addition to the normal superconducting transition, here at 88 K, the data clearly show the presence of two anomalous resistance discontinuities, at 52 K and 83 K, which are presumably associated with the 1-2-3-6.7 oxygen deficient, "60 K bulk" phase of YBCO [3] and the 2-4-6-20 oxygen excess, double-cell phase whose transition temperature is near 80 K.[4] These lower temperature phases would not be apparent with conventional resistance measurements since the transition at 88 K would mask the two lower transitions in a single polycrystalline sample.

The analysis shown by the solid curve is based on the use of the Sridhar "dirty" superconductor modification[5] of rigorous Mattis-Bardeen theory.[6] In addition, the analysis assumes that the presence of a lower transition superconducting phase acts as an additive residual contribution to the surface resistance.

Data such as illustrated in Fig. 1 can be analyzed to estimate important fundamental parameters of specific superconductors such as their coherence length, pair mean free path penetration depth and phase composition. These quantities, in turn, allow scientists and engineers to predict how worthwhile a superconductor will prove in use as microwave circuit elements.

This paper reviews a number of important aspects of the theory of microwave surface impedance of superconductors and presents a new single-integral form of the theory applicable for any range of the ratio of mean free path to coherence lengths in the high temperature superconductors.

2. SURFACE IMPEDANCE ANALYSIS

The analysis of Fig. 1 assumes that there are a number of residual phases which contribute to the surface resistance according to

$$R(T) = R_0 + R_1(T) + R_2(T). \tag{1}$$

Here R_0 represents the residual contribution to superconducting surface resistance which is substantially independent of temperature and commonly found in single phase material[7] while R_1 and R_2 are temperature dependent superconducting contributions.

The use of a single normal state residual resistance in the analysis of microwave surface resistance of superconductors is a long-standing empirical practice. The appearance in multi-phase superconducting material of a additive set of superconducting residual resistances is novel and requires study. Practical devices will require low residual resistance and so understanding the nature of the residual resistance is critical to the microwave use of superconductors. A particular question is whether the residual sum given in Eq. 1 applies also to the surface reactance.

3. "DIRTY' SUPERCONDUCTOR ANALYSIS

The "dirty" superconductor analysis illustrated by the solid curve of Fig. 1 uses the following formula[5] for surface resistance

$$R/R_n = \mathrm{Re}\,\{ 2i / [(\sigma_{1c}/\sigma_n) - i(\sigma_{2c}/\sigma_n)m]\}.^{1/2} \quad (2)$$

The quantities σ_{1c} and σ_{2c} are the real and imaginary Mattis-Bardeen superconductor surface conductivities in the coherent limit as discussed below and σ_n is the normal state conductivity at the same temperature as the superconductor. The complex conductivity is defined by $\sigma = \sigma_1 - i\,\sigma_2$. Because the coherent limit was first applied by Mattis and Bardeen to the case of extreme anomalous surface conduction it is often referred to as the extreme anomalous limit but is equally applicable to the local case characteristic of the high temperature superconductors. The adjustable "dirty" parameter m is assumed in Ref. 5 to be related to the square of the london penetration depth to an effective penetration depth. It is an open question whether the "dirty" parameter follows from rigorous Mattis-Bardeen theory.[8]

4. COHERENCE EFFECTS IN LOCAL LIMIT SUPERCONDUCTORS

The new classes of high temperature superconductors have very large penetration depths, moderate to small mean free paths and small coherence lengths. The ratio of mean free path to penetration is quite small placing these new superconductors well in the local limit regime of microwave surface impedance where the conductivity depends solely on the local electromagnetic fields at a given depth. Mattis-Bardeen theory then

gives the following formula for the complex surface impedance:

$$Z/Z_n = (\sigma/\sigma_n)^{1/2} \tag{3}$$

The real and imaginary parts of the superconducting conductivity are given by

$$\sigma/\sigma_n = (3/4) \iint \exp(-R/L)\ (1 - U\ )\ H\ dU\ d(R/L). \tag{4}$$

The distance from a carrier to the point where the conductivity is evaluated is R and U is the cosine of the angle between R and the electric field which is along the surface. The mean free path is L. The integration on R is from zero to infinity and on U from -1 to 1.

The quantity H represents the Mattis-Bardeen integrals (in their notation, $I/(-\pi ihf)$). $H = H_a + H_b$ where H_a is complex but H_b is real. Then

$$\sigma_1/\sigma_n = \mathrm{Re}\ H_a, \qquad \sigma_2/\sigma_n = H_b - \mathrm{Im}\ H_a. \tag{5}$$

In the new superconductors the energy gap is almost always larger than the microwave wave energy hf at the frequency f and

$$H_a = (1/v) \int du\ [F_p(g \cos au_2 - i \sin au_2)\exp iau_1 \\ -F(g \cos au_1 + i \sin au_1)\exp iau_2] \tag{6}$$

$$H_b = (1/v) \int du\ F(|g| \cos au_2 + \sin au_2)\exp -a|u_1|. \tag{7}$$

The integral H_a represents conduction by normal particles thermally excited across the superconducting gap while H_b describes conduction due the superconducting pairs. The integral for H_a runs from u_o to infinity while that for H_b runs from u_o -v to u_o. H_b is purely real. The double bar I I denotes absolute value.

The u's and v are scaled energies and frequency given by

$u = E/kT, \; u_o = D/kT,$ (8)

$u_1 = (u^2 - u_o^2)^{1/2}, \qquad u_2 = \{(u+v)^2 - u_o^2\}^{1/2}$ (9)

$v = hf/kT$ (10)

E and D are the pair energy and energy gap halfwidth, respectively. Using the fermi function as f(u) we have

$F = 1 - f(u), \qquad F_p = 1 - f(u+v).$ (11)

The density of superconducting pairs is

$g = (u^2 + uv + u_o^2)/u_1 u_2 .$ (12)

Finally,

$a = sR/L, \qquad s = LkT/hv_f$ (13)

with the fermi velocity as v_f. The parameter s is closely related to the ratio of mean free path to coherence length and determines the effects of phase coherence due to both the excited normal particles and the superconducting pairs. When $s \lll 1$ the coherence is full and

$H_{ac} = (1/v)\int (F_p - F)\, g\, du$ (14)

$H_{bc} = (1/v)\int F_p |g|\, du$ (15)

These values of H determine the conductivity ratios to be used on the "dirty" modification.

With other values of s, even small ones, variable coherence makes substantial changes to the conductivity and surface resistance. In these cases Eq. (4) can still be integrated twice to give

$$\mathrm{Re}\, H_a = (1/2v)\int (F_p-F)[(g+1)/(1+s^2 u_d^2) + (g-1)/(1+s^2 u_s^2)]\, du \qquad (16)$$

$$\mathrm{Im}\, H_a = (s/2v)\int [[(F_p+F)(g+1)u_s/(1+s^2 u_s^2)] - [(F_p-F)(g-1)u_d/(1+s^2 u_d^2)]]\, du \qquad (17)$$

$$H_b = (1/v)\int F[(1+s\, u_1)|g| + s\, u_2]/[(1+s\,|u_1|) + s^2 u_2^2]^2\, du. \qquad (18)$$

Here

$u_s = u_1 + u_2 \qquad u_d = u_1 - u_2.$ (19)

5. SOME CONCLUSIONS

The solid curve of Fig. 2 present calculations using the variable coherence formulas given above for the surface resistance as a function of scaled temperature for a wide range of the coherence parameter s. The scaled frequency v is .01 in these figures. The single numerical integrations involved in Eqs. (15) to (17) run rapidly compared to the full Mattis-Bardeen calculations which still require a triple numerical integration after one Turneaure analytic integration.[9] The open circles in Fig. 2 give the results of full Mattis-Bardeen calculations for comparison to the variable coherence calculations. It can be seen from these curves that $s \ll 0.1$ is a reasonable requirement for the coherent limit to be applicable.

Preliminary analysis indicates that a value of s near 4 will matches the dirty multiplier analysis with m =.045 for the data between 54 K and 83 K in Fig. 1. Over a restricted range of temperatures it does appear that the use of a dirty multiplier does mimic the variable coherence formulae derived from Mattis-Bardeen analysis.

REFERENCES

1. Paul H.Carr, "Potential Microwave Applications of High Temperature Superconductors", Microwave Journal, 91-93, December 1988.

2. Peter J. Walsh, John S. Derov, W. D. Cowan, C. Von Benken and A. Drehman, "Measurement and Analysis of the Microwave Resistance in a Mixed Phase Y-Ba-Cu-O Superconductor", to be submitted, Appl. Phys. Letters.

3. R. J. Cava, B. Battlogg, C. H. Chen, E. A. Rietman, S. M. Zhurak and D. Werder, "Single phase 60-K bulk superconductor in annealed Ba2YCu3O7-d(0.3<d<0.4) with correlated oxygen vacancies in the CuO chains ," Phys. Rev. B 36, 2305, 1987.

4. D. D. Berkley, D. H. Kim, B. R. Johnson, A. M. Goldman, M. L. Mecartney, K. Beauchamps and J. Maps, "Preparation of Y2Ba4Cu8020-x thin films by thermal coevaporation," Appl. Phys. Lett. 58, 708-709, 1988.

5. S. Sridhar, C. A. Shiffman and H. Hamdeh, "ElectrodynamicResponse of YlBa2Cu3Oy and Lal.85Sr0.15cuO4-d in the Supercon¡ducting State", Phys. Rev. B 36, 2301 - 2305, 1987.

6. D. C. Mattis and J. Bardeen, "Theory af the Anomalous SkinEffect of Normal and Superconducting Metals," Phys. Rev, 111, 412- 417, 1958.

7. J. I. Gittleman and Bruce Rosenbloom, "Microwave properties of superconductors," Proc. IEEE 52, 1138 - 1147, 1964.

8. Various limiting forms of the expressions for superconducting surface impedance are presented in

Sang Boo Nam, "Theory of electromagnetic properties of superconducting and normal systems. I", Phys. Rev. 156, 470-486, 1967.

9. J. P. Turneaure, Microwave Measurements of the Surface Impedance of Superconducting Tin and Lead, Ph. D. dissertation, Stanford University, Stanford, CA 1967.

FIGURE CAPTIONS

Fig. 1. YBCO absolute surface resistance at 16.6 GHz.Temperatures are acurate within ± 1 K over the region where the points are dense. The solid line represents residual analysis using "dirty" modification of Mattis - Bardeen theory.

Fig. 2. Normalized superconducting surface resistance versus normalized temperature using the variable coherence formulation given here. In this figure v = .01 and the open circles represent full Mattis-Bardeen calculations.

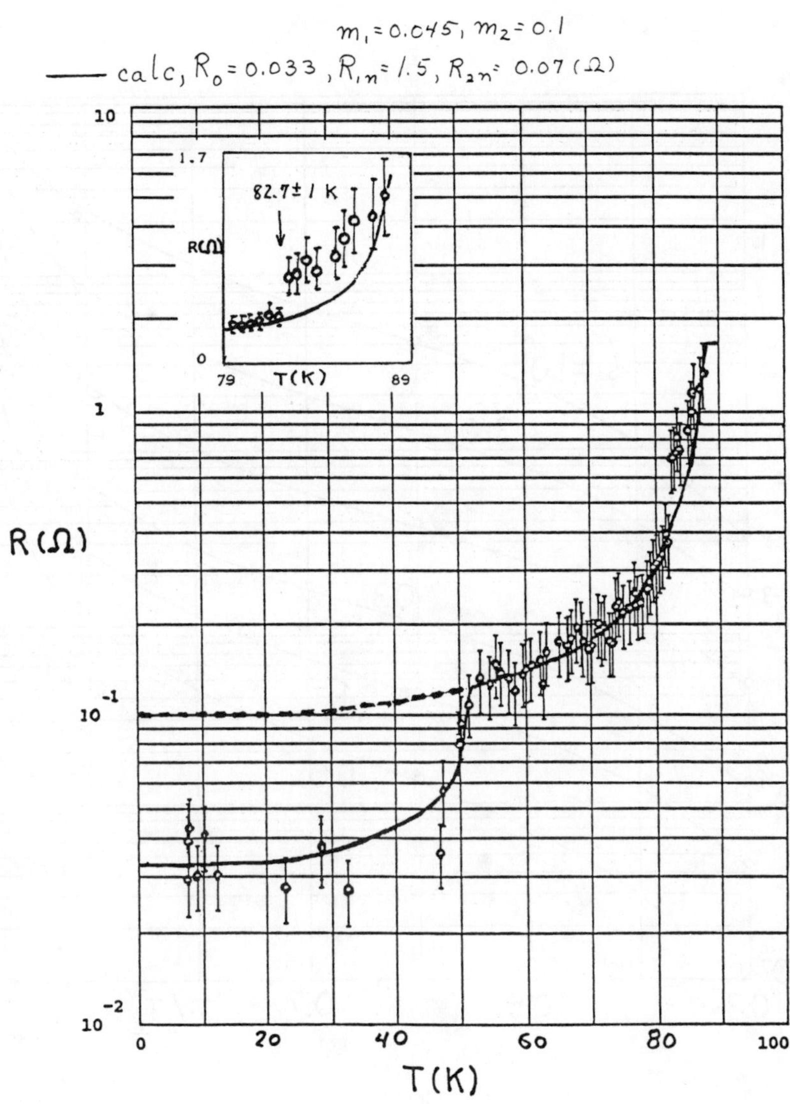

m1=0.045, m2=0.1
calc, R0=0.033, R1n=1.5, R2n=0.07 (Ω)
10
1
10-1
10-2
R(Ω)
1.7
0
82.7±1 K
R(Ω)
79
T(K)
89
0
20
40
60
80
100
T(K)

Fig. 1

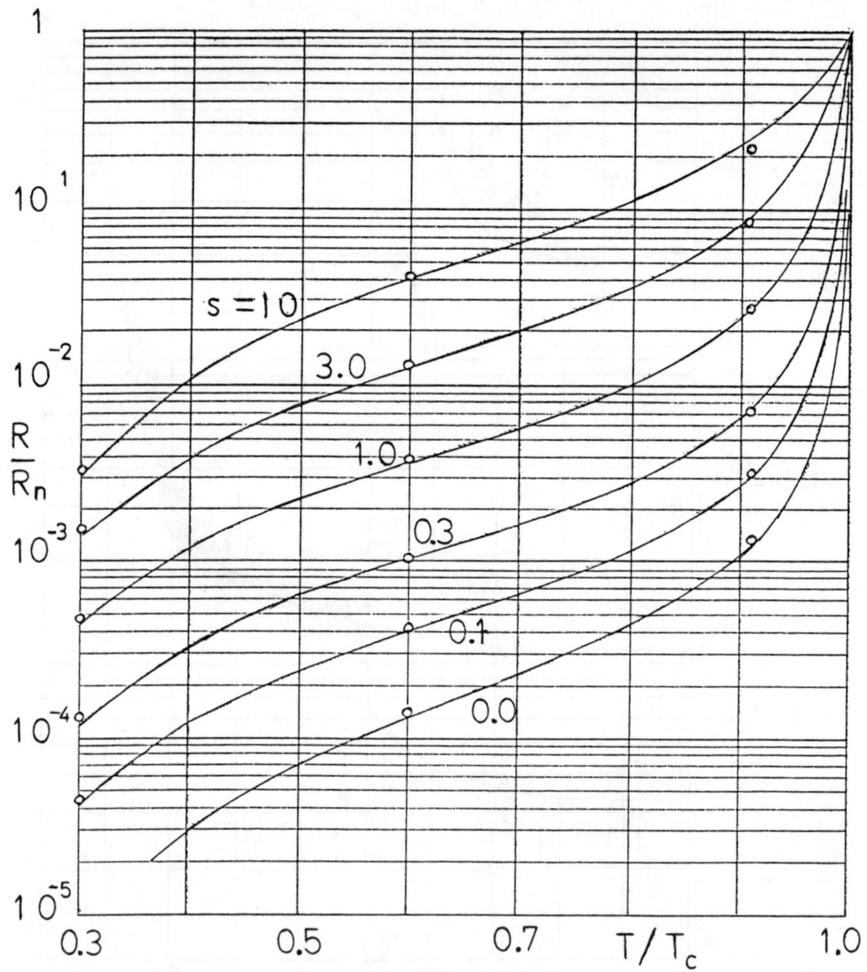

Fig. 2

^{57}Fe MÖSSBAUER STUDY OF A Tl-Ca-Ba-Cu-O HIGH-T_c SUPERCONDUCTOR

A. H. Morrish, X. Z. Zhou* and Y. L. Luo*
Department of Physics, University of Manitoba
Winnipeg, MB R3T 2N2, Canada

ABSTRACT

Samples of the high-T_c superconductor $Tl_2Ca_2Ba_2$-$(Cu_{1-x}Fe_x)_3O_{10}$ were made by a standard ceramic method. Resistivity and ac susceptibility measurements show that the transition temperature, T_c, decreases with increasing x but at a rate that is much smaller than for the iron substituted Y-Ba-Cu-O system. For x = 0.02 the Mössbauer spectra between 10 and 300 K consist of two doublets; their relationship with the crystal structure is proposed. The iron orders in the vicinity of liquid helium temperatures.

INTRODUCTION

The substitution of iron, often enriched in the isotope ^{57}Fe, for copper in $YBa_2Cu_3O_{7-\delta}$, the 123 high-temperature superconductor, has led to Mössbauer studies in a number of laboratories, including our own. Since these investigations have borne considerable fruit, we have commenced experiments on iron substituted $Tl_2Ca_2Ba_2Cu_3O_{10}$. In addition to the use of Mössbauer spectroscopy, these ceramic samples have been characterized by x-ray and electron diffraction and by resistivity and ac susceptibility measurements.

When iron occupies copper sites an additional microscopic local probe is then available to provide information on the superconducting coupling mechanism. This approach will be particularly important if it turns out that antiferromagnetic exchange interactions are involved.

* On leave from Southwest Institute of Applied Magnetics, Mianyang, Sichuan, China.

^{57}Fe Mössbauer spectra of $YBa_2(Cu_{1-x}Fe_x)_3O_{7-\delta}$ between about 10 and 300 K consist of doublets. We have consistently obtained good fits with three doublets[1], although sometimes others have made fits with two doublets[2] or with four[3]. We have also proposed site assignments; alternatives have also been made[4].

In the vicinity of liquid helium temperatures the iron orders in the 123 compound[5]. We have analyzed these spectra with a model that includes relaxation effects[6]. Similar experiments on $Tl_2Ca_2Ba_2(Cu_{1-x}Fe_x)_3O_{10}$ are not only of interest with respect to the new high-T_c superconductor family, but also have the potential to shed new light on the older 123 compounds.

SAMPLE PREPARATION

A ceramic procedure was used to make the samples of the high-temperature superconductor $Tl_2Ca_2Ba_2(Cu_{1-x}Fe_x)_3O_{10}$ (2223). Appropriate amounts of Tl_2O_3, CaO, BaO and CuO were mixed by ball milling with acetone for 5 hours. Then small amounts of a α-Fe_2O_3 corresponding to $x = 0.02$, 0.05 and 0.10 were added and mixed in a mortar. This procedure was made necessary because only a limited amount of Fe_2O_3 enriched to 93% ^{57}Fe was available to us. Small disk-shaped pellets, 8 mm in diameter and about 2 mm thick, were made for each iron concentration, heated at 850° C for 10 minutes, and then furnace-cooled. Oxygen was kept flowing during the entire heat processing in order to prevent Tl_2O_3 from decomposing before reacting and to assist in sample stability on cool-down[7]. Our experience indicates that the 2223 compounds are more sensitive to the preparation conditions than the 123 superconductors. Therefore, in order to ensure essentially uniform reaction conditions, all the samples including those with $x = 0$ were made simultaneously in the same furnace.

The x-ray diffraction pattern of $Tl_2Ca_2Ba_2(Cu_{0.98}Fe_{0.02})_3O_{10}$ is the same as that for $Tl_2Ca_2Ba_2Cu_3O_{10}$. Spectral analysis indicates that the structure is tetragonal with $a = 0.385$ nm and $c = 3.59$ nm[8].

RESISTANCE MEASUREMENTS

The electrical resistance of the samples were measured by the four-probe method. Contact resistance was reduced by depositing silver spots about 4 mm apart by evaporation in vacuum. Indium was used as the solder. Data for x = 0, 0.02 and 0.05 are plotted in Fig. 1. The zero resistance temperature for x = 0 ($T_c \approx 102$ K) is lower than those observed for many of our earlier samples [7], which have been as high as $T_c \approx 113$ K; this serves as an illustration of the sensitivity of the samples to the preparation conditions alluded to earlier.

When iron is present the transition temperature as measured by resistance is lowered, but only by a few degrees Kelvin. This result

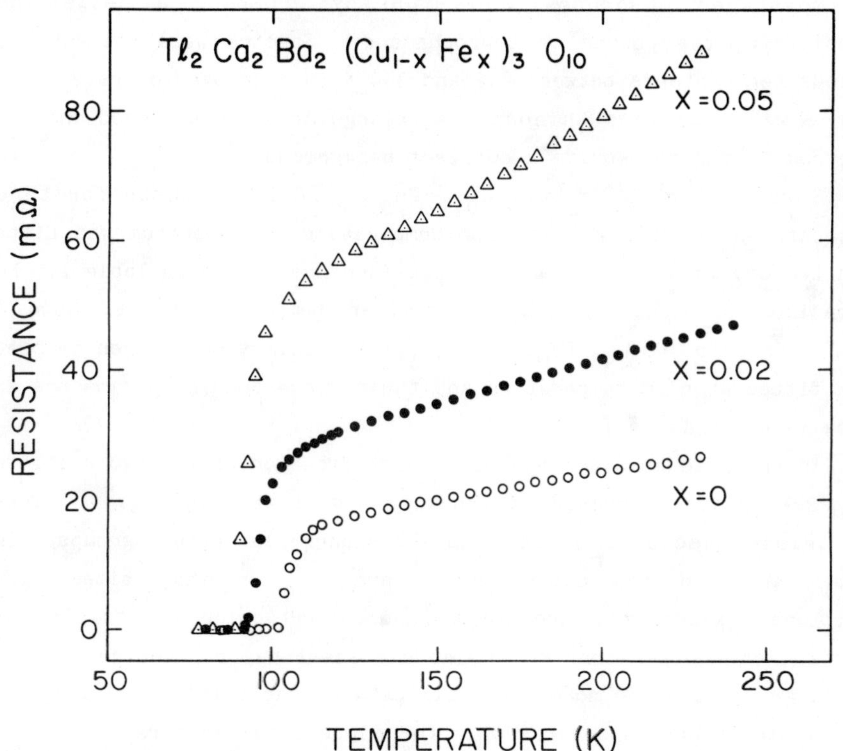

Fig. 1. Resistance versus temperature of $Tl_2Ca_2Ba_2(Cu_{1-x}Fe_x)_3O_{10}$ for x = 0, 0.02 and 0.05.

is in marked contrast to that for the 123 superconductors for which the reduction in T_c is much larger[1]. It will be recalled that the crystal structure of the 123 compound is orthorhombic (x = 0) but gradually approaches the tetragonal structure as the iron concentration is increased[1]. These structural differences between the 123 and 2223 compounds on doping with iron are a possible source of the different behaviors in T_c.

The real and imaginary parts of the ac susceptibility have also been measured for the iron-doped 2223 samples. The transition temperature then occurs below that determined for zero resistance. Similar results have been obtained earlier for the 123 samples[1].

MÖSSBAUER STUDIES

Mössbauer absorbers were made by crushing a pellet and immobilizing the powder in benzophenone. Spectra were collected at various temperatures between 4.2 and 300 K in a cryogenic system. The source was ^{57}Co in a Rh host. A triangular drive was used and each spectrum folded to provide a constant background.

Spectra for $Tl_2Ca_2Ba_2(Cu_{0.98}Fe_{0.02})_3O_{10}$ are shown for three temperatures in Fig. 2. At room-temperature the spectrum was fitted with two doublets; the parameters obtained are listed in Table I. For comparison purposes spectra for similar temperatures are shown in Fig. 3 for $YBa_2(Cu_{0.985}Fe_{0.015})_3O_{7-\delta}$. For this sample three doublets were fitted at room temperature and their corresponding parameters are also listed in Table I.

In order to consider the site identification of the two doublets, the 2223 crystal structure is shown schematically in Fig. 4. There are triple sheets of corner-sharing square-planar CuO_4 groups. The copper ions in the outer sheets are in pyramidal sites since additional oxygen ions are located above and below the triple Cu-O sheets. The copper ions on the middle sheet may be considered to be in planar sites since no oxygens lie between the triple Cu-O sheets[8].

If iron substitutes randomly for copper, the area ratio expected for the two doublets would be 2:1 corresponding to the pyramidal and planar sites, respectively. However, our present computer fits yield

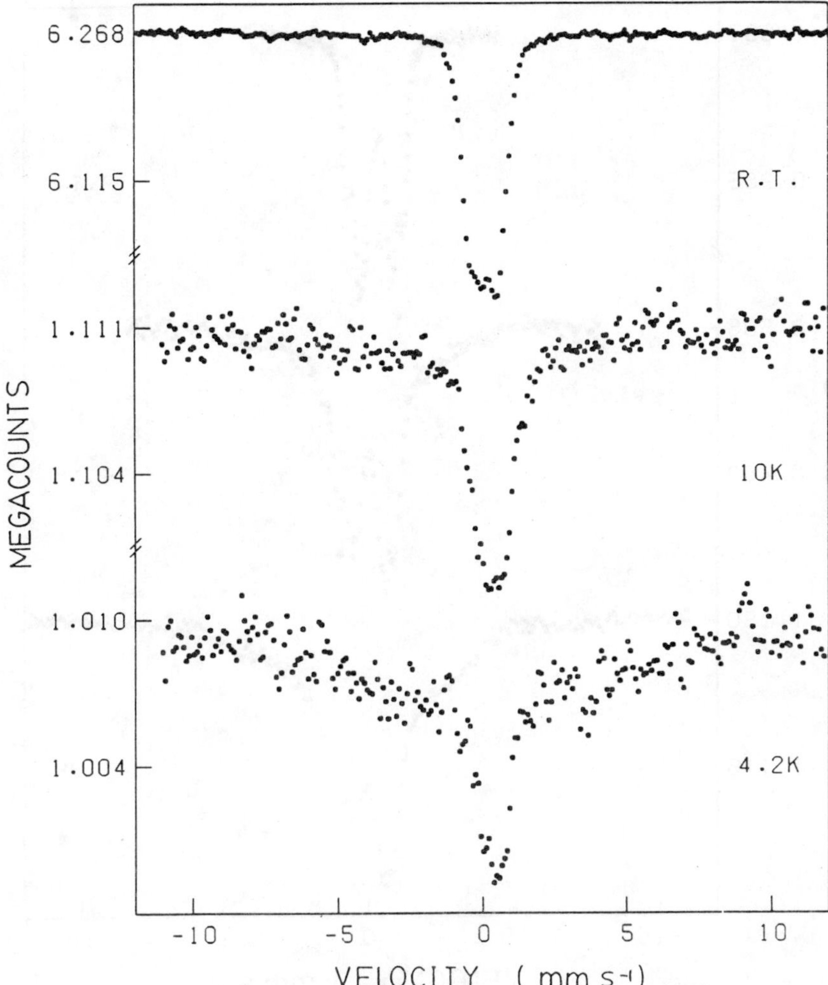

Fig. 2. Mössbauer spectra of $Tl_2Ca_2Ba_2(Cu_{0.98}Fe_{0.02})_3O_{10}$ at various temperatures as indicated.

a 1:1 ratio; this result suggests that the iron prefers the middle copper sheet. Doublet 1, with the large quadrupole splitting, has a value about equal to that for doublet 1 for the 123 compound.

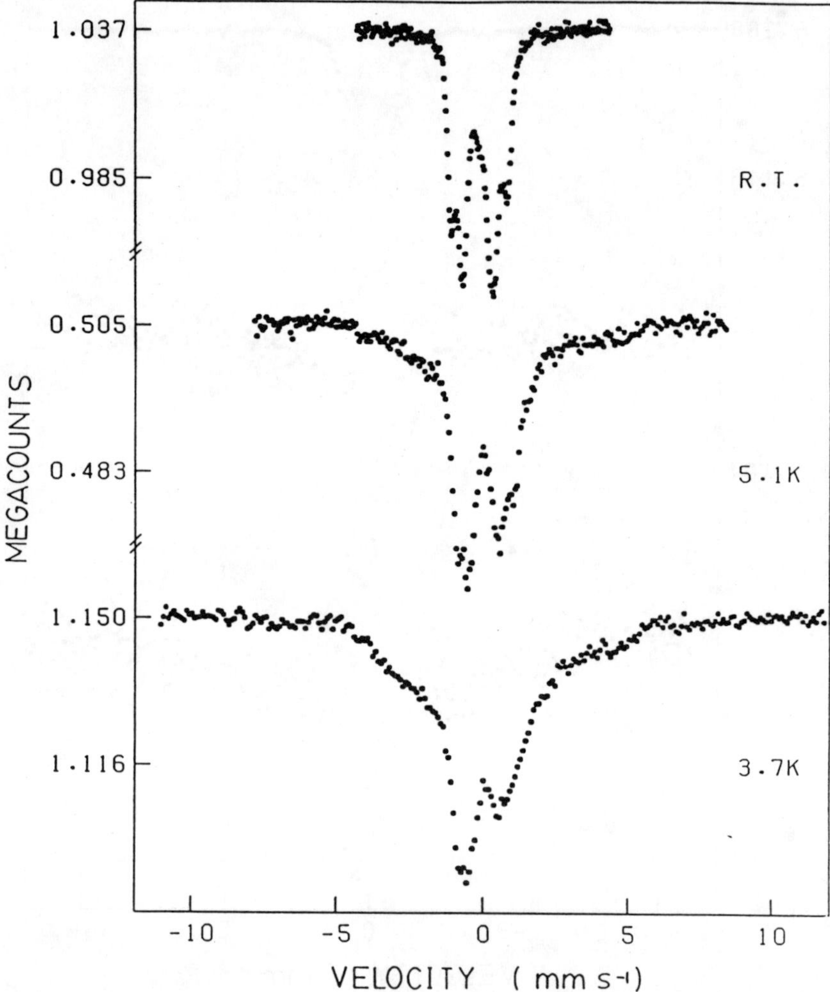

Fig. 3. Mössbauer spectra of $YBa_2(Cu_{0.985}Fe_{0.015})_3O_{7-\delta}$ at various temperatures.

Consequently this doublet may represent iron in the outer copper sheets of the 2223 structure. Doublet 2 then may be associated with iron ions lying on the central copper sheet. Confirmation of these

TABLE I

Mössbauer parameters for $Tl_2Ca_2Ba_2(Cu_{0.98}Fe_{0.02})_3O_{10}$ (2223) and for $YBa_2(Cu_{0.985}Fe_{0.015})_3O_{7-\delta}$ (123) at room temperature. δ is the isomer shift relative to α-Fe, ε is the quadrupole splitting, Γ is the linewidth and A is the relative area of a doublet.

Sample	Doublet	δ mm s^{-1}	ε mm s^{-1}	Γ mm s^{-1}	A %
2223	1	0.053	1.12	0.53	52
	2	0.081	0.40	0.54	48
123	1	-0.03	1.00	0.45	66
	2	0.04	1.93	0.29	25
	3	0.36	0.52	0.29	9

two assignments must await further experimental data, including neutron diffraction results, and theoretical calculation of the electric-field gradients. Doublet 3 of the 123 superconductor corresponds to an octahedrally coordinated iron site; no evidence for a similar doublet is found in the 2223 spectra.

For temperatures between 10 and 300 K the shape of the Mössbauer spectra of the 2223 sample does not change significantly. At T = 4.2 K distinct absorption wings appear on the sides of the central doublets (Fig. 2). A similar behavior was observed earlier for the 123 materials as shown in Fig. 3. Such a Mössbauer pattern is characteristic of magnetic ordering together with relaxation. Further experiments on the low-temperature properties of $Tl_2Ca_2Ba_2(Cu_{1-x}Fe_x)_3O_{10}$ as a function of iron concentration are in progress.

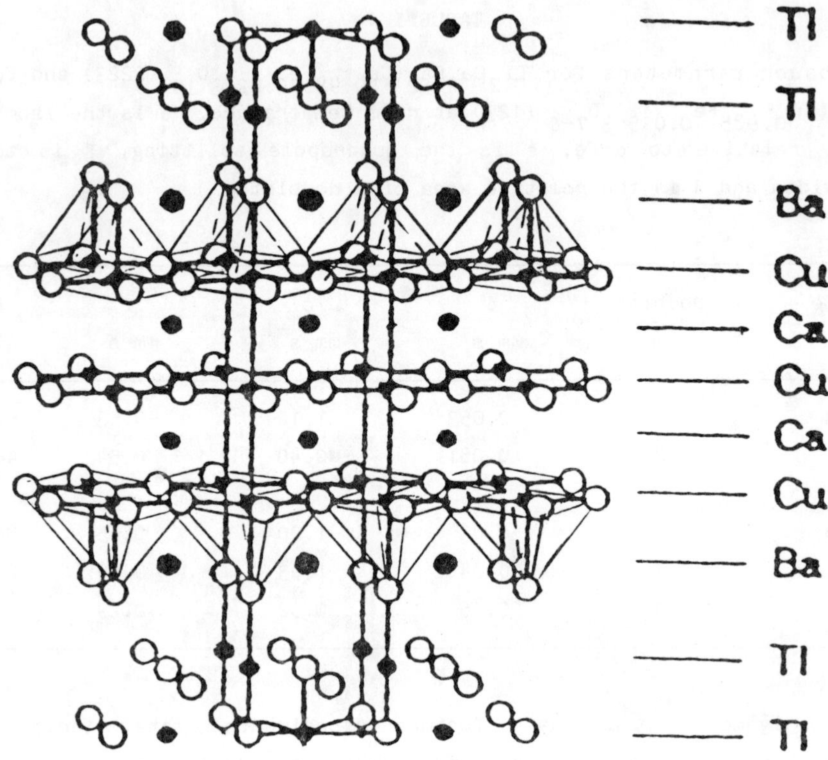

Fig. 4. A schematic of the $Tl_2Ca_2Ba_2Cu_3O_{10}$ structure.

ACKNOWLEDGEMENT

This research was funded by the Natural Science and Engineering Research Council of Canada (NSERC).

REFERENCES

1) Zhou, X. Z., Raudsepp, M., Pankhurst, Q. A., Morrish, A. H., Luo, Y. L. and Maartense, I., Phys. Rev. B36, 7230 (1987).
2) Gómez, R., Aburto, S., Marquina, M. L., Jiménez, M., Marquina, C., Akachi, T., Escudero, R., Barrio, R. A. and Plios-Jara, D., Phys. Rev. B36, 7226 (1987).
3) Blue, C., Elgaid, K., Zitkovsky, I., Boolchand, P., McDaniel, D., Joiner, W. C. H. and Oostens, J., Phys. Rev. B37, 5905 (1988).
4) For example, Tarascon, J. M., Barboux, P., Miceli, P. F., Greene, L. H., Hull, G. W., Eibschutz, M. and Sunshine, S. A., Phys. Rev. B37, 7458 (1988).
5) Pankhurst, Q. A., Morrish, A. H., Zhou, X. Z. and Maartense, I., Hyperfine Int. 42, 1235 (1988).
6) Pankhurst, Q. A., Morrish, A. H., Raudsepp, M. and Zhou, X. Z., J. Phys. C21, 17 (1988); Pankhurst, Q. A., Morrish, A. H. and Zhou, X. Z., Phys. Lett. A127, 231 (1988).
7) Zhou, X. Z., Morrish, A. H., Luo, Y. L., Raudsepp, M. and Maartense, I., J. Phys. D21, 1243 (1988).
8) Torardi, C. C., Subramanian, M. A., Calabrese, J. C., Gopalakrishnan, J., Morrissey, K. J., Askew, T. R., Flippen, R. B., Chowdhry, U. and Sleight, A. W., Science 240, 633 (1988).

MAGNETIC LEVITATION BY ATTRACTION IN SOME HIGH T_C SUPERCONDUCTORS

R. D. Shull and L. J. Swartzendruber
Magnetic Matls. Group, National Institute of Standards and Technology
Gaithersburg, MD 20899

C. K. Chiang
Ceramics Division, National Institute of Standards and Technology
Gaithersburg, MD 20899

M. K. Wu
School of Mines, Columbia University, New York, NY 02988

P. N. Peters
Space Science Laboratory, NASA Marshall Space Flight Center
Huntsville, AL 35812

C. Y. Huang
Research and Development Div., Lockheed Missiles and Space Company,Inc.
Palo Alto, CA 94304

ABSTRACT

Superconducting composites of AgO and $YBa_2Cu_3O_{7-x}$ are found to possess the unusual property at 77 K of being able to levitate by "attraction" (with the composite being suspended in space "below" a permanent magnet). At the same temperature, however, these composites are still able to levitate a magnet above it by means of the better known method of repulsion. Magnetization measurements as a function of field for several composites of AgO and $YBa_2Cu_3O_{7-x}$ are presented to explain the origin and composition dependence of the novel "attractive" force. These data support the previous conclusion that the ability of the material to be suspended below a magnet is dependent upon the degree of magnetic flux trapping in the superconductor and that the ability to obtain a stable equilibrium position depends on the shape of the hysteresis loop.

INTRODUCTION

Conventional demonstrations of magnetic levitation using the large diamagnetism of superconductors have been performed by fixing the superconductor (S), as shown in figure 1a, and levitating a permanent magnet (PM) above S, or vice versa. Using conventional wisdom [1], this levitation occurs because a mirror image (PM') of the magnet is created

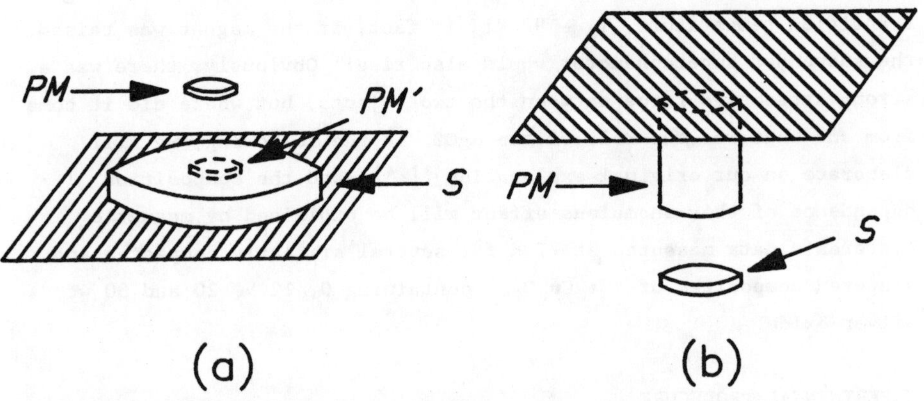

Figure 1. Schematic of (a) conventional and (b) novel methods for the demonstration of magnetic levitation forces existing between superconductors (S) and permanent magnets (PM).

in the superconductor due to the supercurrents induced in S needed to expell the magnetic field of PM. The force between PM and PM′ is repulsive and PM rises above S. The repulsive force decreases roughly as the fourth power of the separation distance (Z); and stable equilibrium is achieved when the negative (defined downward) force of gravity (which does not change noticeably with Z) equals the positive repulsive magnetic force. (Actually, it has recently been shown [2] that stable equilibrium requires a small amount of flux pinning in addition.) If the magnet was placed underneath the superconductor and the experiment repeated, both the magnetic repulsive force and the gravitational force acting on PM would be in the same direction (negative) and no stable equilibrium would be expected. In this case the magnet would simply fall to the ground. Similarly, if in the original experiment the magnet (located above the superconductor) was fixed in position while S was allowed to move, both forces on S would be negative and the superconductor would fall.

Recently P. Peters, et. al [3] prepared a composite material, 80 wt. % high temperature superconductor $YBa_2Cu_3O_{7-x}$ and 20 wt. % AgO, that possessed a very unusual property. If the magnet (PM) was fixed and the superconductor (S) at 77 K was placed underneath the magnet (as shown in figure 1b), S would not fall to the ground, but would become suspended underneath PM until it warmed to near its superconducting transition temperature, $T_c \sim 92$ K! In fact, if the magnet was raised, the suspended superconductor would also rise! Obviously, there was a strong attractive force between the two objects, but where did it come from and what was the role of the AgO? In the following, we will elaborate on our original explanation [3],[4]; and the composition dependence of this anomalous effect will be explained by magnetic hysteresis data measured at 77 K for several similarly compacted and sintered composites of $YBa_2Cu_3O_{7-x}$ containing 0, 12.5, 20 and 50 wt. % silver oxide.

EXPERIMENTAL PRODECURE

The composite samples were prepared by mixing powders of high quality single phase $YBa_2Cu_3O_{7-x}$ (prepared by conventional techniques) and AgO in weight ratios of 1:1, 1:4, and 1:7 respectively [5]. The pressed pellets were subsequently fired at 980 C for 6 hours in oxygen and slowly cooled. Only diffraction lines for pure silver and $YBa_2Cu_3O_{7-x}$ were observed in the X-ray diffraction spectra for these sintered materials. It was also found [5] that composite formation occurred only in a narrow sintering temperature range. Herein, the samples are referred to in terms of their original mixture of silver oxide and $YBa_2Cu_3O_{7-x}$ powders in order to denote weight ratios of the composites, even though the x-ray data, indeed, shows the AgO decomposed during the final heat treatment.

Magnetization measurements were performed using a vibrating sample magnetometer and a 10 kG electromagnet with a bipolar power supply. The superconducting transition temperatures were measured by four-terminal electrical resistivity measurements [3] except for the $YBa_2Cu_3O_{7-x}$ sample containing no silver oxide. In this latter case, the T_c was determined by ac magnetic susceptibility measurements [6] using a Hartshorn-type bridge circuit operated with a 0.4 Oe rms field at 1.68 kHz. T_c = 92 K was approximately the same for all the samples. The magnetic "levitation by attraction" property of these superconductors was judged only qualitatively by cooling the samples (~100 mg each) in liquid nitrogen and observing whether a small (3mm x

3mm x 20mm) rectangular laboratory magnet (~1 kG maximum field) could lift the sample out of the nitrogen, as previously done with a SmCo magnet [4].

RESULTS AND DISCUSSION

For one of the materials that possessed the ability to be levitated easily by a magnet from liquid nitrogen, $YBa_2Cu_3O_{7-x}$ + 12.5 wt. % AgO, the field dependence of its magnetization at 77 K after being cooled to this temperature in zero field is shown in figure 2. Two successive low-field cycles (figure 2a), measured in order of increasing maximum field, and a subsequent 10 kOe high-field cycle (figure 2b) are shown. For this sample (density ≃ 6 g/cm³) the initial small field (-10 Oe < H < +10 Oe) loop (not shown) indicated no magnetic flux penetration into the sample and had an initial slope of -0.078 emu/cm³ Oe (very close to the theoretical $-1/4\pi$ value for perfect diamagnetism). Flux penetration (deviation from M vs. H linearity) began to occur in this sample at fields near 30 Oe ($=H_{C1}$); and from the measurable width of the -100 Oe < H < +100 Oe loop shown in figure 2a, flux pinning may be deduced to occur at slightly higher fields. Unusual about the larger-field hysteresis loops shown in figure 2 is their sizable width at 77 K, the fast <u>sign reversal</u> of the magnetization at the highest fields for relatively small (<25% and 10% respectively in Figures 2a and 2b) field decreases, and the large remanent magnetization (M_r) remaining after a decrease in field to zero following a field excursion in excess of 300 Oe (a field close to that at which the magnetization peaks). These are characteristics of a material that very effectively traps magnetic flux. In addition, the magnetic critical current, J_C, in this sample (0.5 mm thickness) at 77 K is also unusually large: estimated using the phenomenological model of Bean [7] from the -10 kOe < H < +10 kOe hysteresis data to be ~17000 A/cm² (at zero field).

As explained in our earlier paper [4], the novel ability of these composite superconductors to be <u>attracted</u> to a magnet at 77 K is a result of the large positive magnetization (M) possessed by the superconductor at this temperature as the field is decreased from its maximum positive value toward zero, as occurs in the region indicated by point A in figure 2a. The magnetic force (F_m) between the

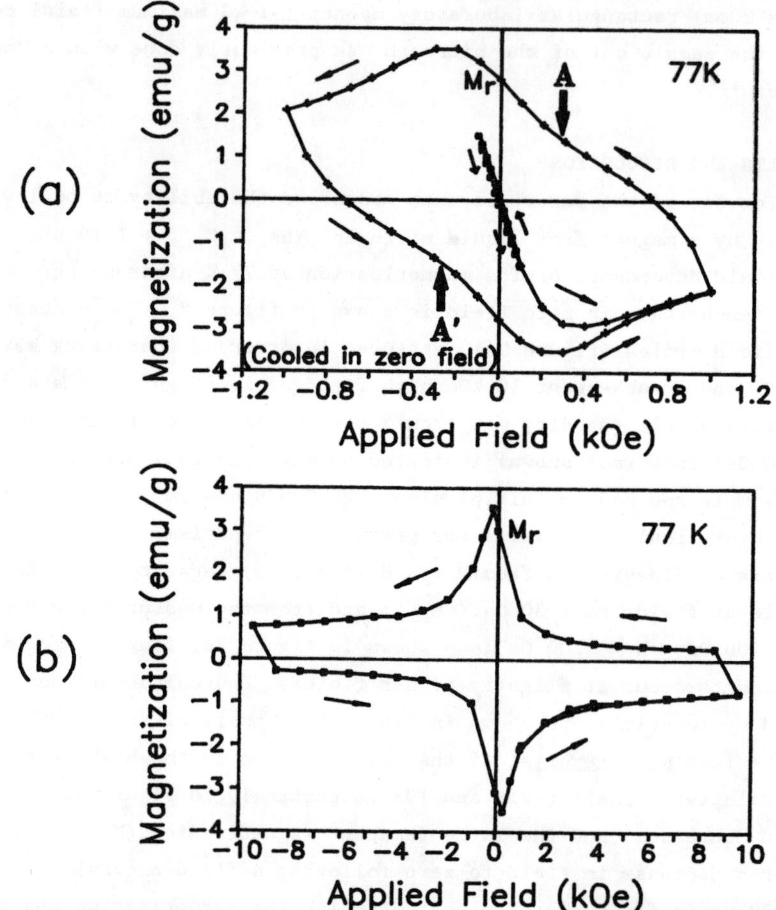

Figure 2. Field dependence of the magnetization for a superconducting $YBa_2Cu_3O_{7-x}$ + 12.5 wt. % AgO composite, which possesses the novel "attractable levitation" effect, as the field was cycled in the sequence indicated by the thin arrows at 77 K. The $|H|\leq100$ Oe (filled symbols) and $|H|\leq1000$ Oe (crosses) loops in (a) and the $|H|\leq10{,}000$ Oe loop in (b) were measured consecutively after cooling to this temperature in zero field.

superconductor and the magnet is described by the relation $F_m=-m[M(H)(dH/dZ)]$, where m is the mass of the superconductor, Z is the distance of separation, and M is in units of emu/g. Since around this location dH/dZ is negative and M is positive, F_m will also be positive (attractive). However, M is only positive around point A because of the trapped magnetic flux in the superconductor. In fact, if the field is decreased to zero, the superconductor becomes a permanent magnet! Stable equilibrium at a non-zero distance of separation between the two objects is due to the negative slope of the M vs. H relationship (M(H)) of the superconductor and the sign reversal property of M at a sufficiently large value of H. Qualitatively, the mechanism whereby the superconductor is kept at a stable location below the magnet may be visualized by referring to figure 2a. Starting at the equilibrium point A, if the superconductor approaches the magnet, H increases in magnitude (roughly proportional to $1/Z^3$), causing M to decrease and, therefore, a decrease in the attractive force F_m. If the superconductor approaches too close to the magnet, the sign of M reverses to negative and F_m becomes negative (repulsive), forcing the two objects apart. Conversely, if at point A the superconductor moves away from the magnet, the field near the superconductor (H) decreases, thereby causing an increase in M. F_m consequently increases and the two objects are pulled back together. The presence of gravity only shifts the location of point A on the M vs. H curve. The negative M(H) relationship observed by the superconductor in establishing this equilibrium position is not identically that shown by the high field hysteresis loop in figure 2a at point A, but is a minor loop (possessing a shape and slope similar to that of the virgin -100 Oe < H < +100 Oe loop) peaked at the equilibrium point A.

The ability to be easily levitated from liquid nitrogen by attraction to a permanent magnet was also demonstrated by the composite superconductor containing 20 wt. % silver oxide. Note from figure 3 that both the low-field and high-field hysteresis loops measured for this material at 77 K also enclose large areas, possess zero magnetization crossings close to the highest applied fields, and show a large remanent magnetization in the material upon removal of the field similar to that shown above for the other "attractable" composite containing 12.5 wt % silver oxide. The effectiveness of trapping magnetic flux in these composites is shown for the 20 wt. % silver oxide material at low fields in figure 4. The vertical shift of the field-cooled loop ($4\pi M$ = 40 Oe) is almost equal to the cooling field,

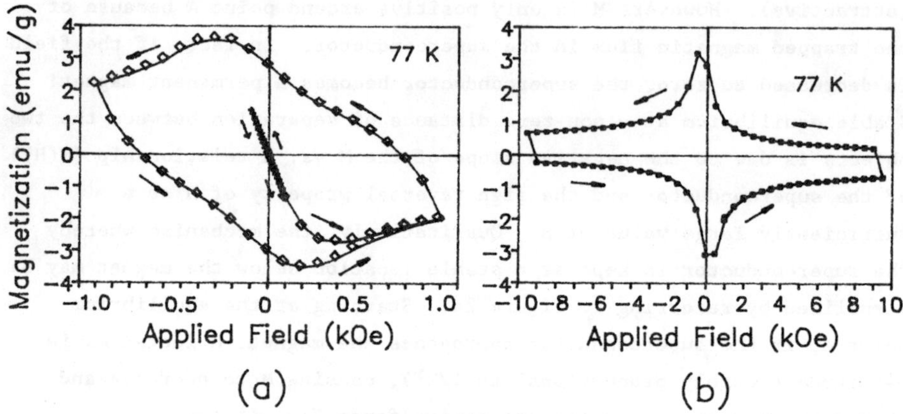

Figure 3. Field dependence of the magnetization for a superconducting $YBa_2Cu_3O_{7-x}$ + 20 wt. % AgO composite at 77 K as the field was cycled in the sequence indicated by arrows. The $|H| \leq 10$ Oe (not shown), $|H| \leq 100$ Oe (crosses), and $|H| \leq 1000$ Oe (open symbols) loops in (a) and the $|H| \leq 10{,}000$ Oe hysteresis loop in (b) were measured consecutively after cooling to this temperature in zero field.

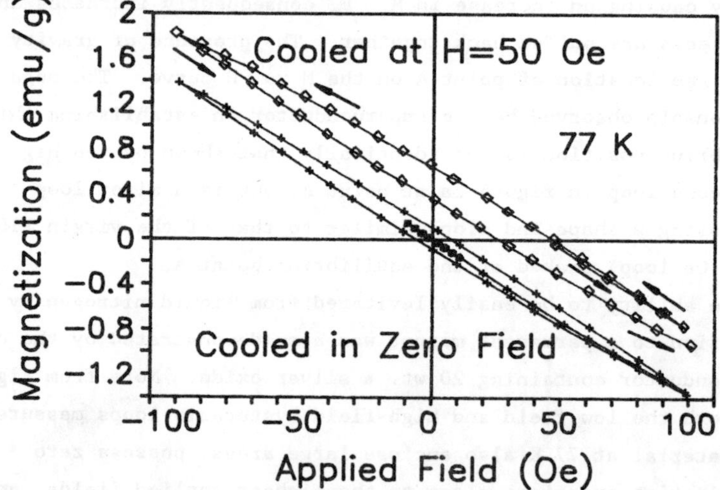

Figure 4. Field dependence of the magnetization for a superconducting $YBa_2Cu_3O_{7-x}$ + 20 wt. % AgO composite at 77 K as the field was cycled following cooling to this temperature from room temperature in zero applied field (filled symbols and crosses for the $|H| \leq 10$ Oe and $|H| \leq 100$ Oe data respectively from the central portion of figure 3a) or in an applied field of 50 Oe (open symbols). The arrows (shown for the field cooled data) indicate the measurement sequence used for both sets of data.

showing that the magnetic flux inside the superconductor above T_c is almost completely trapped during the field cooling through the transition temperature to 77 K. Note from the narrow widths of these loops, however, that these materials still possess good magnetic shielding properties for low fields even when they have trapped magnetic flux inside. The lack of any noticable width in the -10 Oe < H < +10 Oe hysteresis loop for the zero-field-cooled sample also indicates that the lower critical field (H_{c1}) for this material is above 10 Oe.

It was found that in the case of the $YBa_2Cu_3O_{7-x}$ + 50 wt. % AgO superconducting composite, our laboratory magnet was not capable of lifting the sample out of the liquid nitrogen bath. Comparison of the field dependence of the magnetization for this material at 77 K (figure 5) with that shown above for the more silver-oxide dilute composites explains the loss in magnetic "attraction" by this 50 wt. % AgO material. The hysteresis loops for this material are very narrow, possess zero magnetization values only at low fields (regardless of the maximum applied field), and possess order of magnitude smaller remanent magnetization values than do the "attractable" composites. The 50:50 composite does <u>not</u> trap magnetic flux very well and, consequently, cannot be levitated at 77 K by a magnet positioned above it [8]. Consistent with this interpretation, the critical currents in this material were also estimated from the magnetic data to be only ~1500 A/cm². The low-field hysteresis loop measured at 77 K for a typical $YBa_2Cu_3O_{7-x}$ sample (1 mm thickness) containing no silver oxide indicates this material traps even less magnetic flux (figure 6). J_c for this material is only about 600 A/cm². It is, consequently, not surprising that the novel "attractable" levitation effect is also not observed in our typical single-phase high temperature superconducting oxide $YBa_2Cu_3O_{7-x}$ when immersed in liquid nitrogen. The addition of silver to the superconductor is important to the development of the novel magnetic attraction described in this paper because it provides for the pinning of magnetic flux and the enhancement of J_c. Possibly, the flux pinning in these two phase composites is provided simply by the introduction of more normal/superconducting boundaries. Alternatively, the pinning may be provided by the silver modification

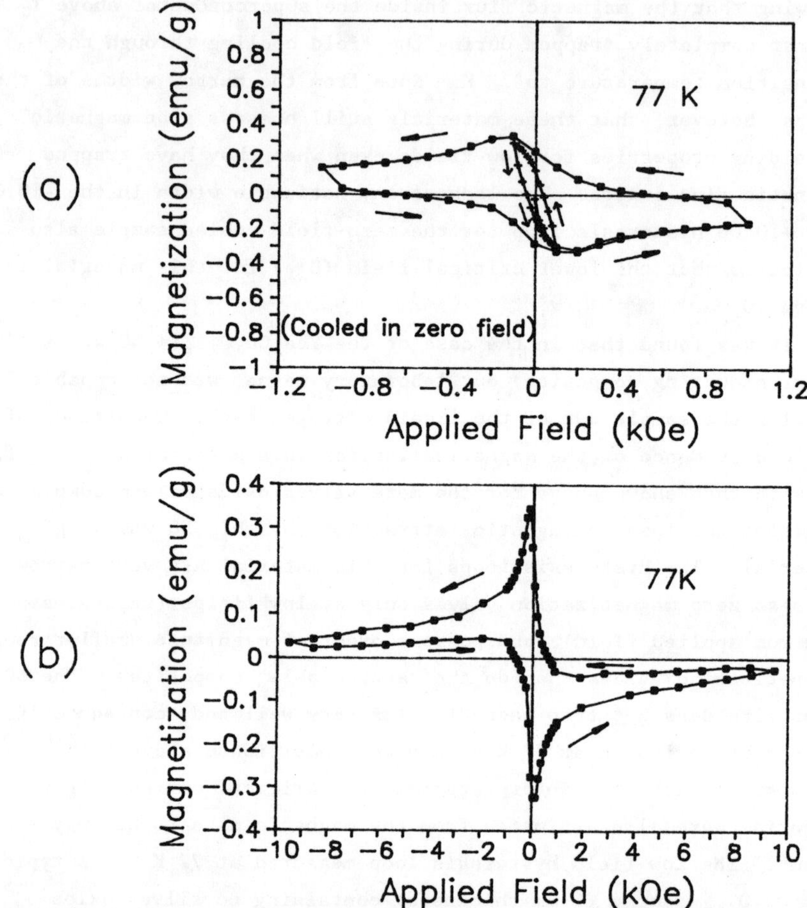

Figure 5. Field dependence of the magnetization for a superconducting $YBa_2Cu_3O_{7-x}$ + 50 wt. % AgO composite, which does NOT possess the novel "attractable levitation" effect, as the field was cycled in the sequence indicated by the thin arrows at 77 K. The $|H| \leq 100$ Oe (crosses) and $|H| \leq 1000$ Oe (filled symbols) locps in (a) and the $|H| \leq 10,000$ Oe loop in (b) were measured consecutively after cooling to this temperature in zero field.

of only specific locations (such as the "weak link" grain boundaries) in the material which are important in controlling the ease of magnetic flux motion through the material. It was explained earlier [3] as being due to an enhancement in the critical currents of the many weak-link inter-grain contacts. The exact nature of this enhanced pinning ability of the composite materials is presently under study and will be the subject of a future publication. Magnetic flux pinning is a temperature dependent effect (becoming stronger at lower T). Consequently, we have predicted [4],[9] that the "levitation by attraction" phenomenon is common to all type II superconductors (not just the high T_c oxides) at sufficiently low temperatures.

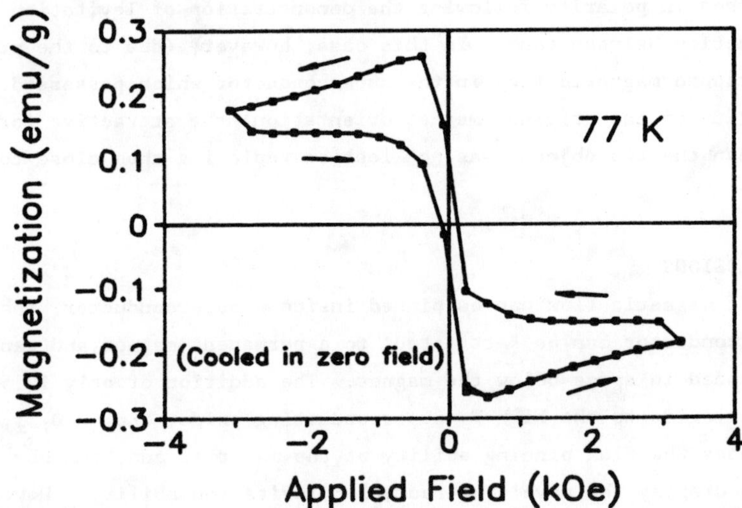

Figure 6. Field dependence of the magnetization for a typical high temperature superconducting oxide, $YBa_2Cu_3O_{7-x}$, at 77 K as the field was cycled in the sequence indicated by arrows after first cooling to this temperature in zero applied field.

If the novel "attractable" levitation ability of the silver oxide doped $YBa_2Cu_3O_{7-x}$ is, indeed, simply due to the pinning of magnetic flux in the material, a couple of additional simple experiments were conceived as a test. (1) First the "attractive levitation property was demonstrated in the composite superconductor (S) at 77 K as described above. The magnet (PM) was removed (while restraining S in the liquid nitrogen), and at this stage any trapped magnetic flux in the superconductor would possess the polarity (positive) of the original magnet orientation. Upon reversal of the magnet's polarity, it was observed that the force between the superconductor and magnet was now repulsive, indeed, indicating the presence of previously trapped flux of the opposite polarity in S [10]. If the magnet was forced close to S, magnetic flux penetrated and was trapped in S with a negative polarity, resulting in a return of the "attractive levitation" ability of the superconductor upon withdrawing PM. The central symmetry of the hysteresis loop shown in figure 2 explains this latter effect as stable levitation was achieved around the image point A'. (2) In this second experiment both the magnet and the superconductor were separated and reversed in polarity following the demonstration of levitation by attraction between them. In this case, however, (due to the presence of trapped magnetic flux in the superconductor which possessed the polarity of the original magnet orientation) the attractive force between the two objects was not lost on replacing them close to each other.

CONCLUSIONS

If magnetic flux can be pinned inside a superconductor, that superconductor can be "attracted" to a permanent magnet and can be suspended in space below the magnet. The addition of only 12.5 wt % silver oxide to the high T_c superconducting oxide, $YBa_2Cu_3O_{7-x}$, enhances the flux pinning ability of the material and, thereby, allows it to display the novel "attractable" levitation ability. However, there is a maximum to the effectiveness of silver oxide additions in enhancing the flux trapping ability of these materials. The 20 wt. % silver oxide composite traps magnetic flux well, but only marginally better than the 12.5 wt. % material. Much further increase in AgO content to 50 wt. percent, however, results in a significant deterioration in the flux pinning ability of the material. In fact, in this study this latter composite was only marginally better in this respect than the undoped superconductor $YBa_2Cu_3O_{7-x}$ at 77 K.

ACKNOWLEDGEMENTS

The work at the National Institute of Standards and Technology (NIST) was supported by the Institute for Materials Science and Engineering (IMSE) Internal Research funds, at Alabama by NASA grants

NCC8-2, NAGW-812, and NAG8-089, and at the Lockheed Missiles and Space Company, Inc. (LMSC) by the LMSC Independent Research fund.

REFERENCES

[1] Arkadiev, V., Nature 160, p. 330 (1947).

[2] Hellman, F., Gyorgy, E. M., Johnson, D. W., Jr., O'Bryan, H. M., and Sherwood, R. C., J. Appl. Phys. 63 (2), p. 447 (1988).

[3] Peters, P. N., Sisk, R. C., Urban, E. W., Huang, C. Y., and Wu, M. K., Appl. Phys. Lett. 52 (24), p. 2066 (1988).

[4] Huang, C. Y., Shapira, Y., McNiff, E. J., Jr., Peters, P. N., Schwartz, B. B., Wu, M. K., Shull, R. D., and Chiang, C. K., Mod. Phys. Lett. B 2, No. 7, p. 869 (1988).

[5] Wu, M. K., Leong, P. T., Chou, H., Curreri, P. A., Peters, P. N., Huang, C. Y., and Shapira, Y., to be published.

[6] Atzmony, U., Shull, R. D., Chiang, C. K., Swartzendruber, L. J., Bennett, L. H., and Watson, R. E., J. Appl. Phys. 63, p. 4179 (1988).

[7] Bean, C. P., Phys. Rev. Lett. 8, p. 250 (1962).

[8] Another 50:50 composite prepared in the same manner after this study, however, did possess the "attractable levitation" ability, indicating the high sensitivity of this property to the processing conditions.

[9] Huang, C. Y., Peters, P. N., Schwartz, B. B., Shapira, Y., and Wu, M. K., Mod. Phys. Lett. B 2, No. 8, p. 1027 (1988).

[10] In reference 3, a Hall probe was used to measure this flux; and C. Y. Huang has made a video tape of this effect.

MAGNETIC MEASUREMENTS ON POLYMER-HIGH T_c SUPERCONDUCTOR COMPOSITES

A.S. De Reggi, C.K. Chiang, L. Swartzendruber and G.T. Davis
National Institute of Standards and Technology
Gaithersburg, MD 20899

ABSTRACT

Composite materials made by dispersing powdered mixed-oxide superconductors in suitable polymers may be usable in applications where contact from one superconducting grain to another is not essential to the function. In this study, permeability data on composite materials of the superconductor $Y_1Ba_2Cu_3O_{7-x}$ and the vinylidene fluoride polymer, PVDF, with volume concentrations ranging from 5 to 25 % superconductor, were obtained. The permeability data were corrected for demagnetizing effects and compared to the predictions of a model analogous to the Lorentz model for dielectrics.

INTRODUCTION

Composite materials made by dispersing powdered mixed-oxide superconductors in suitable polymers may become important new materials in superconducting applications where contact from one superconducting grain to another is not essential to the function, such as electromagnetic shielding[1] or levitation. The principal advantage of the composite over the undiluted superconductor is that it can be processed and fabricated like a polymer using existing polymer processing equipment and methods. Another advantage of the composite is that the polymer, if properly chosen, can protect the surface of the powder grains from exposure to environmental chemicals which might affect the oxides and degrade their superconducting properties. In this study we show that a previously suggested[2] model, analogous to the Lorentz model for dielectrics, provides a precise description of the permeability of superconductor-polymer composites (for compositions up to 25 vol. % superconductor) once demagnetizing factors are properly taken into account and when applied fields are less than H_{c1}.

EXPERIMENTAL

As described previously[2], composite materials of the 1-2-3 mixed oxide superconductor $Y_1Ba_2Cu_3O_{7-x}$ (YBCO), with x<0.1 and critical temperature $T_c \simeq 92$ K, and the vinylidene fluoride polymer $(CH_2CF_2)_n$ (PVDF) with a glass transition temperature in the 230-240K range, were prepared with volume concentrations ranging from 5 to 25 % superconductor. The YBCO-PVDF composites were pressure molded at 175 °C into the shape of small bricks approximately 10 mm long by 6 mm wide by 2.5 mm thick.

For comparison, and to determine the appropriate demagnetizing factors for this shape, a brick of pure lead with the same dimensions was also fabricated. Magnetic measurements were made in a vibrating sample magnetometer with the magnetic field perpendicular to the 10 mm by 2.5 mm face of the brick. The bricks were cooled from room temperature in zero field to 10 K (for the case of the composites) or to 4.5 K (for the case of the lead). The magnetization was measured in intervals as the applied field was increased to 100 Oe, then decreased to -100 Oe, then increased again to 100 Oe. The data obtained are displayed in Fig. 1.

THEORY

In a magnetic field below H_{c1}, each grain of the superconductor behaves as a linear diamagnet and can be assigned a susceptibility. If the YBCO material is completely superconducting this susceptibility is $1/4\pi$ (in CGS units). In the dilute range, i.e. when the distance separating grains is much greater than the characteristic grain dimension, the problem of calculating the local magnetic field in the composite is analogous to that of calculating the local electric field in a dielectric medium containing electric dipoles responding to an applied electric field. Lorentz considered the dielectric problem by introducing a spherical cavity centered around a reference dipolar molecule, a model leading to the Clausius-Mossotti relation[3]. The magnetic analog of this equation is, in CGS units,

$$(\mu-1)/(\mu-2) = (4\pi/3)\Sigma_i N_i \alpha_i , \qquad [1]$$

where μ is the relative magnetic permeability of the composite, N_i is the number of induced magnetic dipoles with effective magnetic susceptibility α_i. If we assume that all the superconducting grains are roughly spherical in shape then they will each have an individual demagnetizing factor of $4\pi/3$. If the grains exhibit a full Meissner effect, then the susceptibility would be $-1/4\pi$. If they are not fully superconducting, then their susceptibility can be taken as $-x/4\pi$, where x is between 0 and 1. Then, taking into account the demagnetizing factor for a sphere, the effective susceptibilities α_i will all be $-(x/4\pi)/(1-x/3)$. After the $\Sigma_i N_i$ is replaced by v, where v is the fractional volume of the superconductor in the composite, the susceptibility, χ, of the composite-superconductor combination will be, in SI units,

$$\chi = -3xv/[3-x(1-v)]. \qquad [2]$$

For x=1 the value of χ will go from 0, for v=0, to -1, for v=1. However the range of validity of Eq. 2 will be limited to the dilute superconductor range because of the Mossotti assumption in evaluating the local field. The values for v we have used experimentally (v=0.05 to 0.25) should fall into the range of validity.

RESULTS AND DISCUSSION

Fig. 1 shows the measured values of $4\pi M$ vs. H for a series of composite bricks at 10 K and for a lead brick of the same size and shape at 4.5 K. The measured values have been least-squares fitted to straight lines. The slopes of these lines give the effective susceptibilities (in SI units) of the samples. Using the fact that the lead sample at 4.5 K is a known perfect superconductor with

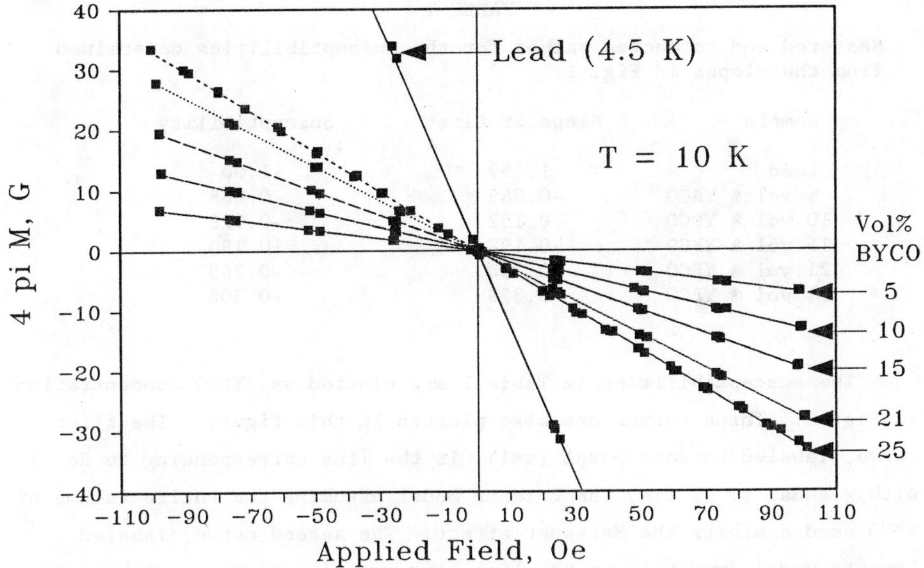

Figure 1. **Measured values of** $4\pi M$ vs. applied field H for a lead brick at 4.5 K and for composites with varying volume percent of YBCO at 10 K. The lines represent linear least squares fits to the data points. The samples were cooled in zero field before beginning the measurement.

susceptibility -1, the slope of the line from lead can be used to determine the demagnetizing factor (D=0.206) for the sample shape and orientation used. Using this demagnetizing factor the true susceptibility (given by s/(1+Ds), where s is the slope and D the demagnetizing factor) of the composites can be determined from the line slopes in Fig. 1. The results are shown in Table I. Note that the correction for sample shape demagnetization ranges from 1.5 % for the 5 vol % sample to about 5 % for the 25 vol % sample.

TABLE I

Measured and corrected values for the susceptibilities determined from the slopes in Fig. 1.

Sample	Slope of Line	Susceptibility
Lead	-1.259	-1.00
5 vol % YBCO	-0.069	-0.068
10 vol % YBCO	-0.132	-0.128
15 vol % YBCO	-0.198	-0.190
21 vol % YBCO	-0.280	-0.265
25 vol % YBCO	-0.329	-0.308

The susceptibilities in Table I are plotted vs. YBCO concentration in Fig. 2. Three curves are also plotted in this figure. The first curve, labeled Lorentz Model (x=1), is the line corresponding to Eq. 1 with x equal to 1, i.e. the Lorentz model assuming the entire volume of YBCO used exhibits the Meissner effect. The second curve, labeled Lorentz Model (x=0.94), is the line corresponding to Eq. 1 with x equal to 0.94, i.e. the Lorentz model assuming only 94% of the YBCO used exhibits the Meissner effect. The curve labeled Linear Model is simply a line which assumes that the susceptibility varies linearly with the volume fraction of superconductor in the composite. We note that assuming that x is less than 1 caused the linear model to move away from the experimental points, producing a poorer fit to the measured values.

It thus appears that the Lorentz model of Eq. 1 gives a good description of the susceptibility for a composite consisting of a finely ground powder of YBCO and a polymer. The results obtained here were for applied fields not exceeding H_{c1} so that no flux is trapped in the superconductor. We also note in passing that the polymer appears to give very good protection to the YBCO since its properties did not change perceptibly over a period of one year even though exposed to a humid atmosphere.

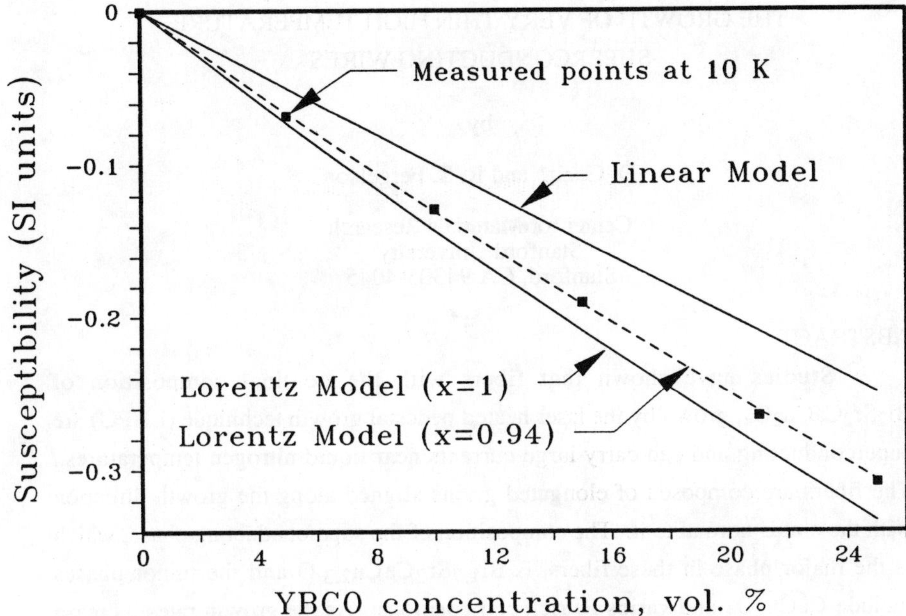

Figure 2. Comparison of the composite susceptibilities (after correction for sample shape demagnetization) with the linear and Lorentz models as described in the text.

REFERENCES

1. E. Tjukanov, R.W. Cline, R. Krahn, M. Hayden, M.W. Reynolds, W.N. Hardy, J.F. Carolan, and R.C. Thompson, Phys Rev B36, 7241 (1987)
2. A.S. De Reggi, C.K. Chiang, L. Swartzendruber and G.T. Davis, to be published.
3. C. Kittel, "Introduction to Solid State Physics", 4th ed., J. Wiley and Sons, Inc. New York, NY, p. 459.

THE GROWTH OF VERY THIN HIGH TEMPERATURE SUPERCONDUCTING WIRES

by

D. Gazit* and R. S. Feigelson

Center for Materials Research
Stanford University
Stanford, CA 94305-4045

ABSTRACT

Studies have shown that fibers with the nominal composition of $Bi_2Sr_2CaCu_2O_x$, grown by the laser heated pedestal growth technique (LHPG) are superconducting and can carry large currents near liquid nitrogen temperatures.. The fibers are composed of elongated grains aligned along the growth direction with the c-axis normal to it. The composition of the superconducting phase, which is the major phase in these fibers, is $Bi_{1.7}Sr_2CaCu_{2.3}$ O and the minor phases include $CaCu_2O_4$ and varios Bi rich compositions. As the growth rates increase from a few mm/hr to 50 mm/hr, the amount of the superconducting phase decreases, and at the higher rates it's composition also changes. Thin fibers 200 μm diameter yielded single crystal sections 8 mm long when the growth rate was 1.5 mm/hr. Single crystal fibers 25 μm diameter and 10 mm long could be produced at 24 mm/hr.. A 110 μm diameter fiber 80 mm long and a 45 μm dia.30 mm long fiber were flexible.

The LHPG technique has also been used to produce metal wires coated with superconducting layers by pulling platinum wire through the superconducting molten material. The wire, which was used as a support for the superconducting material, enabled much easier control of the growth, and yielded uniform layers. Due to platinum incorporation, a 2 μm layer adjacent to the wire was not superconducting. The rest 6 μm of the coating was mostly superconducting.

A bending test has shown that the coating adheres well to the wire and cracks rather than peels off. A cleavage plane revealed by rough handling suggests that the grains grow parallel to the wire axis.

----

* On leave from Nuclear Research Center, Negev, P.O. Box 9001, Beer Sheva, Israel

OPTICAL EXCITATIONS IN THIN FILM $YBa_2Cu_3O_7$

K. Kamarás, S.L. Herr, C.D. Porter and D.B. Tanner

Department of Physics, University of Florida, Gainesville, FL 32611

S. Etemad and Siu-Wai Chan

Bell Communications Research, Red Bank, NJ 07701

ABSTRACT We have measured the optical reflectivity of a $YBa_2Cu_3O_{7-\delta}$ thin film having its surface oriented in the ab-plane. Our data span the far infrared to the ultraviolet at temperatures between 6 K and 300 K . A superconducting gap can not be unambiguously derived from the spectra. In the normal state, we model the reflectivity in terms of two processes: (1) a relatively narrow Drude absorption with a plasma frequency of $\sim$ 1 eV and a relaxation rate which increases with temperature and (2) a broad excitation centered at 0.1 eV, with little temperature dependence. We associate the Drude part with free carriers and the mid-IR absorption with charge transfer connected with the mixed-valent nature of the material.

1. INTRODUCTION

Optical methods have been successfully applied in the past to conventional superconductors.[1] There has been, consequently, very intense activity in this field since the discovery of high T_c materials. However, due to their unusual properties even in the normal state, the results can not be easily interpreted and thus the situation is far from resolved. The properties which make the high T_c superconductors so different from low temperature superconductors are, among others, low carrier concentration and strong anisotropy.

In this paper we report our results on the optical spectra of a c-axis oriented $YBa_2Cu_3O_7$ film. The oriented nature enables us to study primarily excitations in the ab-plane. We have measured optical reflectivity from the far infrared to the ultraviolet at seven temperatures between 6K and 300K.

First we briefly review the optical methods used and the general information one can obtain by these measurements. Then we address the normal-state spectra and try to relate them to other solid-state properties as well as to different, currently used, theoretical approaches. We also examine the superconducting state and the possibility and limitations of extracting a gap from our optical data. Comparisons with conventional superconductors will be given throughout the discussion as necessary.

2. GENERAL OPTICAL PROPERTIES

Conventional superconductors are metals in their normal states. The dielectric function of a metal is given by the Drude form:

$$\epsilon(\omega) = \epsilon_1 + \frac{4\pi i}{\omega}\sigma_1 = \epsilon_\infty - \frac{\omega_p^2}{\omega^2 + i\omega/\tau}$$

with ω_p the plasma frequency and τ the relaxation time of the free carriers. Ideally, the frequency-dependent conductivity, $\sigma_1(\omega)$, approaches the dc conductivity as $\omega \to 0$. Although there were early attempts to fit optical spectra of $YBa_2Cu_3O_7$ by this expression,[2] it is now widely agreed on that the spectra of $YBa_2Cu_3O_7$ clearly deviate from this simple picture. The deviation consists of extra absorption in the mid-infrared region, above 500 cm^{-1}.

The dielectric function for a superconductor has been described by Mattis and Bardeen.[3] The calculation is valid either in the extreme clean limit ($2\Delta \gg 1/\tau$) or the extreme dirty limit ($2\Delta \ll 1/\tau$). In terms of optical properties, the dirty limit approximation leads to reflectivity of unity below the energy gap 2Δ, as was indeed observed for conventional superconductors like lead.[1] In the high T_c superconductors it is not *a priori* clear which limit to assume. The gap value may be large enough to satisfy the clean limit condition, in which case no feature is expected at 2Δ in the reflectivity.[4] The situation may be further complicated if other optical excitations are present close to the gap value. These can give extra absorption at low frequencies and thus cause the reflectivity to drop below unity. For $La_{2-x}Sr_xCuO_4$ it has been elegantly shown by Sherwin et al.[5] that the sharp reflectance drop at 60 cm^{-1} can be caused either by the superconducting gap (in the dirty limit) or by a plasma edge connected with the zero-crossing of the dielectric constant $\epsilon_1(\omega)$ (in the clean limit).

Furthermore, unlike superconducting metals, it is very difficult to prepare samples of sufficient optical quality. So far, only twice has a reflectivity equal to a constant 100 percent (i.e., zero loss) been observed. Thomas et al.[6] did so in an oxygen-deficient $YBa_2Cu_3O_x$ with Tc=68K, leading to a gap estimate of $2\Delta = 3.5k_BT_c$, while Reedyk et al.[7] observed a similar behavior in $Bi_2Sr_2CaCu_2O_8$, but were unable to state with certainty whether the edge seen near $4.8k_BT_c$ was a gap or the onset of mid-infrared absorption.

3. EXPERIMENTAL

The film was prepared on a $SrTiO_3$ substrate through a combination electron-beam/thermal evaporation process starting from Ba, Y_2O_3, and copper, with post-deposition annealing in oxygen.[8] X-ray diffraction confirmed that the orientation of the surface was in the ab-plane. The midpoint of the superconducting transition occurred at 86 K with a width of 10K. Details on the preparation are given in Ref. 7.

Reflectance spectra are shown in Fig. 1. These spectra were taken at near-normal incidence on two different instruments operating in different spectral regions. The visible/ultraviolet part was covered by a home-made grating spectrometer based on a Perkin-Elmer monochromator; in the far- and mid-infrared range a Bruker Fourier-transform interferometer was used. Since the granular texture of the film causes considerable diffuse scattering, thus distorting the data especially at higher frequencies, we adopted the following normalization procedure: After the temperature cycle was completed, a 2000 Å aluminum layer was evaporated onto the surface of the film and its reflectivity taken as reference instead of a flat Al mirror. We believe the values obtained this way to be correct within 1 percent in the mid-infrared to UV range and 3 percent in the far IR.

From the spectra we extracted the frequency-dependent optical constants (optical conductivity, dielectric constant, loss function, etc.) by means of Kramers-Kronig analysis.[9] Since this process ideally requires the knowledge of the reflectivity as a function of frequency from zero to infinity, extrapolations had to be used beyond the measured range. At low frequencies we applied a metallic (Hagen-Rubens) extension in the normal state: $R = 1 - A\sqrt{\omega}$ while in the superconducting state we adapted a different function, derived from the Mattis-Bardeen expression:[3] $R = 1 - B\omega^4$. At high frequencies R was continued smoothly until about 200,000 cm^{-1} and then made to go to zero as $\sim \omega^{-4}$.

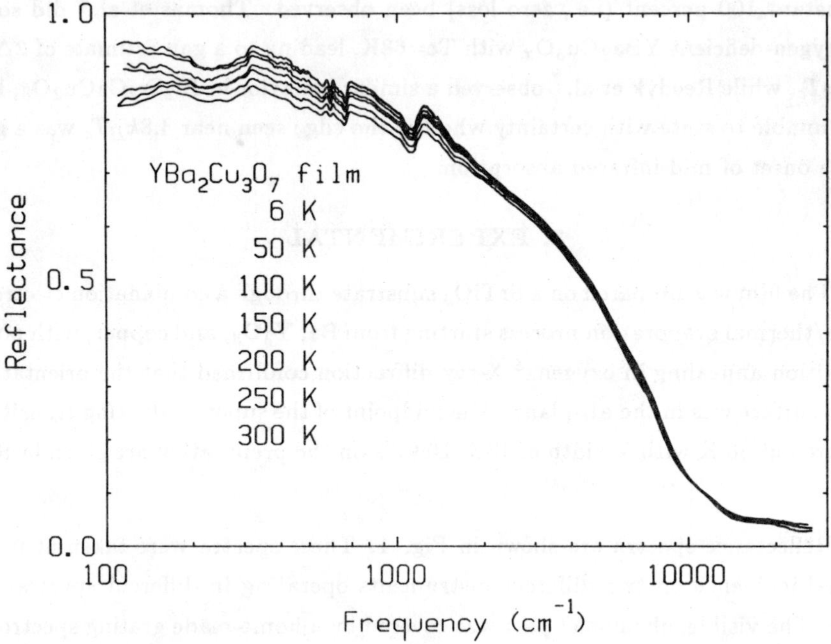

Fig. 1. Reflectivity vs. frequency of $YBa_2Cu_3O_7$ film at different temperatures. Note the logarithmic frequency scale.

Having reflectance data in a range of three orders of magnitude in frequency, we have reason to be confident that the microscopic constants derived this way reflect the true characteristics of the material. However, since the extrapolations necessarily introduce an ambiguity in the analysis, we also used the method of fitting the measured reflectivity data to that derived from an assumed model dielectric constant. The model used was a sum of a Drude term and several Lorentz oscillators, the starting values of which we estimated from the Kramers-Kronig optical conductivity curves.

4. RESULTS

In the reflectivity spectra at various temperatures shown in Fig. 1 we see a strong variation with temperature at low frequencies (below 2000 cm^{-1}), with the high-frequency reflectance being essentially temperature independent. This

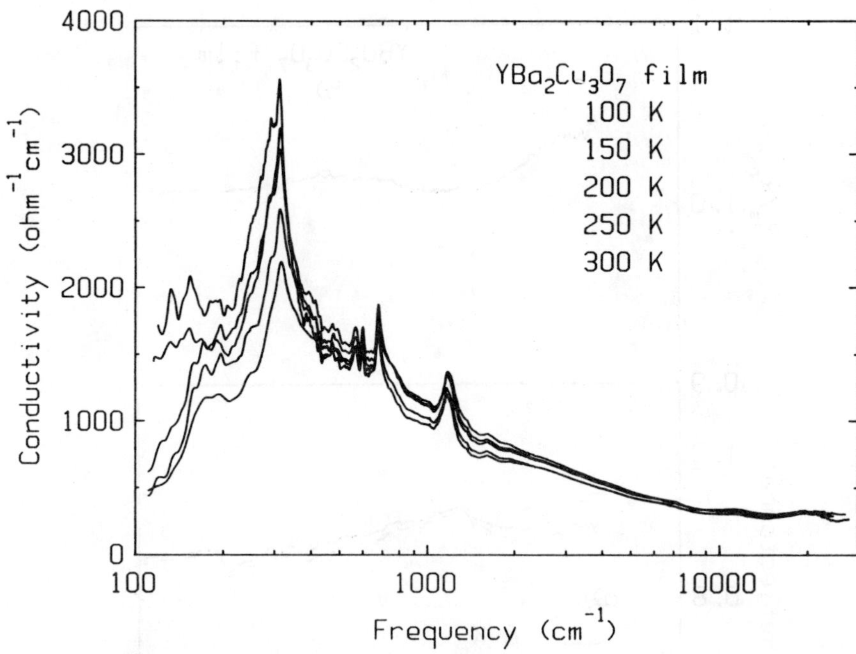

Fig. 2 Optical conductivity at five temperatures in the normal state.

behavior is similar to that found in textured ceramics [10] and oxygen-deficient single crystals. [6] A similar temperature dependence is seen in the Kramers-Kronig derived frequency-dependent conductivity, shown in Fig. 2. The low-frequency conductivity is clearly decreasing with increasing temperature, the curves gradually merging at higher frequencies.

In Fig. 3a we show the change in reflectivity upon going to the superconducting state. Below the transition temperature, the reflectivity is higher, but it does not reach 100 % and is almost parallel to the normal-state curve. This means that a residual absorption is present at low frequencies which can have two origins: either the orientation of the film is not complete, i.e., there is some c-axis polarization, or there is a more intrinsic reason, as we mentioned above. The relaxation rates obtained for the Drude part in the normal state range from 400 to 1600 cm^{-1}; thus the dirty limit approximation is not fully justified for gap values above about 200 cm^{-1}. The reflectivity ratio R_s/R_n, as shown in Fig. 3b, starts to drop

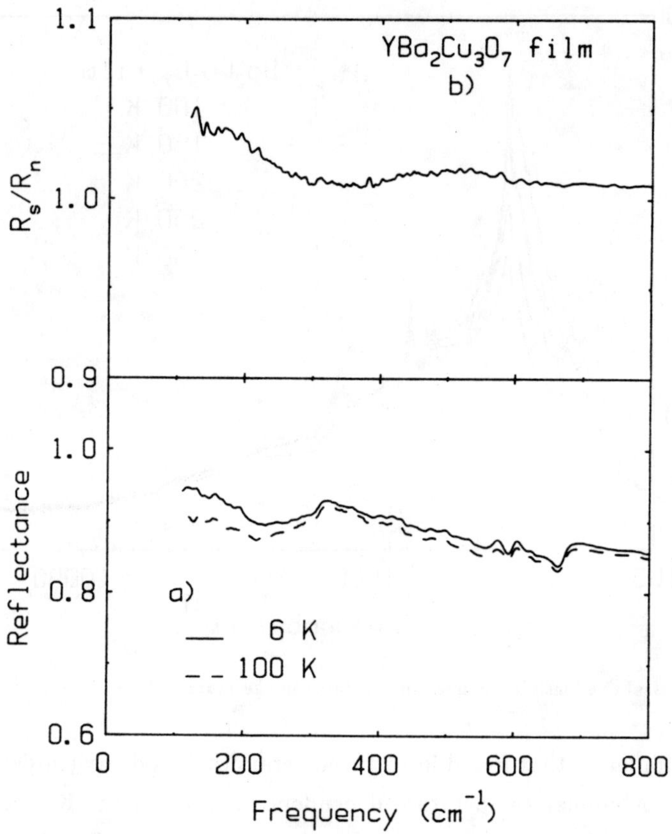

Fig. 3a) Far-infrared reflectivity of $YBa_2Cu_3O_7$ film in the normal (100K) and superconducting state (6K). 3b) Reflectivity ratio R_s/R_n in the far infrared.

at this value, making it impossible to distinguish between a superconducting gap and a plasmon.[5,7]

At this point we feel that because of these difficulties, an attempt to extract a gap value from the present data is not possible. Therefore, in the discussion below, we will limit ourselves to the normal-state properties, which nevertheless tell us very important facts about the electronic structure of the material.

In choosing a model to fit the reflectivity, the observed temperature dependence clearly rules out a simple Drude approach. In analogy with our earlier analysis of textured ceramics and crystal mosaics[10] we model the electronic part of the spectrum with a narrow Drude band and a broad mid-infrared excitation. In addition, several phonon lines and higher energy transitions appear to be present.

Our model dielectric function, which takes into account all of the above, is:

$$\epsilon(\omega) = -\frac{\omega_{pD}^2}{\omega^2 + i\omega/\tau} + \frac{\omega_{pe}^2}{\omega_e^2 - \omega^2 - i\omega\gamma_e} + \sum_{j=1,N+2} \frac{S_j\omega_j^2}{\omega_j^2 - \omega^2 - i\omega\gamma_j} + \epsilon_\infty \quad (1)$$

There are four contributions to this dielectric function: (1) Drude, characterized by a plasma frequency ω_{pD} and a relaxation rate $1/\tau$; (2) mid-infrared, having a strength ω_{pe}, a center frequency ω_e, and a width γ_e; (3) the phonons, represented by N sharp oscillators; and (4) a high frequency part, which includes 2 additional oscillators and a constant ϵ_∞.

Due to the large number of parameters, some restrictions had to be applied when carrying out the fit. After getting a satisfactory fit at 100 K by varying all parameters, we fixed the center frequencies and oscillator strengths of the phonons above 400 cm^{-1}, letting only the width vary. Also, the two high-frequency oscillators remained fixed at all temperatures. We list the values obtained by this fit in Table 1.

Table 1. Drude and mid-infrared-band parameters.
Values in cm^{-1} (8066 cm^{-1} = 1 eV).

	100 K	150 K	200 K	250 K	300 K	units
Drude ω_{pD}	6400	6300	8100	7800	7300	cm^{-1}
$1/\tau$	400	500	1200	1500	1650	cm^{-1}
σ_{dc}	1700	1300	880	680	540	Ω^{-1}cm^{-1}
Mid-IR ω_{pe}	15900	15700	15600	15600	15700	cm^{-1}
ω_e	960	950	940	980	1070	cm^{-1}
γ_e	5600	5400	6200	5900	5800	cm^{-1}

5. DISCUSSION

In accounting for the deviation from the Drude model, an alternative approach is that taken by Webb et al.[11] and Thomas et al.[6] They consider inelastic scattering of the free electrons by optically inactive low-energy excitations (such as magnons), resulting in a frequency-dependent relaxation rate and plasma frequency:[4]

$$\frac{1}{\tau_{\text{eff}}} = -\frac{\omega\epsilon^2}{\epsilon_1 - \epsilon_\infty}$$

$$\omega_{\text{eff}} = -(\epsilon_1 - \epsilon_\infty)(\omega^2 + \frac{1}{\tau_{\text{eff}}^2})$$

In order to calculate these values from reflectance data, we have to obtain ϵ_1 without the phonon contributions. Such curves are shown in Fig. 4. for 100K and 300K. In order to get meaningful τ_{eff} and ω_{eff} values, the condition $\epsilon_1 < \epsilon_\infty$ should hold, which is clearly not the case at 300K. Therefore, we model the optical properties as a superposition of independent processes, which we examine separately in the following subsections.

5.1 Drude Part

The conductivity curves in Fig. 5, at two temperatures and with phonon lines subtracted, illustrate the change in the low-frequency electronic properties. In Fig. 6 we plot the Drude parameters as a function of temperature. Within the accuracy of the fits, these seem to be characteristic of a simple metal, with ω_{pD} constant, approximately 1 eV, and the relaxation rate linear in T. The dc resistivity

$$\rho_0 = \sigma_0^{-1} = \frac{4\pi}{\omega_{pD}^2\tau}$$

shows the familiar linear behavior,[12] with a very small intercept. We note that the intercept obtained by direct resistivity measurements was substantially larger, indicating that we see less of the grain-boundary effects in the far infrared than by dc. The optical absolute values, though, are about a factor of two higher.

The temperature dependence of the fit parameters suggests that the Drude part is representing the free carriers taking part in the dc conduction process. However, a serious discrepancy exists with Hall data.[13] Assuming a one-band model,

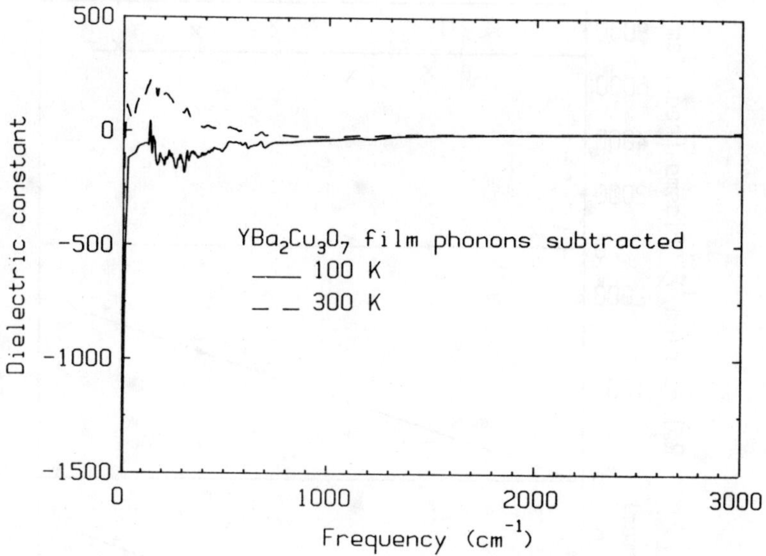

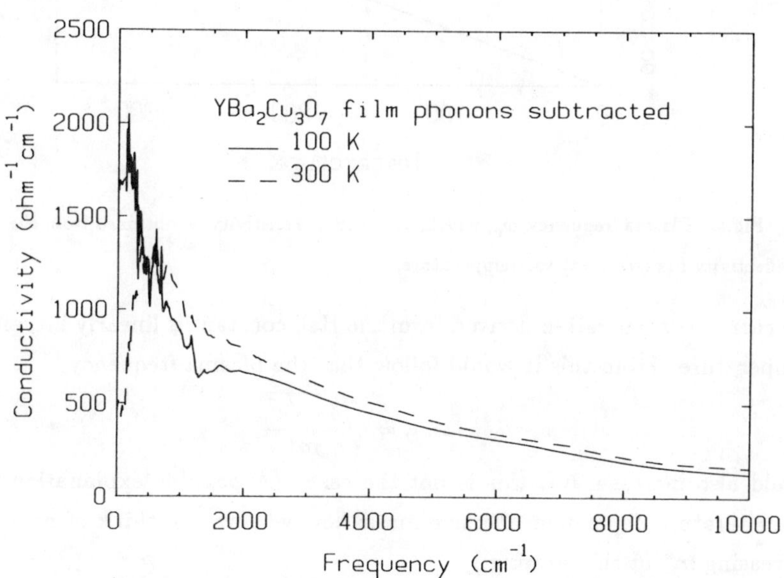

Fig. 5. Optical conductivity, phonon contributions subtracted, at 100K and 300K.

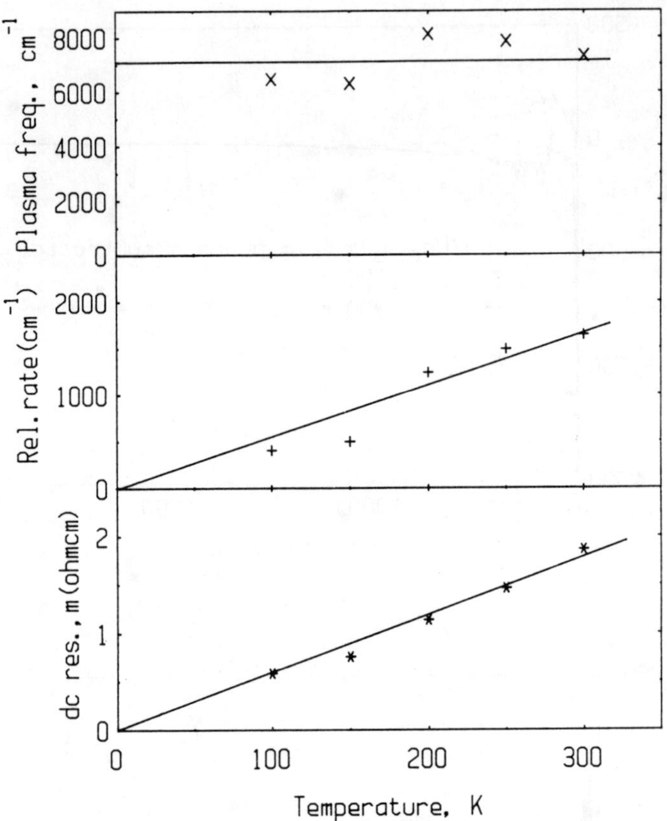

Fig. 6. Plasma frequency, ω_p, width, $1/\tau$, and dc resistivity, ρ, obtained from the Drude part of reflectivity fits (see text) vs. temperature.

the carrier concentration derived from the Hall constant is linearly increasing with temperature. From this it would follow that the plasma frequency

$$\omega_{pD} = \sqrt{\frac{4\pi N e^2}{m^*}}$$

would also increase, but this is not the case. (A possible explanation would be a temperature dependent effective mass; however, we can think of no mechanism increasing m^* upon warming.)

5.2 The Mid-infrared Band

In contrast to the Drude optical parameters, it is obvious from Table 1 that the mid-IR band does not have a marked temperature dependence. As to the origin of this band, several mechanisms have been proposed.

Lyons et al.[14] see a broad feature at about 0.4 eV (3000 cm^{-1}) in the Raman spectra of non-superconducting oxygen-deficient $YBa_2Cu_3O_{6+x}$, which they attribute to two-magnon scattering, based on its temperature and polarization dependence. On this basis, it is tempting to associate the mid-IR band with a direct two-magnon absorption, similar to what has been observed[15] in MnF_2. In our film, though, the center frequency is substantially lower than that of the Raman peak. Also, the Raman intensity decreases with increasing oxygen concentration as the antiferromagnetic order disappears, whereas the IR absorption is absent in $YBa_2Cu_3O_6$ and strong at high oxygen concentration.[16]

We interpret this absorption as a charge-transfer transition arising from the mixed-valent character of the material.This explanation is in accord with the fact that the band completely disappears[16] in $YBa_2Cu_3O_6$, where both in-plane copper ions are divalent. By increasing the oxygen concentration we add extra holes to the system (in other words, increase the formal valence of copper above 2), thus creating a possibility of valence fluctuations. These are most probably not simple $Cu^{2+} \leftrightarrow Cu^{3+}$ charge transfer processes, particularly as no evidence has been seen so far of Cu^{3+} ions by XPS spectroscopy. More likely, the charge transfer occurs between or within relatively small Cu-O clusters, e.g., $Cu^{2+}O^{2-} \leftrightarrow Cu^{+}O^{-}$ as suggested by Varma et al.[17] The correlation of mixed valence and the mid-infrared band is also evident in recent measurements of Kaplan et al.[18] on non-superconducting $La_4BaCu_5O_{13}$ and $La_2SrCu_2O_{6+x}$. They find a mid-IR absorption in the former compound (formal Cu valence: 2.4) and also in the latter series if $x > 0$ (formal Cu valence > 2). This suggests that a one-to-one correlation exists between mixed valence and the infrared band, rather than between the band and superconductivity; mixed valence can be a necessary but not sufficient condition for superconductivity, as also in the case of organic superconductors.[19]

Such particle-hole excitations are often referred to as excitons, meaning that they are treated in the solid as an excitation of the whole system rather than a single cluster; they then can be formally treated as quasiparticles when describing the excitation spectrum. We want to clarify here that the excitation itself is strongly localized, i.e. a tightly bound (Frenkel) exciton. The exact levels of such an exciton are very difficult to calculate, in contrast to a delocalized (Wannier) exciton, which can be treated very similarly to a hydrogen atom (an electron and a hole moving in a Coulomb potential screened by some effective dielectric constant), leading to a Rydberg-like series of narrow absorption lines. In contrast, for Frenkel excitons all the interactions have to be taken into account locally, representing a complicated many-body problem.[17] The calculations of Varma et al. predict a peak in the particle-hole excitation spectrum at 0.5 eV, but they also have a requirement of large momentum transfer which is clearly not satisfied in one-photon processes we are dealing with here.[20]

5.3 Phonons

The center frequencies, strengths and widths of phonons at 100K and 300K are listed in Table 2. Because of the particulars of the fit mentioned above, the 300 K values are to be considered tentative; however, qualitative conclusions can be drawn.

Table 2. Phonon parameters.

Values in cm^{-1} ($8066\ cm^{-1} = 1\ eV$).

100 K			300 K		
ω_j	γ_j	S_j	ω_j	γ_j	S_j
			167	16	10
			192	33	19
268	47	33			
309	44	50	310	75	42
400	60	6	400	60	6
495	42	3	495	42	3
564	13	0.3	564	13	0.3
689	100	5	689	100	5
1190	200	3	1190	180	3

The most striking feature is the extreme strength of some of the lines. S is around 2 for simple optically active phonons, whereas here all the peaks below 400 cm^{-1} show an S value larger than 10. Following the assignment of Crawford et al.,[21] these are all Cu-O vibrations with components parallel to the **ab**-plane, except for the localized Y mode at 192 cm^{-1}. The 155 cm^{-1} feature can be associated with Cu-O chains, and the doublet at 270 and 310 cm^{-1} (merging into one peak at 300 K) with Cu-O in-plane bending. The very small S value for the **c**-axis polarized Cu-O stretching mode at 564 cm^{-1} confirms that very little of that orientation is present. The absence of the peaks below 200 cm^{-1} at 100 K can be explained by screening due to increased free-electron scattering in this region, caused by the narrowing of the Drude peak.

We can not offer any specific assignments regarding the 400 and 495 cm^{-1} lines. The former has been observed occasionally in sintered pellets,[22] and the latter may be part of the 500 - 600 cm^{-1} range where strongly sample dependent structure is known to occur. Two lines are definitely of extrinsic nature, at 700 and 1200 cm^{-1}. These were not observed during the first run and emerged gradually as a function of time. Whether their presence is due to contamination or degradation of the surface is not known. Raman spectra of solid peroxides show[23] a symmetric O-O stretch around 1200 cm^{-1}; we may see such a stretching mode becoming allowed here, probably because of bond asymmetry.

5.4 Bands Above 1 eV

In addition to the mid-infrared band, we see evidence of two more excitations in the 1–3 eV range. Oscillator fit parameters for these bands are shown in Table 3.

Table 3. Parameters for high-energy bands.

Values in cm^{-1} (8066 cm^{-1} = 1 eV).

ω_j	γ_j	S_j
13500	22000	1
19500	40000	0.7

The presence of these additional terms is best seen from the loss function,

-Im(1/ϵ), shown in Fig. 7 for 100K and 300K. (We note that this function, derived from optical reflectance, compares remarkably well with electron-energy loss measurements.[24]) The intensity of these absorptions suggests that they are also of electronic origin, but the underlying mechanism is not clear. Interband transitions are estimated[25] to be above 2.5 eV, leaving the interpretation of at least the 1.5 eV transition open.

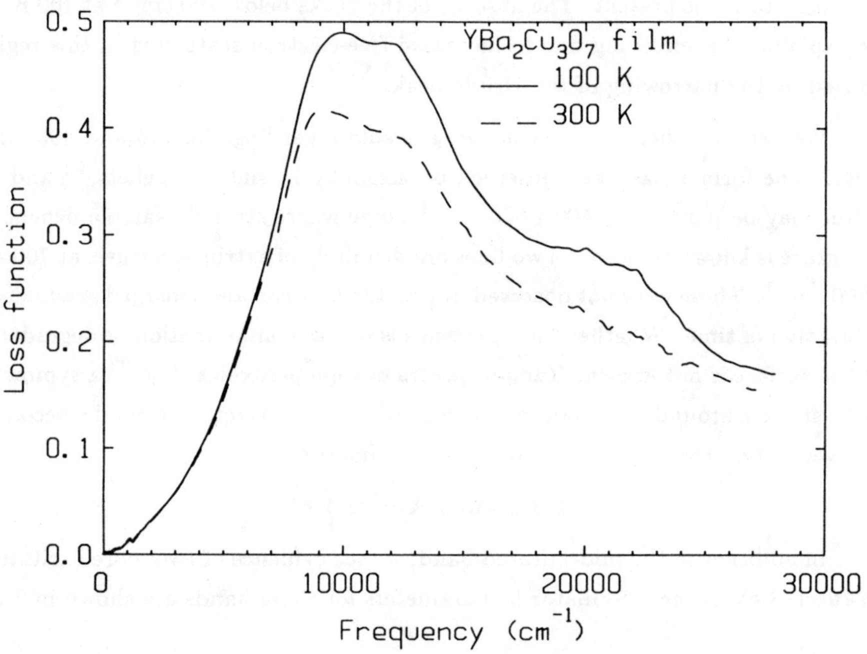

Fig. 7. Loss function Im(-1/ϵ) at 100K and 300K.

6. CONCLUSIONS

Based on our results outlined above, we can draw the following conclusions:

1. The optical spectrum of the $YBa_2Cu_3O_7$ film consists of a low-frequency Drude part and a very intense mid-infrared absorption band centered at 0.1 eV. Superimposed on this electronic part are strong phonons at 167, 192, 268 and 310 cm^{-1}. We associate the Drude part with free electrons responsible

for the dc conductivity and the mid-infrared absorption with charge-transfer processes connected to the mixed-valent character of the material.

2. The temperature dependence of the Drude parameters in the normal state is consistent with a metallic behavior at low frequencies, i.e. constant plasma frequency and increasing relaxation rate with increasing temperature. The mid-IR band, in contrast, is temperature independent.

3. There is a marked change in the low-frequency optical properties upon going to the superconducting state, but a gap value can not be unambiguously determined from the spectra.

Clearly, more experimental data on high-quality, well-characterized samples are needed to confirm and further explore the unusual optical properties of these exciting new substances and such measurements are underway in our laboratory and elsewhere. What we feel we have demonstrated here is that optical investigations are an integral part of the combined efforts of solid-state physicists to elucidate the mechanism of high Tc superconductivity and that any theory concerning this mechanism should be able to explain the optical data as well.

Acknowledgements

Research at Florida supported by DARPA grant MDA972-88-J-1006.

References

1. Glover, III, R.E., and Tinkham, M., *Phys. Rev.* **108**, 243 (1957).

2. Schlesinger, Z., Collins, R.T., Kaiser, D.L., and Holtzberg, F., *Phys. Rev. Lett.* **59**, 1958 (1987).

3. Mattis, D.C., and Bardeen, J., *Phys. Rev.* **111**, 412 (1958).

4. Timusk, T., and Tanner, D.B., World Scientific, to be published.

5. Sherwin, M.S., Richards, P.L., and Zettl, A., *Phys. Rev. B* **37**, 1587 (1988).

6. Thomas, G.A., Orenstein, J.,Rapkine, D.H., Capizzi, M., Millis, A.J., Bhatt, R.N., Schneemeyer, L.F., and Waszczak, J.V., *Phys. Rev. Lett.* **61**, 1313 (1988).

7. Reedyk, M., Bonn, D.A., Garrett, J.D., Greedan, J.E., Stager, C.V., Timusk, T., Kamarás, K., and Tanner, D.B., *Phys. Rev. B* **38**, 11981 (1988).

8. Chan, S.-W., Greene, L.H., Feldmann, W.L., Miceli, P.F., and Bagley, B.G., in 1987 American Vacuum Society series *Conference Proceedings on Thin Film Processing and Characterization of High Temperature Superconductors*.

9. Wooten, F., *Optical Properties of Solids* (Academic Press, New York, 1972).

10. Timusk, T., Herr, S.L., Kamaras, K., Porter, C.D., Tanner, D.B., Bonn, D.A., Garrett, J.D., Stager, C.V., Greedan, J.E., and Reedyk, M., *Phys. Rev. B* **38**, 6683 (1988).

11. Webb, B.C., Sievers, A.J., and Milalisin, T., *Phys. Rev. Lett.* **57**, 1951 (1987).

12. Gurvitch, M., and Fiory, A.T., *Phys. Rev. Lett.* **59**, 1337 (1987).

13. Cheong, S.-W., Brown, S.E., Fisk, Z., Kwok, R.S., Thompson, J.D., Zirngiebl, E., Gruner, G., Peterson, D.E., Wells, G.L., Schwarz, R.B., and Cooper, J.R., *Phys. Rev. B* **36**, 3913 (1987).

14. Lyons, K.B., Fleury, P.A., Schneemeyer, L.F., and Waszczak, J.V., *Phys. Rev. Lett.* **60**, 732 (1988).

15. Allen, S.J., Loudon, R., and Richards, P.L., *Phys. Rev. Lett.* **16**, 463 (1966).

16. Kamaras, K., Porter, C.D., Doss, M.G., Herr, S.L., Tanner, D.B., Bonn, D.A., Greedan, J.E., O'Reilly, A.H., Stager, C.V., and Timusk, T., *Phys. Rev. Lett.* **59**, 919 (1987).

17. Varma, C.M., Schmitt-Rink, S., and Abrahams, E., *Solid State Comm.* **62**, 681 (1987).

18. Kaplan, S.G., Noh, T.W., Sulewski, P.E., Xia, H., Sievers, A.J., Wang, J., and Raj, R., *Phys. Rev. B* **38**, 5006 (1988).

19. Jerome, D., and Schultz, H.J., *Adv. Phys.* **31**, 299 (1982).

20. Varma, C.M., this volume.

21. Crawford, M.K., Farneth, W.E., Bordia, R.K., and McCarron, E.M.III, *Phys. Rev. B* **37**, 3371 (1988).

22. Wrobel, J.M., Wang, S., Gygax, S., Clayman, B.P., and Peterson, L.K., *Phys. Rev. B* **36**, 2368 (1987).

23. Blunt, F.J., Hendra, P.J., and Mackenzie, J.R., *J.C.S. Chem. Comm.* 1969, 278

24. Tarrio, C., and Schnatterly, S.E., *Phys. Rev. B* **38**, 921 (1988).

25. Massida, S., Yu, J., Freeman, A.J., and Koelling, D.D., *Phys. Lett.* **A122**, 198 (1987).

PANEL SUMMARIES

Panel: HIGH-T_c THEORIES – ELEMENTARY EXCITATIONS

Moderator: J. ASHKENAZI
Physics Department, University of Miami, Coral Gables, FL 33124

ABSTRACT - Elementary excitations play a key role in the different mechanisms proposed for high-temperature superconductivity. Three major classes of excitations were discussed in the Panel: "single particle" excitations in the normal state, "collective" excitations in the normal state, and excitations in the superconducting state.

PANEL PARTICIPANTS:

F.J. Adrian, APL, John Hopkins University, Laurel, MD 20707.
S. Berko, Brandeis University, Waltham, MA 02254.
R.E. Cohen, Naval Research Laboratory, Code 4691, Washington, D.C. 20375.
F.J. Crowne, Martin Marietta Laboratories, 1450 So. Rolling Road, Baltimore, MD 21227.
M.J. DeWeert, Naval Research Laboratory, Code 4690, Washington, D.C. 20375.
D.L. Ederer, NIST, Physics A251, Gaithersburg, MD 20899.
G. Fletcher, University of Texas, Department of Physics, Box 19059, Arlington, TX 76019.
A.J. Freeman, Physics Department, Northwestern University, Evanston, IL 60201.
K. Kamarás, University of Florida, 215 Williamson Hall, Gainesville, FL 32611.
V.Z. Kresin, LBL, University of California, Berkeley, CA 94720.
K. Machida, LASSP, Clark Hall, Cornell University, Ithaca, NY 14583-2501.
A. Meulenberg, COSMAT Laboratories, 22300 Cosmat Drive, Clarksburg, MD 20871.
S. Nam, Wright University Center, 7735 Peters Pike, Dayton, OH 45414.
J.J. Rhyne, NIST, Reactor Building A118, Gaithersburg, MD 20899.
R. Shuker, NIST, Physics A251, Gaithersburg, MD 20899.
D. Singh, Naval Research Laboratory, Code 4684, Washington, D.C. 20375.
G.C. Vezzoli, US Army Materials Technology Laboratory, Watertown, MA 02172.
M.A. Wilczewski, US Patent Office, 2021 Jefferson Davis Hwy. CP310E09, Arlington, VA 22202.

1. "SINGLE PARTICLE" EXCITATIONS IN THE NORMAL STATE

The discussion on these excitations addressed several questions:

1.1 What Is The Appropriate Theoretical Approach?

Several theoretical approaches to these excitations were mentioned. One approach is the molecular orbital method, generally used by chemists. Within this approach the single particle excitations of a finite cluster of atoms is calculated. Lattice periodicity is ignored, however, one can in practice introduce better approximations for electron correlations than in an infinite lattice calculation (*e.g.* Configuration Interactions).

Another approach consists of band-structure calculations based on density functional (DF) theory. Within this theory an effective self-consistent single-particle potential, and the resulting band structure, are used to obtain the ground-state density and energy. This band structure has no rigorous correspondence to the physical excitations; however, it was pointed out that it has been successful in predicting certain aspects of single-particle excitations, even in highly-correlated electron systems – for example, it successfully predicted the Fermi surface (FS) of the heavy-fermion system UBe_{13}. R.E. Cohen mentioned that DF calculations predicted transport properties of high-T_c materials.[1] However, though DF calculations yielded the correct FS of UBe_{13}, a mass enhancement of about 20 over the DF results was necessary to explain the experimental data. Also, spectroscopic measurements on high-T_c materials seem to show shifts of ~ 1 eV of the DF density of states (DOS) peaks away from the Fermi level E_F, probably due to many-body effects; however, such effects might still be approximated using the DF scheme in an unconventional manner.[2]

Another very popular theoretical approach is the Hubbard model. High-T_c theories such as the resonating valence bonds (RVB) and the spin-bag models are based on the simple (single-orbital and intraatomic interactions) Hubbard model. Other theories require an extended Hubbard scheme, including interatomic interactions, and a number of orbitals, to account for the complexity of the high-T_c systems. The Anderson model is included in this category. A problem under debate is whether the relevant high-T_c physics could be based on an effective simple Hubbard model derived from an extended one.[3] Also a combination of the DF and the extended Hubbard model approaches was mentioned, in which the model parameters are derived from DF calculations.

1.2 What Is The Energy-Dependence?

There is a debate concerning the nature of the energy dependence of single-particle excitations in high-T_c materials. DF band-structure calculations predict wide metallic bands around the FS. In order to prove the validity of this picture, A.J. Freeman showed spectra of recent photoemission measurements[4] carried out on a single crystal of $EuBa_2Cu_3O_{6.7}$, which was cleaved at a very low temperature (20 K) to avoid oxygen loss from the surface. A Fermi edge was clearly observed in these spectra, but the DOS peak below E_F was still shifted by ~ 1 eV compared to the DF results. Another experimental result shown was angle-resolved photoemission spectra[5] of a single crystal of $Bi_2Sr_2CaCu_2O_8$ (which does not seem to suffer from the problem of oxygen loss from the surface). These spectra showed a number of separated narrow bands (of widths of $\sim 0.1 - 0.5$ eV each), the highest crossing E_F. It was pointed that E_F falling in a narrow band would be consistent with many experimental observations.[6,7] R.E. Cohen demonstrated that the distinction between a wide-band and a narrow-band interpretation could be very minute in these materials. Such narrow bands could be obtained as a result of many-body effects in the high-T_c materials,[8] and the measured bandwidth would then depend on the energy scale of the experimental probe, as had been found before for heavy fermion systems.

Another approach is to analyze the electronic structure of the high-T_c materials in terms of impurity states in a gap (possibly of the Hubbard type). The point of view of the the RVB theory[9] is that these materials are basically Mott insulators, and that the single-particle picture is wrong for them. Instead, they should be analyzed in terms of soliton type chargeless fermions (spinons) and spinless charged bosons (holons).

1.3 What Is The Orbital Character of the States Near E_F?

DF band-structure calculations predict that the electronic states around E_F are composed mainly of strongly hybridized Cu(d) and O(p) orbitals of the CuO_2 planes (where conduction is mainly by to $d_{x^2-y^2}$ and p_x–p_y electrons), and also of orbitals of the inter-plane atoms (called "intercalates" by A.J. Freeman). Spectroscopic data[10–11] tend to indicate that the states around E_F are composed mainly of O(p) orbitals. A number of high-T_c models are thus based on conduction of p-holes; some of them are based on O(p)–Cu(d) hopping, and others on direct O(p)–O(p) hopping, or on both types of hopping. There are also high-T_c models based primarily on Cu(d) orbitals (or on both orbitals).

A.J. Freeman mentioned that the new photoemission data on low-temperature cleaved single crystals[4] show both Cu(d) and O(p) orbitals around E_F. New experimental data on this issue seem to be essential to test high-T_c theories.

1.4 Is The Fermi Liquid Approach Appropriate?

BCS-type high-T_c theories (of any type of pairing) will answer this question by YES, while RVB or bipolaron theories will answer it by NO. A Fermi liquid is characterized by the existence of a Fermi surface. Due to the short mean free path in the high-T_c materials, classical experimental methods to determine a FS, such as the de Haas van Alphen method, do not seem to work so far, and the only method which seems to show some success is the positron-annihilation method (which is based on a direct probing of the electron momentum distribution, and not of electron orbits).

S. Berko mentioned the success of positron-annihilation measurements on single crystals of $YBa_2Cu_3O_x$ in several laboratories to determine a FS resembling the one obtained in DF band-structure calculations. However, he pointed out that the resolution of these measurements is quite poor. Thus it is not possible so far to decide definitely whether they really show a sharp FS, or just a variation in the momentum distribution of the electrons. Further experiments are essential here, too.

1.5 What Is The Effect of Stoichiometry Variation?

The most investigated high-T_c system has so far been $YBa_2Cu_3O_x$. It shows an interesting behaviour when the oxygen-stoichiometry parameter x is changed. For x close to 7, it is a superconductor (SC) with $T_c \sim 90K$. When x is decreased below 7, T_c decreases with it, but there seems to be a plateau of $T_c \sim 55-60$ K for $6.8 > x > 6.5$, and below $x = 6.5$, T_c decreases sharply to zero, and the material becomes non-metallic and antiferromagnetic (AF) at low temperatures (with a possible intermediate x range of spin-glass behaviour). An important problem, which will be discussed in section 2.2, is the existence of AF fluctuations in the SC state.

The non-metallic AF behaviour for $x < 6.5$ cannot be explained on the basis of DF band-structure calculations and electron correlations seem to be essential. It has been difficult to observe a semiconducting gap in this range, and only for $x \sim 6.0$ there seems to be optical and resistivity evidence for a gap ~ 1.5 eV. For higher x-values, the low-temperature non-metallic behaviour seems to be of the variable-range hopping type,[12] which led some to suggest an Anderson-localisation transition. However, it is not absolutely clear whether this non-metallic behaviour is intrinsic, or results from damage caused during the preparation of the oxygen-deficient sample. (They are generally prepared by quenching them from high temperatures.)

X-ray absorption results[13] seem to connect the plateau in T_c for $6.5 < x < 6.8$ with a plateau in the number of holes in the peak near E_F. At least two theoretical explanations to this plateau in T_c have connected it to a plateau in the number of carriers. One of them[14] is based on the assumption that one half of the Cu–O chains are filled for $6.0 < x < 6.5$, and the other half for $6.5 < x < 7.0$. The other explanation[8] connects this plateau with experimental observations that the valence of the Cu atoms in the chains is close to +1 for $x = 6$, and to +2 for $x = 7$, and that most of the valence change from +1 to +2 seems to occur[15] when $6.5 < x < 6.8$.

2. "COLLECTIVE" EXCITATIONS AND SUPERCONDUCTING PAIRING

Two general types of excitations were discussed:

2.1 Charge-Fluctuation Excitations:

Most of the Panel participants seemed to support a BCS-type high-T_c mechanism, where the pairing mediators are excitations of this type. V.Z. Kresin supported such a mechanism based on mixed phonon–plasmon pairing. The plasmons are low frequency acoustic modes, with typical energies somewhat lower than 0.1 eV, which are several times higher than the typical phonon energy, and several times lower than his estimate for E_F; so one can get a fairly high T_c within the BCS theory. His answer to the question whether SC would still exist in these materials if the atoms were nailed to their equilibrium positions was: "NO". Part of the other participants agreed with him on this answer, but others, like G.C. Vezzoli, gave the opposite answer: "YES".

R.E. Cohen supported phonon pairing. He showed the results of his first-principles DF frozen-phonon calculations in La_2CuO_4, which agreed well with the experimental spectrum. He also mentioned he has in progress calculations of the electron-phonon matrix elements. He explained the failure of previous electron-phonon calculations to predict high-T_c in these materials by the fact that they ignored the effect of long-range Coulomb interactions which were important in these ionic crystals (as was found in his calculations). Since the calculated frozen phonon involves considerable electron charge rearrangement, it was pointed out that it should be regarded as a hybridized phonon–plasmon or phonon–exciton, but the relevant energy in pairing through such an excitation is a phonon energy.

K. Kamarás mentioned recent infrared reflectivity results[16,17] on single crystals of $YBa_2Cu_3O_x$. Unlike previous single crystal results, they seem to confirm previous results on polycrystalline samples, showing that the mid-infrared spectrum cannot be fitted with a simple Drude term, and that a broad absorption peak exists in the range above ~ 0.1 eV. Such a peak has been observed also in the La-Sr-Cu-O and Bi-Sr-Ca-Cu-O systems and its composition dependence correlates with high-T_c. This peak has been interpreted as an exciton, or another electronic excitation which is the SC pairing mediator. A mid-infrared B_{1g} Raman peak, observed[18] in single crystals of $YBa_2Cu_3O_x$, has been interpreted to be due to two-magnon scattering. However, very recently[19] a wide A_{1g} Raman peak has been observed in a somewhat higher energy range, and was found to persist and change its energy as x is increased from 6 towards 7. At this time it is not clear whether this peak, and a broad mid-infrared Raman peak observed recently[20] in a polycrystalline sample of $Bi_4(Ca,Sr)_6Cu_4O_{16+x}$, correspond to excitons or other charge-fluctuation excitations.

It seems that further experiments are essential to determine whether there exists an exciton-type excitation responsible for high-T_c. It is also important to determine where in the lattice such a mode is located. High-T_c models have been proposed locating such a mode in the CuO_2 planes or in the intercalates. The existence of such a mode in the intercalates could, in principle, induce pairing between electrons in two neighboring CuO_2 planes.[8] Such pairing has the advantage that the Coulomb repulsion is very low, so that the excitation mediated repulsion does not have to be retarded compared to the Coulomb attraction in order to get net attraction, and the excitation energy (and thus T_c) could be quite high.

2.2 Magnetic-Fluctuation Excitations:

There is evidence on the existence of AF fluctuations in the $La_{2-x}Sr_xCuO_4$ and $YBa_2Cu_3O_{6+x}$ systems.[21,18] These fluctuations are strong close to the non-superconducting AF phase (small x values), but there is evidence that they also exists for larger x values, where SC exists. Thus, it is in principle possible that these magnetic excitations are responsible for high T_c SC. However, it was pointed out that the strength of these magnetic correlations seems to anti-correlate SC, namely the higher T_c becomes (with x) the weaker these fluctuations become. This might imply that the magnetic interactions do not cause SC in these materials, but compete with it. But, as was mentioned, this conclusion is not definite. It is possible that the magnetic fluctuations contribute to SC only when they are not strong. It is also possible that, due to a structural transition, selection rules prevent the full experimental observation of the magnetic correlations in SC samples.

The existence of these magnetic fluctuations motivated a considerable number of high-T_c models based on pairing through magnetic excitations. Among them is the spin-bag model,[22] in which the existence of holes weakens the magnetic correlations by local energy gain; in another model[23] the oxygen p-holes induce ferromagnetic correlations between the Cu magnetic moments, opposing the AF coupling between them, and causing magnetic frustration. Magnetic fluctuations are expected also within the RVB model.[9]

3. EXCITATIONS IN THE SUPERCONDUCTING STATE

The discussion on these excitations addressed two questions:

3.1 Is There A Gap and Are There Electron Pairs in the Superconducting State?

This is a basic question because the validity of BCS-type models depends on a positive answer. No one on the Panel suggested a negative answer to this question. The gathering evidence for the existence of a SC gap and pairing was detailed by V.Z. Kresin who pointed out recent infrared reflectivity data,[17] Andreev scattering, and[24] tunneling results, *etc.* proving the existence of SC pairing, and a gap.

3.2 Is There An Intrinsic Linear Specific Heat Term Below T_c?

One of the puzzling effects in the high-T_c materials was the observation of a linear temperature dependence of the specific heat below T_c. This term has been interpreted as a proof to the existence of spinons, predicted by the RVB theory. There were also other explanations for it, such as the existence of tunneling states, unpaired electrons, *etc.* Thus, the question whether this linear term really originates from the SC material and not from impurities is an essential one.

No one on the Panel had a positive answer to the above question. V.Z. Kresin pointed out that either this linear term is missing, as[25] in Bi-Sr-Ca-Cu-O, or that its presence could be explained as due to an impurity.[26]

ACKNOWLEDGMENTS - The author acknowledges the contribution of the Panel participants, and the comments of G. Dewar and J.C. Nearing on the manuscript.

REFERENCES

1. Allen, P.B., Pickett, W.E. and Krakauer, H. , Phys. Rev. **B37**, 7482 (1988).
2. Redinger, J., Freeman, A.J., Jaejun Yu and Massidda, S., Phys. Lett. **A124**, 469 (1987).
3. Zhang, F.C. and Rice, T.M., Phys. Rev. **B37**, 3759 (1988).
4. Arko, A.J., List, R.S., Fisk, Z., Cheong, S-W., Thompson, J.D., O'Rourke, J.A., Olson, C.G., Yang, A-B., Tun-Wei Pi, Schirber, J.E. and Shinn, N.D., J. Mag. Mag. Mater., in press.
5. Takahashi, T., Matsuyama, H., Katayama-Yoshida, H., Okabe, Y., Hosoya, S., Seki, K., Fujimoto, H., Sato, M. and Inokuchi, H., *Proceedings of 1st International Symposium on Superconductivity*, Nagoya, August 1988, in press.
6. Kresin, V.Z., Deutscher, G. and Wolf, S.A., *High T_c Superconductivity World*, in press.
7. Fisher, B., Genossar, J., Lelong, I.O. , Kessel, A. and Ashkenazi, J., J. Supercond. **1**, 53 (1988); Physica **C153-155**, 1349 (1988); Bar-Ad, S., Fisher, B., Ashkenazi, J. and Genossar, Physica C, in press; Genossar,J., Fisher, B., Lelong, I.O. , Ashkenazi, J. and Patlagan, L., Physica C, submitted.
8. Ashkenazi, J. and Kuper, C.G., Physica **C153-155**, 1315 (1988); Ann. Phys. (France) **13** (1988).
9. Anderson, P.W., Science **235**, 1196 (1987).
10. Bianconi, A., De Santis, M., Di Cicco, A., Clozza, A., Congiu Castellano, A., Della Longa, S., Gargano, A., Delogu, P., Dikonimos Makris, T., Giorgi, R., Flank, A.M., Fontaine, A., Lagarde, P. and Marcelli, A., Physica **C153-155**,

115 (1988).

11. Nücker, N., Fink, J., Fuggle, J.C., Durham, P.J. and Temmerman, W.M., Physica C**153-155**, 119 (1988).

12. Yu Mei, Jiang, C., Green, S.M., Luo, H.L. and Politis, C., Z. Phys. B**69**, 11 (1987).

13. Kuiper, P., Kruizinga, G., Ghijsen, J., Grioni, M., Weijs, P.J.W., de Groot, F.M.F., Sawatzky, G.A., Verweij, H., Feiner, L.F. and Peterson, H., Phys. Rev. B**38**, 6483 (1988).

14. Zaanen, J., Paxton, A.T., Jepsen, O. and Andersen, O.K., Phys. Rev. Lett. **60**, 2685 (1988).

15. Oyanagi, H., Ihara, H., Matsubara, T., Tokumoto, M., Matsushita, T., Hirabayashi, M., Murata, K., Terada, N., Yao, T., Iwasaki, H. and Kimura, Y., Jpn. J. Appl. Phys. **26**, L1561 (1987).

16. Timusk, T., Herr, S.L., Kamarás, K., Porter, C.D., Tanner, D.B., Bonn, D.A., Garret, J.D., Stager, C.V., Greedan, J.E. and Reedyk, M., Phys. Rev. B**38**, 6683 (1988).

17. Thomas, G.A., Orenstein, J., Rapkine, D.H., Capizzi, M., Millis, a.J., Bhatt, R.N., Schneemeyer, L.F. and Waszczak, J.V., Phys. Rev. Lett. **61**, 1313 (1988).

18. Lyons, K.B., Fleury, P.A., Schneemeyer, L.F. and Waszczak, J.V., Phys. Rev. Lett. **60**, 732 (1988).

19. Lyons, K.B., Fleury, P.A. and Sulewski, P.E., private communication.

20. Néstor E. Massa, Jorge Güida, Oscar E. Piro and Pedro J. Aymonino, *Proceedings of 1st Latin-American Conference on High-T_c Superconductors*, Rio de Janeiro, May 1988 (World Scientific, Vol. 9, Oct. 1988).

21. Shirane, G., Edndoh, Y., Birgeneau, R.J., Kastner, M.A., Hidaka, Y., Oda, M., Suzuki, M. and Murakami, T., Phys. Rev. Lett. **59**, 1613 (1987).

22. Schrieffer, J.R., Wen, X.-G. and Zhang, S.-C., Phys. Rev. Lett. **60**, 944 (1988).

23. Aharony, A., Birgeneau, R.J., Coniglio, A., Kastner, M.A. and Stanley, H.E., Phys. Rev. Lett. **60**, 1330 (1988); Birgeneau, R.J., Kastner, M.A. and Aharony, A., Z. Phys. B**71**, 57 (1988).

24. Hoevers, H.F.C., van Bentum, P.J.M., van de Leemput, L.E.C., van Kempen, H., Schellingerhout, A.J.G. and van der Marel, D., Physica C**152**, 105 (1988).

25. Fisher, R.A., Kim, S., Lacy, S.E., Phillips, N.E., Morris, D.E., Markelz, A.G., Wei, J.Y.T. and Ginley, D.S., preprint.

26. Sasaki, T., Nakatsu, O., Kobatashi, N., Tokiwa, A., Kikushi, M., Liu, A., Hiraga, K., Syono, Y. and Muto, Y., preprint.

Panel: SPACE APPLICATIONS OF HIGH-Tc SUPERCONDUCTORS

Moderator: Y. FLOM
NASA-Goddard Space Flight Center
Code 313, Greenbelt, MD 20771

ABSTRACT

NASA has identified a number of potential space applications for HTSC materials in the areas of sensors, cryogenics, power and propulsion, communications and data systems. Due to the time constraint the scope of the panel discussion was limited to only two applications related to cryogenic technology (an area actively pursued at GSFC), namely: 1) current leads and ground straps and 2) magnetic bearings.

PANEL PARTICIPANTS:

A. Banerjea, NASA - Lewis Research Center, M.S. 302-1, 21000 Brookpark Rd., Cleveland, OH 44135.
L.H. Bennett, NIST, Bldg. 223, Room B152, Gaithersburg, MD 20899
J.C. Brasunas, NASA - GSFC, Code 693, Greenbelt, MD 20771.
S.R. Breon, NASA - GSFC, Code 713, Greenbelt, MD 20771.
D. Chung, Howard University, Dep. of Physics, Washington D.C.
A. DeReggi, NIST, Polymers Bldg., Room B320, Gaithersburg, MD 20899
N. Eror, Oregon Graduate Center, Beaverton, Oregon 97006.
H. Ettedgui, NIST, Bldg. 223, Room B150, Gaithersburg, MD 20899.
D. Gazit, Stanford University, Center for Materials Research, Stanford, CA 94305
S.L. Herr, University of Florida, Dept. of Physics, 215 Williamson Hall, Gainesville, FL 32611.
M. Levinson, GTE Labs, 40 Sylvan Rd., Waltham, MA 02254.
D. Lunoy, NIST, Gaithersburg, MD 20899
R. Ono, NIST, Boulder, Colorado
E.K. Smith, Martin Marietta Labs, 1450 S. Rolling Rd., Baltimore, MD 21227
L. Swartzendruber, NIST, Bldg. 223, Room B150, Gaithersburg MD 20899
M. Vlasse, NASA-Marshall Space Flight Center, Huntsville, AL 35812.
F. Wimenitz, Kaman Science Corporation, 2560 Huntington Ave, Suite 100, Alexandria, VA 22303.

A total of 18 people attended this panel, including representatives from three NASA centers: MSFC, LeRC and GSFC. During an introduction it was stated that the connection between the basic research and applications is vital to the progress in the area of high T_c superconductors (HTSC). Also, it was pointed out that this panel was an important step towards improvement in cooperation between the scientific, industrial and application communities.

CURRENT LEADS AND GROUND STRAPS.

At the present time, metallic current leads are responsible for as much as 20 to 30% of the parasitic heat load on cryogenic systems. Replacing the conventional materials-copper, manganin, stainless steel-with HTSC ceramics would greatly reduce this heat load due to the low thermal conductance of the ceramics. This could lead to a dramatic improvement in performance of cryogenic coolers, greatly enhancing long term flight missions to the outer planets.

In one particular application, HTSC material could be used as the current leads for a low-current superconducting magnet which will be used in the Advanced X-ray Astrophysics Facility (AXAF)/X-ray Spectrometer. AXAF is scheduled for launch in 1996. Currently, manganin leads are used to carry the 1.7 A magnet current. These leads, which are exposed to a maximum field of approximately 300 gauss, result in a heat leak of 1.8 mW from the inner vapor cooled shield, which is kept at a temperature of 20 K to the 1.5 K helium bath. Thus, for the HTSC ceramic to replace manganin leads, its thermal conductance should be less than 10^{-4} W/K.

In a similar vein, HTSC ground straps would serve as thermal insulators and at the same time as perfect electrical conductors. Presently, no ground straps are provided for electronics inside helium dewars due to the large heat leak down a metal strap. The ground straps would have to carry a current of 3-5 mA and have a thermal conductance less than 10^{-4} W/K.

In the panel session, it was pointed out that these applications do not require high values of critical current (typically, Jc values of 10^5 - 10^6 A/cm2 are specified for electronic applications) and recently reported current densities for the bulk polycrystalline HTSC ceramics are quite sufficient.

What does present a problem, however, is the routing of HTSC "wires" within the existing structures. This will require a number of sharp turns and bends which are rather difficult to accommodate considering the brittle nature of the HTSC ceramics.

To overcome this problem, various ideas and HTSC manufacturing methods ranging from the explosively made silver/high T_c superconductor composites to the single fibers grown from the melt were suggested during the panel discussion.

Also it was mentioned that the "internal" noise in polycrystalline HTSC materials could be an important factor in ground strap application. This noise is due to the grain boundaries acting as Josephson junctions in the presence of an external electric field.

PASSIVE MAGNETIC BEARINGS

At the present time, mechanical refrigerators are being developed to support long term missions in space. Stored cryogen systems are not suitable for this since they have not demonstrated lifetimes of more than 3 years and are heavy. Mechanical refrigerators, however, require large input power to produce refrigeration and compensate for friction and heat leaks.

Present state-of-the-art bearings (electromagnetic, gas and flexure) employed in long-lifetime mechanical coolers have significant shortcomings reducing their reliability and efficiency.

Due to the Meissner effect, bearings made with HTSC ceramics and permanent magnets, unlike permanent magnets alone, can achieve a stable equilibrium when cooled below T_c (Earnshaw's theorem does not apply to diamagnetic materials). Therefore, a passive, non-contact, frictionless bearing can be constructed.

Initial results indicate, however, that the stiffness of a permanent magnet/HTSC ceramic pair is several orders of magnitude smaller than that of electromagnetic bearings. This could limit the application of the HTSC passive bearings to miniature machinery only.

The panel also discussed the effect of flux trapping on HTSC bearings. At the present time, this subject is not understood and both theoretical and experimental work is required before any substantive comments be made.

The more general question of which ceramic oxide (Y-Ba-Cu-O, Bi or Tl compound) is preferred for the subject applications was also offered for discussion. No specific compound was singled out as the best candidate. Two approaches were mentioned. One suggested Y-Ba-Cu-O oxide as a starting material for the application work since more data exist on this compound than any others. However, protective coatings have to be developed to protect the oxide from the moisture degradation. The Bi and Tl compounds would be employed later when their technology matures. It was indicated that the substitution of one compound for another will only require engineering changes (that could be a mighty big one).

The second approach advocated the use of the Bi compound due to its inherent stability.

Finally, it should be mentioned that all attendees actively participated in the discussion and left the panel with the hope that similar meetings will take place in the future to advance the effort in applications of HTSC ceramics.

Panel 1B- Thin Film Applications

Moderator: J. Brasunas, NASA Goddard Space Flight Center, Greenbelt, MD

ABSTRACT: The panel on thin film applications considered the impact of theory and experiment on applications, inquired whether there are preferred HTS materials or deposition techniques, discussed the promise of RF properties, and overviewed the progress of infrared detection with HTS materials.

Discussion during this panel centered around the following questions.

1. What is the impact on applications of theory or experiment designed to study the basic phenomena of high temperature superconductivity?

It was recognized that researchers doing applications, and researchers performing basic studies, have to some extent been working apart. Therefore it was felt to be a strength of this particular conference that not only were basic studies presented, but also there were some presentations covering applications. The reasonableness of the applications presented was felt to be a good sign. While the question stated above does not admit an easy or final answer, it was felt that the bringing together of researchers in basic studies and in applications for mutual discussions was an important thing in itself, and that this bringing together should be continued in future conferences.

2. Is there a preferred (application specific) HTS material?

E. Venturini (Sandia) presented some results on polycrystalline thallium thin films prepared by e-beam coevaporation on ZrO2. Since a rather high transport critical current of 10**5 A/cm**2 at 77K had been achieved, the material shows promise for pulsed, high current applications. However, the flux vortices are weakly pinned, and so the material is not yet as promising for applications requiring field exclusion.

Comments were also made on the oxygen stability and moisture degradation of YBaCuO. While there has been a recent report of reduced moisture senstivity due to high deposition temperature or inclusion of a silver layer (Chang, A.P.L., 53, p. 1113, 1988), it was speculated that genuine stability will require encapsulation. It was also noted that higher quality films appear to be more stable.

3. Is there a preferred (application specific) thin-film deposition technique?

While a preferred technique was not singled out, comments were made on the versatility of the co-evaporation technique, allowing for independent control over a number of target materials, compared with, say, sputtering. With respect to laser ablation, some excellent results have been recently achieved with bismuth films.

4. How promising are the RF properties?

It was noted that the surface resistance (Rs) of a single crystal can be quite good, although the 1-2-3 material can degrade in about 30 days.

Wayne Cooke (Los Alamos) presented some results from a variety of groups on Rs of HTS materials from 1-100 GHz. It was reported that cleaning up the grain boundaries appears to improve performance. At 4K, HTS is superior to copper and approaches niobium. It was noted that this 4K result is very important to the various RF applications currently being anticipated and needs to be confirmed; this gets back to question 1 above. The data appear to indicate the superiority of HTS to copper at 77K, probably to frequencies of 10 GHz; at 100 GHz it is not clear there is superiority.

There is need for more study of this key property of HTS materials to scope out the useful RF applications.

With respect to specific applications, one area in which there has been appreciable progress is direct detection, infrared sensors. Leung et al. (A.P.L. 51, p. 2046, 1987) have demonstrated optical detection in thin films of YBaCuO via a phase-slip mechanism. Forrester et al. (A.P.L., 53, 1332, 1988) have demonstrated a bolometric effect in epitaxial films of YBaCuO. Enomoto et al. (Physica C, 153-155, p. 1592, 1988) have also shown a bolometric effect in a YBaCuO thin film.

During the panel J. Brasunas (NASA/GSFC) showed recent results on bolometric detection with a patterned thin film of ErBaCuO on SrTiO3 grown at NIST/Boulder. 1/R dR/dT, Icrit, and R already appear adequate to make a nearly phonon-noise-limited thermal detector at 77K. Responsivity, NEP and D-star derived from load-curve analysis at the transition indicate very promising performance comparable to 300K commercial detectors, without yet having optimized the thermal isolation of the HTS film. There does appear to be excess noise below 200 Hz.

In summary, the panel considered some difficult questions that do not admit easy or definitive answers. The need was recognized to bring together at conferences the people doing basic studies and the people doing or overseeing applications. The dialog between these two groups has begun, but they are not yet speaking the same language.

PANEL ON PATENTS AND SUPERCONDUCTIVITY

Speaker: Gerald Goldberg, U.S. Patent and Trademark Office,
Washington, D.C. 20231

ABSTRACT

A brief introduction to the role of patents is given. The response of the Patent and Trademark Office to the special challenge posed by the new field of high T_c superconductivity is outlined.

INTRODUCTION

Of all the technologies encompassed by practice before the Patent and Trademark Office (PTO), none appears faster-growing or is more exciting, and holds a greater promise of benefit to mankind than superconductivity. The aim of PTO is to see that the United States patent system continues to encourage new technology and that it does not get in its way.

Along the pathway between inventive conception and commercialization, the patent system has the potential for being either a bridge or a roadblock. We want to ensure that all lanes remain clear. The following discussion is intended for those working on high T_c superconductors who might need a quick introduction to the patent process.

BASIC INFORMATION ABOUT PATENTS

Let me start by explaining in very general terms just what the patent system is about. The concept of rewarding inventors and authors with exclusive rights to their works is centuries old. The framers of America's Constitution adopted this general concept and refined it within Article I, Section 8, to reflect the intrinsic right of the people to their respective discoveries and writings. To "promote the progress of science and useful arts," is what our constitution says.

The patent system promotes progress of science and useful art in at least two principal ways. The first has to do with the limited exclusive right to the invented technology which is conveyed to the patent owner through the patent grant. By this I mean that for a limited period of time (17 years) a patentee has the right to exclude others from making, using, or selling the invention. This presents investment security for those who place valuable resources into the otherwise risky process of research, new product development, and commercialization.

A second way that the patent system promotes progress is in its role as a powerful generator of information. To obtain a patent, an inventor must present a full disclosure of the invention. Some refer to this as the "contact theory" of Patent Law. It is sort of a quid pro quo -- disclosure of the full details of the new technology in exchange for its temporary exclusive control. Here, underscore the term "disclosure." Throughout the world, more than a quarter million unique inventions are patented each year. You can imagine what a wellspring of new knowledge they will provide to future inventors.

ARE PATENTS VALUABLE?

Press reports, which had been quite unkind to the U.S. Patent system in the past, have been far more favorable in recent articles. Fortune reported recently on the "new power of patents," noting the on-going opinion of research and development managers that "the increased value of patents should eventually translate into a greater willingness to spend money on research."

An article last September noted that recent patent system changes hold hope for increased R&D spending, and that the positive trend in rewarding patentees is "at last starting to give people confidence they can protect their investments." However even though the past few years have seen a slight recovery in patenting rates by U.S. inventors, the influx of applications from foreign shores has overshadowed this gain. The top 10 corporate patentees formerly consisted mostly of U.S. companies. But the times have changed. In 1987 the top companies were primarily foreign. Applications filed in the U.S. Patent and Trademark Office in 1987 were about 49% foreign origin, up from 20% twenty years ago. This, as much as anything else, reflects a sobering picture of the changing face of global competitiveness. After all, the filing of a U.S. Patent application almost always signals an intent to enter a new product or process into the American market place. And those "signals" are coming in greatly increasing numbers from across the oceans.

PRESIDENT'S INITIATIVE

On July 28, 1987 President Reagan presented his superconductivity initiative. This eleven-point initiative aims to promote further work in the field of superconductivity and to ensure U.S. readiness in commercializing technologies resulting from recent and anticipated scientific advances. This initiative has several objectives. One of these objectives was directed at the prime function of the PTO to wit.

> "Better protecting the intellectual property rights of scientists, engineers, and businessmen working in superconductivity".

The initiative directed the Patent and Trademark Office, when requested by applicants who have filed applications dealing with superconductivity, to

1. accelerate the processing of patent applications,
2. accelerate the adjudication of disputes involving superconductivity technologies.

In response to the President's Initiative, Donald J. Quigg, Assistant Secretary of Commerce and Commissioner of Patents and Trademarks, took several actions:

1. Establishing a special task force to prepare for the processing of patent applications for higher temperature superconductor technology.
2. Marking patent applications relating to this technology special on request of the applicant.
3. Renewing PTO support for legislation providing for an expanded scope of protection for process patents to cover products made by the patented process.

With respect to item 2 above, on September 1, 1987, a notice entitled "Special Status for Patent Application Relating to Superconductivity" was published in the Official Gazette of the Patent and Trademark Office.

What does this mean? A special status! As within any office, work flow depends on resources, number of items to be processed, and timeliness. Examiners generally give priority to the application which has the oldest effective United States filing date. However being granted special status jumps the application out of line (so to speak). When an application is made special and advanced out of turn, the application will continue to be special throughout its entire course of prosecution in the PTO.

HOW DOES AN APPLICATION BECOME SPECIAL?

If the inventions involve superconductive materials, processes or apparatuses the process is simple:

> The applicants or their attorney identify the application and request in writing that it be accorded special status and submit a statement under 37 CFR 1.102 that the invention involves superconductivity.

The statement submitted must be verified if made by a person not registered to practice before the PTO. By verified is meant the applicant signs the statement and states that he or she is aware that willful false statements and the like are punishable by fine or imprisonment or both under 18 U.S.C. 1001 and may jeopardize the validity of the application or any patent issuing thereon. Applicant must state that all statements made on information and belief are believed to be true.

Decisions whether or not to grant "special" status will be made by a Group Director. Again, once made special, applications remain in special status throughout its processing by the PTO.

Can you rely upon examination at PTO to be of highest quality even if the application is made special? Yes!!! The special task force geared up to insure that the PTO would be adequately prepared to "ensure that all lanes" were open. This meant having the best examining personnel and providing them with all information available.

PTO STAFF

The examining staff of the PTO is highly educated. All of our 1400 examiners have as a minimum a Science or Engineering degree. Many have a Masters or Ph.D. and about one-third have Law degrees. To keep examiners abreast of current technology, they are sent to scientific meetings, conventions, and conferences. Special superconductivity courses, both in house and at various universities have been arranged. Video tapes of meetings and courses are available at our training facility called, appropriately, the Patent Academy.

RESOURCE TOOLS FOR PATENT EXAMINERS

PTO began its collection the old fashion way. First we assessed our <u>own records</u> and created a profile of U.S. Patents from 1963 to 1986. It showed patent activity in the pre-high T_c superconductor technologies. This report was forwarded to each examining group. It served to alert all examiners to areas which had past superconductor activity.

Next we created a multi-discipline classification team to develop a manual search file of U.S. Patents and Foreign Patents. The completed manual file is available in the PTO's Public Search Room, this search room is open to the public. We welcome you to use it. No fee charged. The PTO maintains its own Scientific Library. The purpose of this library is the same as other libraries ... to supply its customers with the resource materials they need. Our library has arranged for subscriptions to various superconductivity technical journals, newsletters etc. These materials are maintained in our library or forwarded to appropriate examining groups. The library has developed an on line collection of non-patent literature (primarily bibliographic records of journal articles, preprints, monographs and newspaper articles.) This collection is available on line through the library's automated cataloging system to both the examining corps and the public. (By the way, if you have any literature you would like us to have, we would appreciate receiving it.)

As you can see, the PTO is doing its best to collect as much echnical information as possible. This will ensure that the patent examination will be based upon the most recent information permitted by patent statute.

TRACKING OF APPLICATIONS

The PTO has its own internal system to track applications based upon bar code technology. We need one. After all, in FY 1988, 140,000 applications were filed. All applications filed in the PTO are kept secret. Even if you the inventor or assignee, i.e., the owner of the application, publicly tell of your invention, we in the PTO cannot publicly discuss your application. Only when the application matures into a Patent, does anyone outside of the PTO know of your invention. Even when you place "Patent Pending" on your goods, we won't give out any information.

The PTO is permitted to give out ball park figures about filings. So far, as of October 1988, we have received approximately 400 applications dealing with the new high T_c superconductivity. As a service to the public, each Tuesday we publish our Official Gazette which profiles the patents issued that week.

SPECIFIC QUESTIONS

I will address several questions posed by the organizers of this conference.

<u>Question 1.</u> When is a discovery or an invention considered patentable?

To be patentable the invention must be new, useful and unobvious. The examiner reviews all prior art, i.e., U.S. Patents, literature, foreign patents, etc., to determine if the invention is patentable. The examiner is guided by the patent laws, patent rules, PTO procedures and guidelines.

<u>Question 2.</u> What difficulties arise when so many are working in the same field at the same time and simultaneous discoveries are quite likely?

The U.S. Patent system is a full examination system and grants a patent to the original inventor or inventors. Other countries have their own patent systems. Many use "the first to file" procedure. That procedure results in a race to the Patent Office. In the United States system, if the invention is patentable and the examiner cannot readily ascertain the true inventors, a process called "interference" is instituted. Then "shop notes", conception of the idea, reduction to practice, diligence, and timeliness are used as evidence. A Board requests, reviews and renders a decision as to the true inventors.

Question 3. What is the situation when one attempts to patent work published in the open literature or work presented at a conference or workshop?

In order to answer this question, I must make reference refer to the patent laws particularly section 102b-"Conditions for Patentability." This deals with the so-called one-year grace period. You are given one year from the time you publish, publicly use, or offer your invention for sale to file your patent application. Failure to file timely, may bar the issuance of a patent. Remember, a patent is a contract. Your knowledge in exchange for an exclusive right. If this knowledge is available, the contract will not be enacted.

Now a legal question is: What is a printed publication, what is public use, what is on sale? These questions can be the subject of many separate lectures by patent lawyers. Key points are (1) public accessibility, (2) what is printed, (3) what is published. For example, items can be printed but not distributed or published. Printed ordinarily means that a large number of items could be copied and made public. However, a typewritten patent application laid open for public inspection in foreign patent offices is not usually held to constitute a printed publication.

A printed document may qualify as a publication, even when given to a small group at a meeting, so long as accessibility is sufficient "to raise a presumption that the public concerned with the art would know of the invention". Preprints of scientific papers presented at a technical or scientific meeting may become a printed publication when distributed. The concept of "printed publication" is not easy to grasp or to apply. Many cases on both sides exist. Your patent attorney is the best source for guidance because he will be aware of your particular circumstances and the current case law.

CONCLUSION:

We in the Patent and Trademark Office stand ready to help you gain an economic foothold in exchange for knowledge.

Panel: MEASUREMENT METHODS

Moderator: N.S. DALAL
Department of Chemistry
West Virginia University, Morgantown, WV 26506

ABSTRACT - It became apparent during the Workshop that new experimental data are needed as clues to further theoretical understanding as well as the design of new high-T_c superconductors with improved characteristics. This panel discussed several techniques including low-field microwave absorption, electron and muon spin resonances, positron annihilation, Mossbauer spectroscopy and NMR.

PANEL PARTICIPANTS:

S.M. Bhagat, Department of Physics, University of Maryland, College Park, MD 20742.

R.M. Catchings, Physics Department, Howard University, Washington DC 20059.

M. Chase, NIST, Office of Standard Reference Data, Physics Building, Room A323, Gaithersburg, MD 20899.

W. Cook, Los Alamos National Laboratory, MP-14, MS H844, Los Alamos, NM 87543.

N.S. Dalal (Moderator), Chemistry Department, West Virginia University, Morgantown, WV 20606.

J. Gennosar, Technion, Israel Institute of Technology, Haifa 32000, Israel.

B.F. Kim, Johns Hopkins University Applied Physics Laboratory, Laurel, MD 20707.

K. Moorjani, John Hopkins University Applied Physics Laboratory, Laurel, MD 20707.

J.R. Morton, Division of Chemistry, National Research Council Canada, Montreal Road, Ottawa, Canada KIAOR9.

M. Rubinstein, Naval Research Laboratory, Washington DC 20375.

R.D. Shull, NIST, Building 223/B304, Gaithersburg, MD 20899.

S. Tyagi, Drexel University, Philadelphia, and the University of Maryland, College Park, MD 20742.

1. LOW-FIELD MICROWAVE ABSORPTION

The technique of microwave absorption under magnetic modulation (MAMMA) developed by Moorjani and coworkers at Johns Hopkins, or its variant, the magnetically modulated microwave reflection (MMR) as introduced by Dalal and coworkers at West Virginia University, by Rao and coworkers at Bangalore, India, Smyko and coworkers at the University of Utah, Müller and coworkers at IBM Zürich, and by several other groups more recently, appears to have high promise for become a sensitive, contactless and completely non-abrasive methodology for characterizing the high-T_c superconductors. However, theoretical guidance is needed as to what the technique is actually measuring. In particular, the estimates of the rate of change of T_c as a function of an applied magnetic field H, i.e. dT_c/dH appears to be an order of magnitude larger than that found by other, more standard methods (such as resistivity). Similarly, the low-field absorption from single crystals of Y-Ba-Cu-O reveal an entire new phenomenon — a series of very sharp (~ milligauss) lines, spaced a few milligauss apart. These lines are highly sensitive to temperature, microwave power, and the crystal orientation relative to an externally-applied magnetic field. Additional investigations of this phenomenon, on single crystals of the Bi- and Tl-based high-T_c materials would be very valuable. It is also desirable to extend the measurement frequency from the current range of X-band (9-10 GHz) to both lower and higher frequency ranges. The data would be useful to the construction of devices such as high-Q accelerating cavities, as well as for the theoretical understanding of this novel phenomenon.

2. THERMOLUMINESCENCE MEASUREMENTS

Wayne Cooke of the Los Alamos National Laboratory presented a highly detailed and illuminating discussion of this recent studies of thermally-stimulated luminescence from the surfaces of the high-T_c ceramic samples. This technique measures both the type and the concentrations of the defects and strains within a thin layer of the surface, and its sensitivity far exceeds that of the standard techniques of investigating defects (such as ESR and optical absorption). The technique is inherently very simple and inexpensive but appears to hold high promise for new data on the surface of the high-T_c materials.

3. POSITRON ANNIHILATION SPECTROSCOPY (PAS)

An important conclusion of this panel was that positron annihilation spectroscopy (PAS) measurements are urgently needed for obtaining a clear picture of the properties relating to the Fermi surface of these materials. This is significant because during several of the theoretical talks and discussions in the workshop, it became clear that some of the new theoretical models hinge critically on the details of the Fermi surface, even to the extent of whether or not these materials exhibit any Fermi surface at all. It also became apparent that PAS is perhaps one of the most accurate methods for characterizing the Fermi surface in the high-T_c superconductors. In particular, PAS will complement low temperature photoemission studies and the assignment of the electronic energy levels band structure parameters in these materials, as discussed by Berko and others at this workshop.

4. ELECTRON SPIN RESONANCE (ESR) AND MOSSBAUER MEASUREMENTS

Electron spin resonance (ESR) measurements of the high-T_c materials have shown that as far as Cu^{2+} resonance is concerned, the signals arise essentially from the impurity (non-superconducting) phases. An important question for the Cu^{2+} in Y-Ba-Cu-O is: why is Cu^{2+} ESR-silent? Recent ESR work by Mehran (IBM, Yorktown Heights) on YBaCuO and its non-superconducting variants, and its critique by P.W. Anderson (Princeton) suggest that the absence of Cu^{2+} ESR resonance implies a strong delocalization of the unpaired spins, somewhat in line with the resonating valence bond (RVB) description. It also became clear from the ESR results on Zn-doped YBaCuO by Crowe that systematic ESR studies of Zn-doped samples could yield direct information on the proposed spin delocalization and hence a check on the variously proposed theoretical models. Additional information on this aspect of the effect of impurity substitution and spin delocalization could also come from Fe, Pb and Sn-doped materials, as being carried out in NIST and other laboratories.

5. MUON SPIN RESONANCE (μSR)

This technique provides very critical data on the London penetration depth and local magnetic correlations in the superconducting state of the high-Tc materials. Budnick (University of Connecticut) presented μSR results which showed that strong magnetic correlations are indeed present with onset temperatures well below the superconducting transition temperature for $La_{2-x}Sr_xCuO_4$, for x greater than 0·07. The drawback of μSR, that a cyclotron is needed for these measurements, is well offset by the unique and sensitive data on the magnetic correlations, London penetration depth and the evolution of superconductivity in these materials, not easily obtainable by other methods. Thus this panel urges that the National Cyclotron Facility allocate special funds for μSR and related measurements on high-Tc materials.

CONCLUDING REMARKS

Concluding Remarks

Vladimir Z. Kresin
Materials and Chemical Sciences Division
Lawrence Berkeley Laboratory
1 Cyclotron Road
Berkeley, California 94720

Lately, there have been many conferences of high T_c superconductivity. As a result, it is not easy to organize a successful meeting, but the Workshop on Materials Science (NIST, October 11-13, 1988) definitely stands out. The organizers of the conference, Drs. L. H. Bennett and G. Vezzoli, have displayed a lot of initiative and imagination, which resulted in a very interesting and informative meeting with a well-balanced program. The panels turned out to be an important part of the program. It was nice that they took place on the last day of the Workshop, after all the other presentations.

The workshop has covered a broad range of problems. I think that the problem of critical experiments was the major aspect. As is known, the cuprates display exotic behavior in their normal, as well as superconducting, state. The situation is in some sense opposite to that before the appearance of the BCS theory. At that time, we did not have a real understanding of the phenomenon of superconductivity, but were well informed about the normal properties of conventional superconducting materials. At present, we have knowledge about the superconducting state, but we do not have a reliable and generally accepted picture of the normal state of the cuprates.

The question of the applicability of the usual Fermi liquid approach is a crucial one. The positron annihilation technique is a unique method allowing one to detect the presence of the Fermi surface. The conclusion presented at the Workshop that the Fermi surface exists is very important and imposes many constraints on the theory of high T_c.

The description of other critical experiments such as infrared techniques, neutron scattering, Hall effect, etc., also was very informative. The very important materials science problems such as material preparation, substitution and its effect on the magnetic properties also were addressed at the Workshop.

The problems of the critical fields, pinning force, etc., have been widely discussed. It is clear that past analysis underestimated H_{c_2}, and its real value remains an open question.

Many experimental constraints were discussed in a keynote lecture by Professor W. Little who is pioneer in the field of high T_c superconductivity.

Two years ago our information about the cuprates was very limited. Many theoretical models have been developed, but some of them were excluded by the following critical experiments. The final judgment will be made by the future experiments. Of course, if the elegant and rigorous theory appears to be irrelevant to the oxides, it will not make the efforts worthless. Such a theory can guide a search for new superconducting systems. At present, we are witnessing an intensive development of the theory of superconductivity. I think that one of the major lessons of the discovery by J. Bednorz and A. Muller is connected with their innovative approach. While many scientists focused on Nb-based materials, they made the breakthrough in new direction. Of course, the new oxides are unique, but high T_c superconductivity can be, probably, discovered in other, entirely different materials.

On the whole, the Workshop has displayed several trends. The experimental data on infrared spectroscopy, positron annihilation, microwave absorption, etc., showed that we are dealing with exotic systems. On the other hand, even the interpretation of the data is often controversial and requires a nontrivial theoretical approach which should be developed.

Different mechanisms of high T_c were discussed. Among them were the phonon mechanism along with electronic (excitons, plasmons) as well as magnetic interactions. It is clear that further progress depends strongly on the future critical experiments and their proper interpretation. In addition, nontrivial theoretical predictions should be tested experimentally.

I think that all these trends were presented at the Workshop, along with many interesting results and insights. The potential to meet the challenge posed by the discovery of high T_c was greatly enhanced by the Conference at NIST.

AUTHOR INDEX